湿地恢复手册 原则·技术与案例分析

The Wetland Restoration Handbook Guiding Principles and Case Studies

湿地是极其重要的生态系统，人们通常都把湿地与森林、海洋并称为地球三大生态系统。湿地拥有独特的生态功能，并且在抵御洪水、蓄洪防旱、降解污染、净化水质、涵养水源、调节气候、保护生物多样性等方面有着其他生态系统不可代替的作用和功能。湿地还是自然界中生物多样性最丰富的生态景观，也是各类珍稀濒危水禽的繁殖地和栖息地。因此，湿地被誉为"地球之肾"，又被称为"生命之源"。健康的湿地生态系统，是国家生态安全的重要组成部分和经济社会可持续发展的重要基础。保护好湿地对于维护生态平衡、改善生态状况、促进人与自然和谐、实现经济社会可持续发展，都具有十分重大的现实意义和深远的历史意义。

据专家研究，历史上，我国的湿地总面积曾经达到过6570万hm²，占我国国土总面积的7%。全国首次湿地资源调查表明，今天，我国湿地总面积为3848万hm²（未包括水稻田3800万hm²）。其中，自然湿地面积3620万hm²，库塘湿地面积228万hm²。自然湿地在国土面积中的比例仅为3.77%，远低于全球6%的平均水准。我国现存湿地的生态状况不容乐观，还有约40%的湿地面临着严重退化的危险；全国湿地面积减少、功能下降的趋势仍在继续，湿地生态系统面临着严重的威胁。据专家测算，50年来我国因围垦、改造等各种人为活动，丧失了至少40%以上的自然湿地，其中绝大部分是生态功能最为强大的沼泽湿地、湖泊湿地和滨海湿地。由于湿地面积减少和功能下降，我国湿地在保持水源、净化水质、蓄洪防旱、调节气候、维护生物多样性、抵御海啸和风暴潮危害等重要生态功能大大下降，一些重要的湿地文化和美学价值正逐步丧失，经济效益明显降低。从全球湿地保护的经验来看，要做好湿地保护，必须恢复和扩大湿地保留面积、逐步恢复湿地生态功能。

湿地生态恢复在我国尚处于起步阶段，目前缺乏十分成功的技术和实例，这严重制约了我国湿地生态恢复工作的推广。一些科学家根据中国湿地生态系统的状况，提出应该通过努力将自然湿地在国土面积中的比例从现在的3.77%提高到6%左右，再加上人工湿地，这一比例将提高到10%左右，这样才能满足生态环境改善的需要，实现人与自然的和谐相处，提供良好的生态服务。要实现这个目标，湿地保护和恢复的任务非常艰巨，需要各级政府、各有关部门、国内外自然资源保护组织和所有关注湿地保护的人们共同努力。

开展湿地恢复，就要选择合理有效的技术和方法，以达到在最短时间内，用最少的资金和最合理的技术手段恢复最多的湿地。基于这种考虑，国家林业局湿地保护管理中心（中华人民共和国国际湿地公约履约办公室）和美国易道公司（EDAW），决定合作编写《湿地恢复手册》，在介绍全球湿地恢复成功案例的基础上，结合中国国情，为中国湿地生态的恢复提供技术指导与帮助。

国家林业局是中国组织协调全国湿地保护工作的政府机构，全面掌握中国湿地保护和恢复，具有丰富的湿地保护管理经验。美国易道公司是全球最大的土地和环境规划设计公司之一，在全球范围内接受过诸多景观设计、生态环境和湿地规划项目，实施了许多湿地恢复的成功案例。双方联合编写《湿地恢复手册》是强强联合。《湿地恢复手册》详细介绍了湿地恢复的关键技术和操作方法，汇集中外湿地保护恢复专家的科研成果，对中国及美国、欧洲、东南亚的湿地恢复成功案例进行分析，并根据不同类型湿地提出具有针对性的恢复和修复方案，科学性、专业性较强，对规范和指导各地开展湿地保护和恢复工程具有较强的指导作用，对我国湿地保护恢复工作具有重要意义。

《湿地恢复手册》是中外专家智慧的结晶，凝聚了众多湿地保护工作者的辛勤成果，在此我代表国家林业局对中外专家付出的努力表示衷心的感谢，同时也希望此手册为我国湿地生态恢复工作起到应有的作用。

赵学敏

国家林业局副局长
2006年8月

Wetlands, together with forests and oceans, are one of the Earth's three great ecological systems. The diversity of their utility can hardly be overstated. Wetlands play an indispensable role in fighting floods, balancing hydrologic and climatic cycles, protecting biological d versity, controlling pollution, minimizing the impacts of drought, and conserving water supplies. As one of the most biologically productive ecosystems on the planet, they are a paradise to rare waterfowl and are often referred to as "the cradle of life", and "the kidneys of Earth".

Given this context, it is easy to see why wetlands are a critical part of China's diverse ecology, one which can aid in the sustainable development of the nation's economy and contribute to the society's well being. Wetland conservation has a great contemporary significance for a country developing as rapidly as ours. Within them can be found factors that have long-term implications for maintaining ecological balances, improving environmental quality, and enhancing the harmony between mankind and nature.

The country's first survey of national wetland resources has shown that historically there was a total of 65.7 million hectares of wetlands covering, or 7% of the total area of China. It is estimated that today the total area of China's wetlands is 38.48 million hectares (this excludes 38 million hectares of rice fields), with natural wetlands covering 36.2 million hectares, and constructed wetlands covering 2.28 million hectares. Yet natural wetlands account for only 3.77% of China's total land area, well below the global average of 6%. The current ecological condition of China's wetlands is far from optimistic: 40% of remaining wetlands face a serious threat of degradation. The retreat of wetlands and the subsequent decline of their natural functions on a national scale present grave challenges to China. According to expert surveys, mining, construction and other human activities, have caused the destruction of more than 40% of the natural wetlands over the past 50 years, including many with the highest ecological values, such as swamp, lake and coastal wetlands. While critical ecological functions and economic benefits remain in decline as a result, important culture and aesthetic

values have already vanished. Drawing upon the growing variety of experiences of global wetland conservation, it is imperative that China restore and expand its wetland reserves, through a concerted attempt to restore their diverse ecological functionality.

Wetland restoration in China remains in the early stages. A lack of approved technology and successful case studies have hindered the promotion of wetland restoration. Considering the present condition of China's wetland ecosystems, some scientists have recommended that in order to attain the kind of environmental improvements that offer a better balance between humans and nature, the percentage of natural wetlands covering China's total land area should be increased from 3.77% to approximately 6%, or up to 10% if constructed wetlands are included. To reach this goal, there is a great deal to be done in the conservation and restoration of wetlands. Governments and their departments at all levels, along with international and domestic organizations that concern themselves with natural resources protection, and all who deeply value wetland conservation should join forces in a concerted, consistent effort.

China will be able to make the most of an "economy of protection" in recovering wetlands only if it chooses effective restoration technologies and approaches. This publication has that aim in mind. The State Forestry Bureau of China along with environmental planners from the firm EDAW conceived the idea of co-publishing the Wetland Restoration Handbook to introduce successful case studies on global wetlands restoration, and to offer the best technical guidance and solutions, with a book that is additionally sensitive to the particular needs of China.

The State Forestry Bureau of China is the official organization coordinating nation-wide wetland conservation activities. With the best management experience in wetland conservation and restoration, it controls detailed data and information in this field. EDAW is one of the largest land and environmental-based planning and design firms in the world. It has engaged in landscape designs, ecological environment

and wetland planning projects, and has implemented successful wetland restoration projects in China and across the globe.

The Wetland Restoration Handbook explains in detail key technologies and approaches to wetland restoration, bringing together scientific research and outcomes from foreign and domestic experts. It examines successful cases on wetland restoration in America, China, and Southeast Asia, and offers professional restoration and recovery technologies for various types of wetlands. This handbook aims to act as a definitive reference for our national wetland conservation and recovery effort.

A handbook of this kind has been a long time coming. As a representative of the State Forestry Bureau, I would like to thank all the professionals and their great efforts and contributions to this book. It is my hope that the handbook fulfills its purpose and produces the kind of results that will make China an example of a thoughtful stewardship over its vital natural resources.

赵学敏

Zhao Xuemin
Deputy Director General
State Forestry Bureau of China
August 2006

国家林业局和易道公司首开先河，联手打造出这部《湿地恢复手册》，旨在为中国的湿地保护事业提供实用的蓝本和参考。本《手册》反映了全球对可持续发展的环境设计与规划的呼吁，也表明了中国信守此原则的决心，同时见证了中国正越来越多地加入到关于有效环境管理的国际性对话中。

《湿地恢复手册》是一部专业著作，涵括了全球关于湿地资源保护和修复的重要论述；同时它也是一部符合时代要求的作品—— 将西方和亚洲的广泛经验与中国可持续发展的目标结合起来。这个宏大的目标定位需要政府、社会和个人的努力和支持，同时也给我们这个行业提供了绝佳的机遇。值得骄傲的是，易道公司与国家林业局的合作最终促成了《湿地恢复手册》的诞生。

本《手册》得以成书是许多人精诚努力的结果，限于篇幅，这里不能一一尽数。谨借此机会特别感谢国家林业局野生动植物保护司司长卓榕生和国家林业局湿地保护管理中心全体同仁，感谢他们在湿地保护事业上的持续努力，为中国的环境发展所做出的贡献。此外，还要感谢易道的亚洲区主席乔全生先生(Sean Chiao)，他的远见促成了本书的出版；还有亚洲区环境规划总监艾思龙先生(Stephane Asselin)，他在创建和恢复湿地方面的丰富经验为本书提供了宝贵的素材。

最后，谨将此书献给戴立亚先生(Sandy D'Elia)，以缅怀他为易道亚洲区的成长作出的宝贵贡献，以及他对中国的可持续发展的深厚信心。

乔·布朗
易道公司总裁CEO

The Wetland Handbook is the result of a ground-breaking collaboration between the State Forestry Administration (SFA) of the People's Republic of China and EDAW. That this joint effort was undertaken to create a practical blueprint and reference for wetlands conservation in China testifies not only to the global call for environmental and sustainable design and planning, but to China's commitment to those principles. The handbook is further evidence of China's deeper participation in the critical international dialogue for a responsible environmental stewardship.

The Wetland Handbook is a collection of some of the best examples of global expertise on conserving and rehabilitating the Earth's wetland resources. It also represents a unique moment where lessons learned in the West and Asia intersects with China's goals for a sustainable future. This rare alignment of goals requires political, social and individual champions and provides a great opportunity for our profession. We are proud that this opportunity with the SFA has led to a definitive statement through the Wetland Handbook.

This book is the result of the hard work of many people, too numerous to name. I would like extend special thanks though, to Zhuo Rongsheng, Director General for the Department of Wildlife Conservation, of the SFA of the People's Republic of China and to the staff of the Conservation and Management Center for Wetlands of the SFA, People's Republic of China, for their dedication to advancing an environmentally responsible vision for China through their efforts on wetland conservation. I would also like to thank EDAW's regional director Sean Chiao, whose vision provided the instrumental spark for the publication, and to EDAW Asia's environmental director Stephane Asselin, whose years of experience in creating and restoring wetlands can be found in the pages of this book.

Finally, the Wetland Handbook is dedicated to Sandy D'Elia, an early proponent for EDAW in Asia who brought his considerable energy to building our regional practice, and who was a dedicated believer in China and its vision for a more sustainable future.

Joe Brown, FASLA
President and Chief Executive Officer, EDAW Inc.

目　录
Contents

第 一 章
Chapter 1

湿地及湿地恢复
Wetlands and Their Restoration

1.1 引言

湿地是位于水域和陆地之间的生态交错区，可以控制水域对陆地的侵蚀，对化学物质具有高效的处理与净化能力，还能够提供滨海咸水、河口或淡水栖息地。湿地是一道天然屏障，也是多种生物的避风港，其价值远远高于人们最初的认识。例如，尽管淡水湿地仅占地球表面的1%，其中的生物物种却占地球上总量的40%，湿地的价值由此可见一斑。湿地不仅具有不可替代的生态效益和防洪功能，也是环境保护中不可或缺的一个组成部分。

然而，当今全世界各地的湿地都处于不断退化或即将退化的厄运中（见图1-1,1-2）；与此同时，科学家和专业人士也正在为湿地的保护、恢复与建设而坚持不懈地努力着。

近年来，湿地恢复领域发展迅速，这主要归功于一些大型湿地恢复工程的实施，例如美国佛罗里达州大沼泽湿地(Everglades Wetland)项目和旧金山市湾盐湖(San Francisco Bay Salt Pond)转化工程。在相对较短的时间内，中国也向保护和恢复湿地生态系统迈出了第一步。虽然湿地恢复在美国和大多数西方国家已经是成熟的做法，并继续向复合型专家协作模式发展，但是在其他的地方，湿地恢复尚处于探索的新兴阶段。值得一提的是，尽管湿地恢复在美国不断增长的环境事业中已经是其中常见的部分，然而十年前的情况却远非如此，并且，美国的湿地恢复发展到今天也曾经走过不少弯路。

从非洲到东南亚、美国、欧洲，世界各地的湿地都面临着相似的问题，同时，每个地区又有不同的情况和背景，有许多不同的影响因素。在中国，随着经济的迅猛发展和城市化进程急剧推进，许多地方已然出现自然环境急剧恶化的问题。

尽管如此，湿地恢复正迅速成为中国许多新开发专案和环境工程的重要战略，这一趋势还将延续数十年。而且，湿地数量和品质的恢复有望成为衡量中国环境工程的指标，不仅是开发过后的环境补偿，更是可持续发展进程的重要部分。

1.1 Wetland Restoration: An Introduction

Wetlands are natural water systems that contain erosion and prevent the degradation of tidal ecosystems. They also act as highly effective treatment and cleansing harbors for chemical runoff. Wetlands can be saline and coastal, estuary, or freshwater habitats. As natural barriers and a critical harbor for extensive biodiversity, they are invaluable. A testament of their worth can be found in the fact that while freshwater wetlands cover approximately one percent of the Earth's surface, they hold over 40% of the world's species. In addition to being flood barriers with irreplaceable ecosystems, wetlands are also, as many have come to agree, an essential feature in the business of environmentalism.

However, the story today of wetlands the world over is one of consistent degradation or the danger of degradation almost everywhere they exist. At the same time, significant talent and industry is being applied by scientists and professionals towards the restoration, protection and engineering of wetlands.

The field of wetland restoration has developed rapidly in recent years, most notably in the United States, due in large part to the implementation of ambitious restoration efforts such as the Everglades restoration programs in Florida and the San Francisco Bay Salt Pond Conversion in California. In a relatively short period, China is also undergoing a nascent move towards protecting and creating wetland habitats. While wetland restoration is a more mature practice in the US and most Western countries, one which continues to evolve into a complex collaboration of specialists, outside the US it is an emerging practice in search of greater structure. It should be noted that while restoration projects in the US are now common features of a growing environmental business, this was not the case even a decade ago, and achieving this status has been a process marked by its share of missteps.

While wetlands from Africa to Southeast Asia, to the US and Europe face similar challenges, each region has its own unique context that creates different factors. In China, an ascendant power undergoing one of the most intense urbanizations and economic growth, one unwelcome result has been the rapid degradation of the natural environment.

Wetland restoration though, is fast becoming a critical strategy for many new developments and environmental projects across China. This trend will evolve for decades to come and wetlands are likely to become standard features in environmental projects in China, serving not only as environmental compensation in the wake of development, but serving the larger end of a progress that is sustainable.

◄ 图 1-1：湿地遭到破坏的原因经常是
人为的干预，例如：围垦（摄影：何文
珊）
Figure1-1: Wetland degradation often results
from human disturbance, such as this land
reclamation operation in Hangzhou Bay,
China (Photograph by Wenshan He)

◄ 图 1 - 2：受到侵蚀的滨海湿地（摄影：Rowan
Roderick-Jones）
Figure 1-2: Erosion of a coastal mudflat in
Hangzhou Bay, China (Photograph by Rowan
Roderick-Jones)

湿地的影响意义深远，能够改善生活品质，提供生态、休闲娱乐和社会文化价值。中国对湿地价值的认识日渐形成共识，而且，在湿地保护与恢复领域拥有广泛的国际基础经验作后盾——有越来越多的专业技术和相关立法可以借鉴，许多学者、规划师和专业人士都曾访问过国外许多著名的湿地恢复项目。

迄今为止，介绍湿地恢复的书籍不计其数，但是这些书籍主要针对美国的情况，几乎没有一本结合国际背景进行实用案例研究。本《手册》概述了湿地恢复的理论，辅以正在或已经在美国、欧洲、东南亚及中国实施的案例来加以说明。

湿地恢复这门学科自出现以来，业界和学术界从未停止对其实际作用的关注，不断分析总结20多年来湿地恢复项目中的成败经验教训。这些专案提供了很多宝贵资料和资讯，但是，中国和美国过去都未曾给予足够重视。因此，本书借此契机整合利用世界上主要的湿地项目，从方法入手，找出其中的共性，确立适用于不同国家相似情况的合理框架。

尽管生态的、工程的和湿地研究方面的出版物越来越多，但是专业人士发现在理论和实践之间还存在着一定的差距。希望本书能填补这个空缺，能够为决策者、工程师和工商界，甚至为研究者们提供一本权威、实用的手册。

本书的理念是给国内或其他国家和地区的读者，提供实用的基础知识，并结合全球重要的湿地恢复工程进行个案研究。本书作者中有生态学家、工程师、水文学家、水质专家、经济学家和湿地景观专长的设计师等，他们的理论知识与实践经验被综合进这本实用手册中，因此，本书不仅能为业内专业人士和政府部门提供参考，也能为大众更多地了解湿地与湿地保护、湿地的生态功能及社会功用提供帮助。书中收录了国际上有关湿地恢复的实践经验，为中国的湿地恢复事业提供了一个通用的结构框架，旨在履行将先进的理念及技术应用于保护

Wetlands create far-reaching opportunities that touch on areas such as improving the quality of life, and providing ecological, recreational, social and cultural benefits. The emerging consensus on the value of wetlands in China is abetted by a broader international awareness: Chinese in the field are increasingly exposed to international restoration projects and related legislation. Many top academics, planners and professionals have had access to some of the world's best known wetland restoration sites.

Numerous books have been written as "how to" guides on wetland restoration, primarily in the US. Few of these though, have included practical case studies in an international context. The Wetlands Handbook was conceived along the latter lines, aiming to provide an overview of the theory of wetland restoration, complete with illustrated case studies as they are being or have been implemented in the US, Southeast Asia and China.

Wetland restoration remains a discipline in which professionals and academics are monitoring project performance, analyzing what has succeeded and failed for projects that date up to 20 years ago. These cases provide invaluable data and information that has been largely overlooked not only in China, but also in the United States. As such, there is an opportunity to aggregate and leverage this knowledge using key wetland projects from around the world, to find consistencies in methodologies, and to create a streamlined framework that suits similar situations in different countries.

Yet even those most committed to that task can still find an unfortunate gap between theory and practical learning gained from the implementation of wetland projects, despite the greater availability of published material on ecology, engineering, and advances in related wetland research. That there remains a need for an authoritative, practical manual serving the broad interests of policy-makers, engineers and the business community, is a chasm that this book hopes to fill.

The idea is to present a common, critical foundation of knowledge, regardless of national or regional differences. The Wetlands Handbook integrates theory and practice, explained through case studies on important wetland restoration projects from around the world. While this book has technical aspects that are intended for professionals, scientists, academics and officials with at least some knowledge of wetlands, it will also appeal to the lay reader who would like to know more about wetlands and their preservation, their diverse ecological functions, and the roles they play in society.

By combining the ideas and experiences of ecologists, engineers, hydrologists, water quality specialists, economists

自然环境的责任。

谨向多位湿地恢复领域内的专家致谢，感谢他们无私的参与，提供了湿地恢复中的丰富经验，使本书最终得以成稿。

and landscape architects on wetland restoration, the authors have compiled a highly practical reference book that contains the best international practices of wetland restoration, providing a common framework for wetland restoration efforts in China. While its approach is innovative, it is one attempt in the greater goal of applying better science and methodologies towards the duty that is the protection of the natural environment.

Finally, this book was made possible by the generous participation of many experts in the field of wetland restoration and relies on their documentation of these experiences – both successful and unsuccessful.

▼ 图 1-3: 湿地分类的不同依据（摄影：何文珊）
Figure 1-3: Factors influencing wetland character (Photographs by Wenshan He)

周期性
Periodicity

短暂性（高山草甸）
Ephemeral (alpine meadow)

永久性（木本沼泽）
Permanent (swamp)

www.photos.com

盐度
Salinity

淡水（淡水沼泽）
Freshwater (freshwater marsh)

咸水（海草床）
Saltwater (seagrass)

植被
Vegetation type

无植被覆盖的（潮汐滩涂）
No vegetation (tidal mudflat)

植物繁茂（红树林）
Dense vegetation (mangrove)

水流
Water Flow

静态水体（高山湖泊）
Still water (upland lake)

流动水体（高地河流）
Running water (upland stream)

北部区域防洪工程
美国加州萨克拉门托市
业主：萨克拉门托防洪部门 (SAFCA)
North Area Flood Control
Sacramento, California
Client: Sacramento Area Flood Control Agency (SAFCA)

1.2 湿地的分类

从高原到河谷，从内陆到近海，湿地在地球上几乎无处不在。如果从字面上解释的话，无论是英文还是中文，"湿地"（wetland）一词都很容易被认为是"潮湿的土地"。事实上，潮湿的土地只是众多湿地类型中的一种，一些季节性湿地在旱季时可能非常干旱，一年中有几周或几个月的干涸期；而湖泊和海草床等湿地生态环境则终年被淹没在水下。不同的国家和专业组织对湿地有不同的定义，目前全球广泛引用的是1971年在伊朗拉姆萨尔小城签订的《湿地公约》中第1.1条款关于湿地的定义：

"湿地系指天然或人工、永久或暂时之静水流水、淡水、微咸或咸水沼泽地、泥炭地或水域，包括低潮时水深不超过6m的海水区。"

根据此定义，湿地包括多种类型，泥炭地、沼泽地、河流、湖泊和红树林生态系统，甚至珊瑚礁都属于湿地。和所有自然生态系统一样，湿地不属于任何现成类别，湿地的许多特征不符合环境梯度，使其类别很难界定。影响湿地特征的最重要因素是水文和水文周期；其他有助于湿地分类的重要环境参数有：盐度、水流状态和植被类型（见图1-3）。

除了天然形成的湿地外，还有人工湿地。在《湿地公约》分类系统中，人工湿地也叫人造湿地，包括养殖池塘（见图1-4）、潮湿的农田（如水稻田）（见图1-5）、盐业用地、蓄水用地、运河和废水处理区。

历史上，人工湿地是对自然生态环境的改造，目的是为满足一些明确需求，如食物生产、饮用水供应或交通需求，从而为当地群众带来切实的利益。近来，无论是专业人士还是普通大众，在建设用作其他用途的湿地方面都具有越来越强的意识，如用于提供生态价值、洪水控制功能乃至文化和宗教用途。湿地建造与恢复项目越来越多地被用于提供各种环境和社会功能，而不是简单的经济效益。

1.2 Wetland Classification

From high plateaus to river deltas, to continental interiors and coasts, there are few places on Earth where wetlands do not exist. In both English and Chinese, the word 'wetland' is often taken literally to mean wet landforms. However, 'wet lands' as such are only one of many kinds of wetlands. There are ephemeral or seasonal wetlands, which are dry for all but a few weeks or months during the year. Other wetland habitats such as lakes and seagrass beds are consistently inundated with water. Moreover, wetlands are defined differently in various countries and organizations. Currently, the most commonly adopted definition is that given in Article 1.1 of The Convention on Wetlands (Signed at Ramsar, Iran in 1971, and commonly known as the Ramsar Convention):

"For the purpose of this Convention wetlands are areas of marsh, fen, peatland or water, whether natural or artificial, permanent or temporary, with water that is static or flowing, fresh, brackish or salt, including areas of marine water the depth of which at low tide does not exceed six metres."

As noted above, wetlands can include peatland, marsh, river and mangrove habitats, and even coral reefs. As with all natural habitats, wetland types do not fall into convenient discrete categories, but instead have characteristics that vary along environmental gradients, making them difficult to classify. The most important factor influencing wetland characteristics is hydrology and hydroperiod. Other critical environmental parameters that help classify them include salinity, water flow, and vegetation types (Figure 3).

In addition to naturally occurring wetland habitat types, there are also man-made habitats that are defined as wetlands. Man-made wetlands classified under the Ramsar Convention include aquaculture ponds, 'wet' agricultural land (such as rice fields), salt pans, water storage facilities, canals and wastewater treatment areas.

Historically, man-made wetlands have been modified natural habitats, created in order to meet a clear demand such as food production, drinking water supplies, or transportation requirements, bringing tangible benefits to local communities. More recently though, there has been a growing awareness on the part of both professionals and the general public in creating wetlands that serve other purposes, such as providing ecological value, enhancing flood control abilities, and even serving cultural and religious purposes. Wetland creation and restoration projects are increasingly being designed to provide a whole range of environmental and social benefits, rather than simple economic gain.

▼ 图1-4: 养殖鱼塘（图片来源：www. photos.com）
Figure 1-4: Fishery ponds (Source: www.photos.com)

▼ 图1-5:亚洲南部常见的稻田（图片来源：www. photos.com）
Figure 1-5: Rice fields are common sites in southern Asia (Source: www.photos.com)

1.3 湿地的价值

1.3.1 生态与环境价值

生态价值

湿地是非常重要的生态资源，具有极其丰富的生物多样性。在美国，仅占国土总面积9.5%的湿地滋养着美国1/3的濒危物种，另有1/5的濒危物种在它们生活中的某个阶段会利用湿地。美国的湿地保护源于保护水禽及其栖息地的需要，因为许多禽类都在湿地里筑巢、觅食和栖息。很多大型河口湿地处于海陆交界、咸淡水交汇处，因此生境异质性高，生物多样性也更加丰富，是多种候鸟迁徙途中的重要驿站。

以位于长江口的崇明东滩鸟类自然保护区为例。近年来，每年冬季在崇明东滩越冬的雁鸭类达5万~6万只(徐宏发和赵云龙，2005)。其中珍稀保护鸟类有白鹳(Ciconia ciconia)、白头鹤(Grus monacha)(见图1-6)、小天鹅(Cygnus columbianus)、黑脸琵鹭(Platalea minor)和小青脚鹬(Tringa guttifer)等。湿地也是濒危鱼类的洄游必经之处，如中华鲟(Acipenser-sinensis)等洄游鱼类，对长江及其河口具有极高的依赖性，避免进一步的栖息地丧失和破碎化是其种群恢复的关键因素(见图1-7)。

水文价值

水文价值是湿地生态价值的重要组成部分。在我国的东部省份，洪水是导致人员死亡和财产损失的主要自然灾害，据报

1.3 Wetland Values

1.3.1 Environmental Values

Ecological Value

Wetlands are an extremely important ecological resource, a fact demonstrated by their exceptional levels of biodiversity. In the US, for example, wetlands only cover 3.5% of the total land area, but support a third of all endangered species found in that region. Protection of wetlands originated in the US from the conservation of waterfowl and its habitats, with many bird species using wetlands for nesting, hunting food and shelter. Located at the interface between ocean and land, fresh water and salt water, large estuary systems often boast a rich biodiversity and a high level of biological productivity, and thus serve as important stopovers on the migration paths of many bird species.

The Natural Reserve for birds on the eastern end of Chongming Island for example, the total number of birds wintering at the site is in the range of fifty to sixty thousand, including many endangered species such as the White Stork (Ciconia ciconia), hooded crane (Grus monacha), Bewick's swan (Cygnus columbianus), the black-faced spoonbill (Platalea minor) and Nordmann's greenshank (Tringa guttifer). Wetlands are also vitally important for certain endangered fishes such as the Chinese sturgeon (Acipenser sinensis), which is highly dependent on the Changjiang River and its estuary. Avoiding further degradation and fragmentation of the habitat is essential for the conservation of these species.

▼ 图1-6: 在崇明东滩湿地越冬的白头鹤 (Grus monacha) 种群
（摄影：章克家）
Figure 1-6: Wintering population of hooded crane at Chongming Dongtan wetland
(Photograph by Zhang Kejia)

▼ 图1-7: 渔业产量常取决于湿地的品质
（摄影：章克家）
Figure 1-7: Many fishery products depend on wetlands
(Photograph by Zhang Kejia)

导，2004年的洪水灾害导致1280人死亡、930000户家庭受灾，直接经济损失达714亿人民币，而2000年至2004年间，我国平均每年洪涝受灾面积1280万hm²，受灾人口1.61亿人，死亡1510人，直接经济损失1006亿元(水利部，2004；中国政府门户网站，2005)。同年，我国政府在洪水的控制方面投入了380亿人民币的经费，在水利项目上(包括水坝建设)上还另外投资了240亿人民币。

湿地对区域乃至全球的水循环起了重要的调节作用，其中，防洪是最为显著的功能之一。湿地可以通过截留、储存雨水来降低洪水的危害，然后再通过地表径流或地下渗透缓慢释放到河流中。通过这种形式，湿地能使在较长的洪水期内洪峰流量不同步，降低下游的洪峰峰值，避免(或减轻)下游流域受灾程度。图1-8比较模拟了流域有湿地调节时与没有湿地调节时的水文过程。

湿地的蓄水容量决定着降水以后流入河流的水量和时间，地下水水位、土壤特性以及地表高程都能影响一块湿地的蓄水量。如果湿地附近或下面的土壤处于非饱和状态，将发生地下水补给，水在湿地中停留足够时间，向土壤中慢慢渗透。湿地的地理位置和形状往往有利于其迅速积蓄雨水和地表径流，然后通过一个地表出口慢慢排出，或下渗到渗透性土壤中。

湿地通常和大型河流的漫滩直接联系，或者位于河流分水岭范围内。当大型和小型湿地都能提供洪水控制功能的时候，保护上游多个小型湿地要胜于保护下游的单个大型湿地（见图1-9）。如果只是保护上游湿地的话，洪水控制功能益于整个流域，乃至下游的地区。如果只保护下游湿地的话，则其上游仍然无法控制洪水的发生。因此，当在全流域的范围内规划湿地的保护和恢复时，从空间上考虑是非常重要的。

利用湿地来控制洪水，要比传统的工程控制方法花费少。美国陆军工程兵团(USACE)在马塞诸塞州的查理斯河盆地进行过一项很著名的研究工作，结果显示丧失3400hm²(约51000亩)的淡水湿地会导致下游发生洪水，以及每年600万美元(按1976年的美元价值)的财产损失。对于整个社会来说，综合各方面的损失则更为巨

Hydrological Value

Hydrological benefits, especially in terms of flood attenuation, also contribute to the environmental value of wetlands. Floods are major natural hazards causing deaths and property damage in most of China's eastern provinces. In 2004, 1,280 deaths and 930,000 homes were reported destroyed, with direct economic losses from flood damages estimated at RMB 71.4 billion (Ministry of Water Resources 2004). In that same year, China spent RMB 38 billion on flood control improvement projects and an additional RMB 24 billion on water resource projects, including the construction of dams.

Wetlands reduce flooding by intercepting and storing stormwater and releasing it by increments into stream channels via surface or groundwater flow. In this way, wetlands desynchronize peak discharge flows, allowing areas downstream from wetlands to benefit from reduced peak flood water levels during prolonged flood periods. Figure 1-9 illustrates this concept, showing the hydrology for a hypothetical river with and without the flood control benefits of wetlands. The alteration of fundamental river hydrology can translate into reduced flood damage.

The water storage capacity of a wetland determines the volume and timing of precipitation induced water flows to receiving rivers. A wetland's storage capacity is affected by groundwater elevations, soil characteristics and surface contours. Groundwater recharge occurs when soil is unsaturated, adjacent to or beneath the wetland and water is retained for long enough to allow it to percolate into the soil. Often, a wetland basin is located and shaped as such that it can be rapidly filled by precipitation and surface runoff that is released slowly through a restricted surface outlet, or infiltrates permeable soils.

Wetlands sometimes connect directly to the floodplains of large rivers, or are located in pockets throughout a river's watershed. While both large and small wetlands provide flood control benefits, there are implications to preserving many small upstream wetlands compared to preserving a single large downstream wetland. In the former scenario, the flood control benefit is spread throughout the watershed to all areas downstream. In the latter instance, the upper watershed is without flood control benefits and thus less efficient. As the above illustrates, when planning for protection and restoration of wetlands at the watershed scale, it is important to consider the implications of spatial strategies.

Wetlands provide flood control services at what is often a lower cost than conventional engineering approaches. A well known study by the United States Army Corps of Engineers (USACE) in the Charles River Basin in Massachusetts estimated that the loss of 3,400 ha (approximately 8,100 acres) of forested wetlands

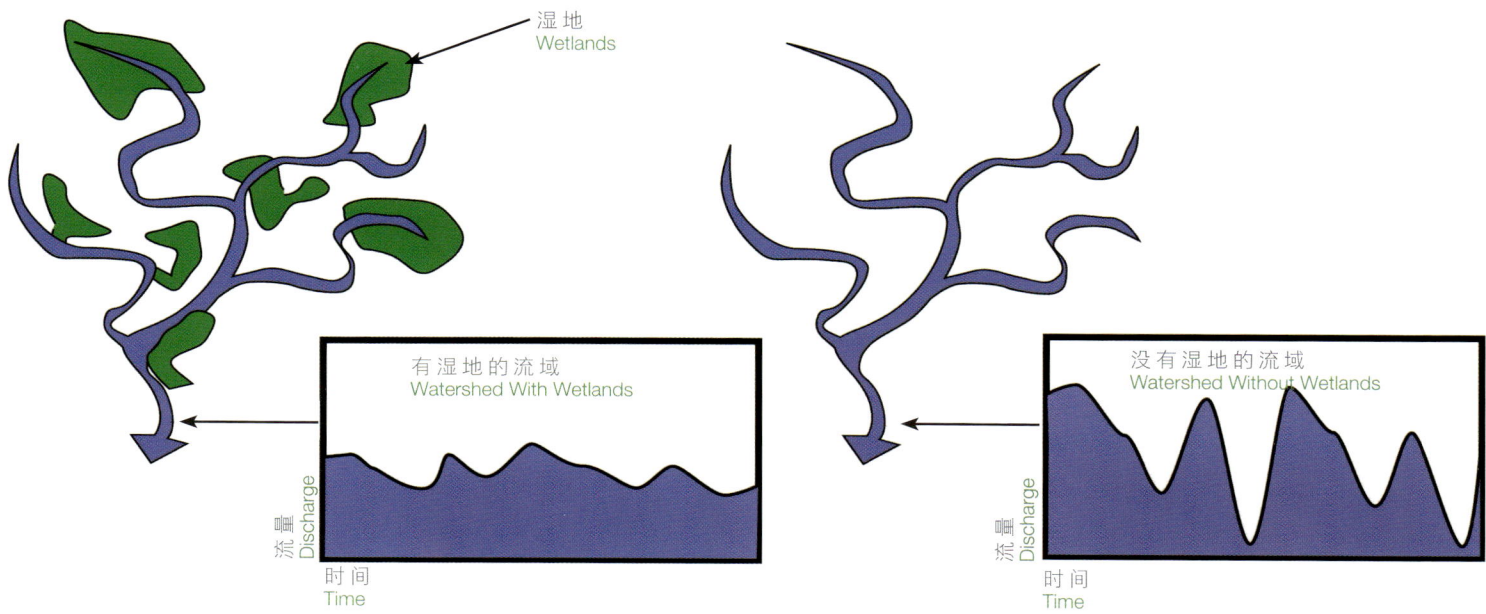

湿地
Wetlands

有湿地的流域
Watershed With Wetlands

流量
Discharge

时间
Time

没有湿地的流域
Watershed Without Wetlands

流量
Discharge

时间
Time

▲ 图1-8: 湿地的缓冲作用能够在较长时期内储存与释放雨水，从而降低水位和减小洪灾发生的可能性（绘制：易道）
Figure 1-8: Hydrologic Buffering by Wetlands-Watersheds with wetlands tend to store and distribute stormflows over longer time periods, resulting in lower levels of streamflow and reduced probability of flooding (Copyright, EDAW)

▲ 图1-9: 沿河(江)滨岸湿地在防洪中起着重要的作用（图片来源：www.photos.com)
Figure 1-9: Riparian wetlands along rivers play an important role in flood control (Source: www.photos.com)

大。这就很容易理解，美国陆军工程兵团为什么会决定购买并保护价值730万美元的湿地，来取代要花费3000万美元的洪水控制工程(Thibodeau & Ostro,1981)。此外，根据经济价值的估算，自然湿地的功能还胜于一些工程结构，例如防洪堤、水库、河渠，因为湿地还能提供一些附加的功能，如休闲、美学及生态学上的价值，而通常传统的工程学方法则无法赋予。

与此相对照，2006年我国国家林业局（湿地主管部门）在湿地保护和恢复上的总预算为20亿人民币（国家林业局，2006），几乎是1995年的4倍。但是，与此同时，湿地仍在不断丧失。自19世纪50年代以来，我国超过50%的湿地丧失了，其中包括超过12000km²的长江流域的湿地（白军红和王庆改，2003）。

改善水质

湿地的第三大环境价值是其改善水质的能力。湿地生态系统的生物产量仅次于热带雨林，高于其他生态系统类型(Campbell & Ogden, 1999)。正因为湿地具有这种高生产力，自然湿地才能够改善水质。湿地中复杂的物理、化学、生物过程相互结合，形成一个强大的可吸收、转化并固定污染物质的环境。因此，湿地常常被称作地球的"自然之肾"。确实，正如人体的肾脏清除出血液中的有害物，湿地可以净化地球上的水环境。

因为湿地有着极大的自然净化能力，所以人们往往误以为这种能力是无限制的。事实上，湿地的水质净化能力是有限的，而且不同类型的湿地具有不同的水质净化能力。例如，淡水沼泽湿地比有林湿地、泥炭藓湿地、盐沼湿地和草本泥炭湿地的净化功能强，所以，人们倾向于使用淡水沼泽湿地处理城市污水、工业废水和农业废水。这种高度工程化的处理系统被称为"人工湿地"。表1-1列出了人工湿地的净化过程和被处理的主要污染物。

would increase downstream flooding and cause US$ 6,000,000 (1976 US dollars) annually in property damages. Compounded annually, this is an enormous cost to society. It is easy to understand why, based on that study, the USACE decided to purchase and preserve the wetlands for US$ 7,300,000 in lieu of building a US$ 30,000,000 flood control structure (Thibodeau and Ostro, 1981). In addition to their economics, wetlands are often preferred over engineered structures such as levees, reservoirs, and channels, because they provide additional amenities such as recreational, aesthetic, and ecological values that conventional approaches do not offer.

Despite the recognized value of wetlands in flood attenutation, the total budget from the SFA (State Forestry Administration) for wetland preservation and restoration in China was only RMB 2 billion in 2006. Meanwhile, wetland loss has continued unabated. Over 50% of China's wetlands have disappeared since the 1950's, with over 12,000 square kilometers of wetland destruction in the Yangtze River basin alone (Bai and Wang, 2003).

Water Quality

A third major environmental benefit of wetlands is their ability to improve water quality. Wetland ecosystems are more productive in terms of biomass production than any other ecosystem except for tropical rainforests (Campbell and Ogden, 1999). This productivity is the driving force behind the well known ability of some natural wetlands to improve water quality. The complex physical, chemical, and biological processes in wetlands combine to create a powerful environment for the recycling, transformation and immobilization of pollutants. Wetlands have frequently been referred to as earth's natural kidneys. Indeed, just as kidneys cleanse the blood of toxins, wetlands can clean Earth's aquatic environments.

Their extensive natural capacities though, have led to a common misconception that natural wetlands have an inexhaustible ability to treat polluted waters. Wetlands have tangible treatment limitations, and not all wetland types have the same capacity to improve water quality. Fresh water marshes offer better treatment capabilities compared to other types such as swamps, bogs, salt marshes and fens. For this reason, marsh wetlands have become one preferred tool for water treatment engineers who design systems for polishing municipal, industrial, and agricultural wastewaters. The highly engineered treatment systems are known as "constructed wetlands" and are designed to magnify and enhance the treatment processes occurring in natural wetlands (see Table1-1).

处 理 过 程 Treatment Process	被 处 理 的 主 要 污 染 物 Major Pollutants Affected
影响化学转化过程的生物作用 Biologically mediated chemical transformation	氮，BOD，有机物 nitrogen, BOD, organics
沉降和掩埋 Sedimentation and burial	磷化物，悬浮固体，碳化物 phosphorous, suspended solids, carbon
沉积 Precipitation	金属，硫化物 metals, sulfides
絮凝和螯合作用 Flocculation and chelation	金属，溶解有机碳，病原体 metals, DOC, pathogens
植物吸收 Plant Uptake	营养物，微量营养素 nutrients, micronutrients
紫外光降解 UV-photodegradation	病原体，COD pathogens, COD
挥发 Volatilization	有毒物质，有机化学物 toxic substances, organic chemicals

1.3.2 社会经济价值

鉴于当前围绕中国的城市化、有限的土地资源和环境保护状况的争论，与湿地相关的环境经济学的重要性与日俱增。湿地可给当地人们和广大社会带来巨大的经济效应，包括湿地本身的使用和非使用价值以及影响更为广泛的或正或负的外部效应。由于有关环境保护和经济发展的争论仍在影响着中国的社会经济的发展进程，决策者必须全面考虑与湿地有关的各种经济成本和效益。

使用与非使用价值

分析湿地经济效益的传统方法是看它的直接使用价值。直接使用价值是指与各种生产和消费活动相关的价值。例如，在东南亚的许多国家，湿地仍然是农业生产的主要基础和农户收入的重要来源。在中国，湿地产品包括烧柴、淡水、稻米、水产品和泥炭等，这些物质对经济发展都很重要。此外，湿地的休闲旅游产业是迅速发展起来的直接使用价值。中国自古就重视身心健康与精神安宁，在现代社会，远足、垂钓和野营等生态旅游越来越受欢迎。从这方面讲，湿地既为游客提供体验自然的机会，又为当地社会增加收入，提供更多的就业机会。可见，湿地的直接使用价值丰富多样，并且每个地区各有不同。易道调查了中国湿地的多样化直接使用价值，表1-2的第一栏列出了这些直接使用价值。

湿地具有很多间接使用价值，更多的人群能够受益于这种间接价值。湿地是一个高效的生态系统，能够起到保持水质、水流与蓄水、防洪、吸收营养物和调节气候等作用。为此，经济学家们制定出多种方法，用于评估间接使用价值。研究表明，间接使用价值远远高于直接使用价值。例如，据估计，在斯里兰卡和荷兰的沿海地区，湿地的使用每年可分别节约500万美元和1.89亿美元的防洪支出。尽管准确衡量湿地间接使用价值的难度越来越大，但这种价值的重要性日益凸现。

湿地的第三类使用价值是社会对其未来使用价值的重视，例如，湿地能够保护将来可用于医学研究的物种和基因遗传资源。经济学家将这类价值称为湿地的

1.3.2 The Social and Economic Value of Wetlands

The economics of wetlands are an increasingly significant issue in light of the current debate involving China's urban expansion, limited land resources and the state of environmental preservation. Wetlands are a source of tremendous economic impact for both local populations and broader communities. Their impact includes basic use and non-use values in the wetland itself, as well as positive and negative externalities with wider effects. As the debate around environmental preservation and economic growth continues to shape China's agenda, decision-makers must thoroughly consider the full spectrum of economic costs and benefits associated with wetlands.

Use and Non-Use Benefits

Wetland economic benefits are traditionally analyzed in terms of their direct uses. Direct uses are values based on a range of production and consumption activities. For example, wetlands continue to be an important source of agricultural production and are responsible for a majority of all rural household income in most South Asian countries. Wetland products in China include fuel wood, fresh water, rice and fish – all valuable inputs to an expanding economy. Furthermore, wetland tourism recreation is a rapidly growing direct use. Nature-based and eco-tourism activities – such as hiking, fishing and camping – have all grown in popularity in a country that has historically placed significant value on health and medicine. In this regard, wetlands provide both a natural getaway for tourists as well as revenue for local communities. The overall direct use of wetlands is widespread and unique to each locality. Table 1-2 lists the various productive wetland direct uses that EDAW has observed in China.

Wetlands also have numerous indirect uses that impact the greater community. As efficient ecosystems, wetlands sustain ecological functions that protect important systems that include the maintenance of water quality, flow and storage; flood and storm control protection; nutrient retention and micro-climate stabilization. Economists have developed various methods to value indirect uses, and studies have revealed that the value of indirect uses often vastly outweighs direct uses. In coastal Sri Lanka and the Netherlands, for example, flood control and prevention has been estimated at US$ 5 million and US$ 189 million a year in reduced expenditures, respectively. While measuring the indirect uses of wetlands presents more challenges, their value is very often more critical.

A third category of use value that wetlands provide are the premiums society places on possible future uses. Preserving a pool of wetlands species and genetic resources, for example, enables the wetland to endure as

使用价值 Use Values	直接使用价值 Direct	• 农业(稻米、蔬菜、树和其他作物）Agriculture (rice, vegetables, trees, other crops) • 渔业 Fishing • 食品 Food • 休闲 Recreation • 建筑材料 Building material • 烧柴 Fuel • 交通 Transport • 其他原材料 Other raw materials
	间接使用价值 Indirect	• 防洪减灾 Flood/storm prevention and protection • 有机物和营养物的储存与循环利用 Storage and recycling of organic matter and nutrients • 腐蚀控制与生物防治 Erosion and biological control • 废水处理 Wastewater treatment • 为当地居民和（或）庄稼灌溉提供水源 Water supplies for local population and/or crop irrigation
	可供选择的价值 Optional Values	将来可以用湿地造福社会的方案：药物研究、工业、休闲、纯净水等 Future potential applications of wetlands that will benefit society: pharmaceutical research, industrial, leisure, clean water, etc.
非使用价值 Non-Use Values		• 文化价值 Cultural value • 美学价值 Aesthetic value • 文物价值 Heritage value • 遗产价值 Bequest value
正外部效应 Positive Externalities		• 房地产价值的提升 Increased real estate values

选择价值。选择价值体现出人们期望能够将自然资源和大自然赋予人类的这些不确定益处留待将来使用。虽然选择价值难以衡量，却是湿地内在价值中的重要组成因素。

经济学家不仅研究了湿地的使用价值，还对与之相对的非使用价值进行了研究（非使用价值也叫特种自然资源的存在与传承）。无论湿地当前或未来如何使用，湿地物种及地域本身毫无例外地具有文化和美学方面的价值。例如，湿地中存在的濒危野生动植物和自然现象可能在当地习俗或宗教中具有特别的意义。尽管湿地的非使用价值不能直接产生价值，但在中国农村和山区可能特别重要，因为当地居民非常重视社会、文化及传统惯例。

积极的外部效应：影响深远的经济增值
湿地的使用价值和非使用价值是湿地综合经济效益中的核心内容。同时，湿地对周边社会和地域具有乘数效应，但是这一效应却常常被忽视。积极的外部效应中最重要的是房地产的增值和税收的增加。美国做过多个案例研究，结果表明，缩小与最近湿地的距离或扩大最近湿地的面积都会使房地产价值随之增高（Mahan等，2000）。其他相关研究也得出了量化结果：房地产与沿海湿地的距离每近3km，就会增加1010美元的价值，比距离湿地9km的房地产价值约高出32%。房地产价值升高后，房地产的税收也随之增加。

房地产的增值来自居民从附近高品质的开放空间获得的满足感和愉悦感。据调查，居住在湿地附近的居民认为居住地空气新鲜，水质更好，户外娱乐设施近在咫尺，他们能享受自然之美，体会到生活品质的提高等。虽然具体的提升幅度随当地条件和居民的偏好不同而有所不同，但是事实清楚地表明，靠近湿地的房地产价值都有很大的提升。

维护与管理成本
与其他复杂的事业一样，湿地在带来效益的同时，也会产生各种费用。湿地成本通常以用在维护和管理上的现金支出来衡量，包括购买设备、建设基础设施，以期维持一个可持续的环境，将旅游活动对湿地的影响降至最小。成本还

a source of medicinal research. Economists describe this type of benefit as the wetland's option value. Option values reflect the desire to maintain access to uncertain future benefits from natural resources and systems. While difficult to measure, option values are an important factor in measuring the intrinsic value of wetlands.

In contrast to use values, economists also examine wetland non-use values, or the existence and inheritance of exceptional natural resources. Wetland species and areas have intrinsic value in cultural and aesthetic terms, irrespective of their current or future use. For example, some individuals feel strongly about protecting endangered wildlife species found in wetland parks. Although not directly productive functions, the non-use value of wetlands is particularly important to consider in rural China, where considerable significance is placed on social, cultural and heritage practices.

Positive Externalities: Far-Reaching Economic Value-Added
Use and non-use values are essential to the overall economic impact of wetlands. Often overlooked, however, are the multiplier effects wetlands have on surrounding communities and land areas. The most significant of these positive externalities are increased property values and tax revenues. Studies in the US have been conducted on numerous cases studies to statistically illustrate that (i) decreasing the distance to the nearest wetland and (ii) increasing the size of the nearest wetland result in a subsequent increase in property values (Mahan, Polasky, and Adams, 2000). Other relevant studies have quantified this increase to be US$ 1,010 for every 1,000 feet (304.8 meters) closer a property is to a coastal wetland, and possibly as much as 32% higher than property located 3,200 feet (975.4 meters) away. The rise in property values likewise leads to higher property tax revenues.

The rise in property values is attributed to the satisfaction and enjoyment residents attain from high-quality open spaces near their homes. Surveys of local residents reveal multiple and related perceptions, such as better air and water quality, nearby outdoor recreation opportunities, natural aesthetics, and an enhanced quality of life. The precise increase varies according to local conditions and preferences, but the evidence clearly indicates a significant increase in property values for communities with good proximity to wetlands.

Wetland Maintenance and Management Costs
As with almost any complex undertaking, in addition to benefits, wetlands incur various costs. Wetland costs are

▼ 表 1-3: 湿 地 的 社 会 经 济 成 本 一 览
Table 1-3: Socio-Economic Costs of Wetlands

管理与维护成本 Management and Maintenance Costs	设备 Equipment
	车辆 Vehicles
	基础设施 Infrastructure
	维护保养 Maintenance
	人员配置 Staffing
	培训 Training
机会成本 Opportunity Costs	就业、收入和其他土地利用方法创造的收入：房地产、工业、写字楼等 Employment, income and revenues from alternative land uses: property, industrial, office, etc.
负外部效应 Negative Externalities	害虫：损害庄稼 Pests: Crop damage
	人类疾病 Human diseases
	牲畜受伤/疾病 Livestock injury/disease

包括建立一个管理团队，以确保长期管理湿地所需的劳务、设备和基础设施（见表1-3）。

机会成本

湿地机会成本是指以其他方式利用湿地可获得的价值。在农村地区，湿地的其他利用方式通常就是农业和畜牧业生产；城市及城市郊区的机会成本一般为工业开发和住宅开发。经济学家采用精确的方法从土地、就业、收入和效益等方面评估机会成本。在中国城市地区，全面、客观地理解湿地的机会成本格外重要，因为经济的发展必须与城市稀缺的土地资源和环境资源相协调。

负面外部效应

除了直接运营成本和机会成本，湿地可能还会给社会带来负面影响。湿地作为可充分进行自身调节的生态系统，本身就具有生态风险。有时，湿地也可能成为传播疾病的环境。常被提到的便是湿地害虫（如蚊子）和鸟类对家畜、人类和庄稼的祸害。虽然在规划湿地保护的成本/效益时很难评估负面外部效应，但仍需充分考虑控制和降低负面外部效应所需要的经济成本。

typically measured in terms of cash expenditures for maintenance and management. This includes equipment and infrastructure for building a sustainable environment wherein visitors have a minimal impact on the wetland. Costs include and are measured by a management team that would identify the required labor, equipment and infrastructure to operate the wetland over a sustained period.

Opportunity Costs

The opportunity cost of a wetland refers to the value that may be gained through an alternative use of the wetland. In rural areas, alternative wetland land use generally refers to crop and livestock production, while industrial and residential developments are typical opportunity costs for cities and urban fringes. Economists use stringent valuation techniques to estimate opportunity costs in terms of land, employment, incomes and profits. A realistic and thorough understanding of a wetland's opportunity cost is particularly important in Chinese cities, where land scarcity and environmental resources must be balanced with economic growth.

Additional Negative Externalities

In addition to direct operating and opportunity costs, wetlands have the potential to negatively impact the larger community. As fully functioning ecosystems, wetlands by nature generate unique environmental threats. Most typically cited is crop and livestock damage by wetland pests and birds. Occasionally, wetlands may also foster an environment that allows certain human diseases to proliferate. While often difficult to value, negative externality costs must be considered in comprehensive cost-benefit discussions of wetland preservation (see Table 1-3, previous page).

1.4 湿地恢复的历史

恢复生态学的概念最早由J.D.Aber和W.Jordan两位英国学者于1985年提出，他们认为使生态系统恢复到先前或历史上（自然的或非自然的）的状态即为生态恢复。随后成立的国际恢复生态学会（Society for Ecological Restoration)先后提出了以下三个定义：

1. 生态恢复是修复被人类损害的原始生态系统的多样性及动态的过程；

2. 生态恢复是维持生态系统健康及更新的过程；

3. 生态恢复是帮助研究生态整合性的恢复和管理过程的科学，生态整合性包括生物多样性、生态过程和结构、区域及历史情况、可持续的社会实践等广泛的范围。

最终国际恢复生态学会以及其他生态学家们采纳了第三个定义作为生态恢复的最终定义，因为大多数被破坏的生态系统已无法恢复到原来的状态，尤其是社会、气候和水文条件都已发生了变化；此外，生态恢复过程并非只有一个改善自然环境的目的，而是一个跨行业、跨区域的过程，在此过程中必须考虑到社会和经济背景。

湿地恢复一般旨在改善湿地主要的环境功能，通过恢复生态结构与功能，实现水文调节、净化水质、创造和提供文化和休闲娱乐价值。在恢复工作开始前，通常需要依据湿地原来的性质或特征制定出一套相应的管理原则。

过去，湿地恢复项目一般采用建立自然保护区和限制人为影响等方法进行管理。现在的湿地恢复项目中，大多数都制订了与恢复地区社会经济发展一致的目标，这完全符合国际生态恢复学会的生态恢复概念。此方法比传统的湿地管理技术（如建立自然保护区和对人为干预的控制等）更具主动性。许多至今仍然有效的传统方法完全依靠湿地生态系统内在的发育和演替机制，因而需要很长时间才能看出效果

1.4 History of Wetland Restoration

Restoration ecology was first defined by two British scholars, J.D. Aber and W. Jordan in 1985 as 'returning an ecosystem to its previous or historical (natural or unnatural) state'. Jordan later founded the Society for Ecological Restoration, which offered three successive refinements of this original definition:

1. Ecological restoration refers to the process of renewing the diversity and dynamic of original ecosystem, which has been damaged by man;

2. Ecological restoration is the process of maintaining and renewing the exuberance of an ecosystem;

3. Ecological restoration is the process of assisting the recovery and management of ecological integrity. Ecological integrity includes a critical range of variability in biodiversity, ecological processes and structures, regional and historical context, and sustainable social practices.

The third and final definition is today universally accepted, since most damaged ecosystems cannot be restored to their original state, especially when societal, climate and hydrological changes are taken into account. Furthermore, it has been recognized that ecological restoration is not a process with the sole purpose of natural enhancement, but one that includes multi-industrial, cross-regional processes wherein social and economic backgrounds must be considered.

Wetland restoration commonly aims to improve principal environmental services such as the restoration of ecological structures and functions, hydrologic regulations, water purification, heritage and recreational values. Restoration usually requires a set of administrative guidelines that correspond to the preservation of its original wetland nature or character.

Most current wetland restoration projects prescribe goals that conform to the social and economic development of restoration sites. This practice is in accord with the concept of ecological restoration adopted by the International Society on Ecological Restoration. This approach is more proactive than traditional wetland management techniques such as the establishment of natural reserves and limitation on human interference. While still effective, more traditional methods rely solely upon the natural succession and evolution of wetland ecosystems, and can therefore take many years to demonstrate any discernable effects (Hu, 1999).

（胡聃，1999）。

虽然关于生态恢复的理论探讨出现在20世纪80年代后，但是，对湿地的生态恢复可追溯至1977年。当时，美国通过立法（美国《清水法案》第404节）控制湿地的开发与利用。自20世纪90年代开始，湿地恢复在全球范围内得以广泛开展和发展，湿地生态环境的价值逐渐被广泛认可与重视，全球湿地退化与丧失等问题也受到普遍关注。

1.5 中国的湿地恢复

中国幅员辽阔，地貌多样，拥有大量的湿地，面积可观。根据国家林业局发布的全国湿地调查数据，截至2004年，中国拥有天然湿地3620万hm²。中国的湿地是全球范围内具有重要意义的生态资源，在保护湿地所在区域的生物多样性、蓄洪防旱、发展农业和水产业以及提供工业原材料方面起着重大作用。

20世纪以来，全球许多地区出现了史无前例的湿地退化和丧失现象（参见表1-4）。与此相似，中国的湿地也在以惊人的速度消失，黑龙江省三江平原湿地就是一个令人痛心的例子。根据《湿地国际》的统计资料，截止2002年，这块湿地从1975年的500万hm²缩小到21世纪初的113万hm²。长期以来，由于缺乏对湿地生态系统价值的理解，中国的湿地面临着工业化、经济的高速发展和因人口压力导致的快速城市化进程带来的直接威胁。目前，中国湿地毁坏的问题仍在继续，剩余的湿地也因为污染和湿地资源开采过度而不断退化。内陆地区湿地的丧失和退化主要受洪水和水质下降影响，沿海地区则受到咸水倒灌、台风危害、土地侵蚀和赤潮的影响。根据官方统计，1949年以来，共有219万hm²的潮汐湿地消失，约占中国全部沿海湿地面积的一半。

同样，由于围垦、砍伐和海岸带开发，红树林湿地也丧失严重。中国的红树林已从20世纪50年代的5万hm²，下降到至今仅有的1.4万hm²，即72%的原始红树林已经消失（湿地国际，2002）。红树林在抵御风暴潮、提供近岸鱼类栖息地方面具有不可替代的作用。近年来，红

In the 1980s, theories about ecological restoration first began to be discussed, yet restoration of wetlands dates back to 1977, when America passed a law to control the exploitation and utilization of wetlands (Section 404 of the Clean Water Act). The 1990s witnessed an expansion of wetland restoration projects and the field has experienced considerable growth since then, as the value of wetland habitats has become far more widely known and appreciated, one that has unveiled a historic backdrop of degradation and loss.

1.5 Wetlands in China

China's vastness includes geographically diverse regions, each supporting a multitude of wetland habitats that are often enormous in scale. According to the China's State Environmental Protection Agency's (SEPA) website, China has 36.2 million ha of natural wetlands as of 2004 (SEPA, 2005). China's wetlands form a globally powerful environmental resource that plays a major role in the protection of the region's biodiversity, its flood and drought prevention capacities, the sustenance of agriculture and aquaculture, and in the provisioning of industrial raw materials.

But like many regions across the globe since the beginning of the 20th century, which experienced an unprecedented degradation and loss of wetlands, wetlands in China have disappeared at an alarming rate (Table 1-4). The Sanjiang Plains Wetlands in Heilongjiang Province is a somber reminder of this trend, which shrank from the original 5 million ha in 1975 to its 1.13 million ha (Wetlands International, as of 2002). Due to a general lack of understanding on the overall value of wetland ecosystems, wetlands in China face immediate threats from rapid industrialization and the urbanization associated with the country's booming economy and high-density populations.

China today faces the continued destruction of its wetland habitats, together with the degradation of remaining wetland areas due to pollution and over-exploitation of wetland resources. Wetland loss and degradation in China has been largely affected by flooding and water quality issues inland, along with salt water intrusion, typhoon hazards, shoreline erosion, and algal blooms along China's coastal zones. According to official national statistics, a total of 2,190,000 ha of tidal marshlands (about half of China's entire coastal wetland area), has disappeared since 1949.

▼ 表 1-4: 全球湿地退化和丧失的比率 (资料来源 : Mitsch & Gosselink, 2000)
Table 1-4: Global Percentage of Lost Wetlands (Source: Mitsch & Gosselink, 2000)

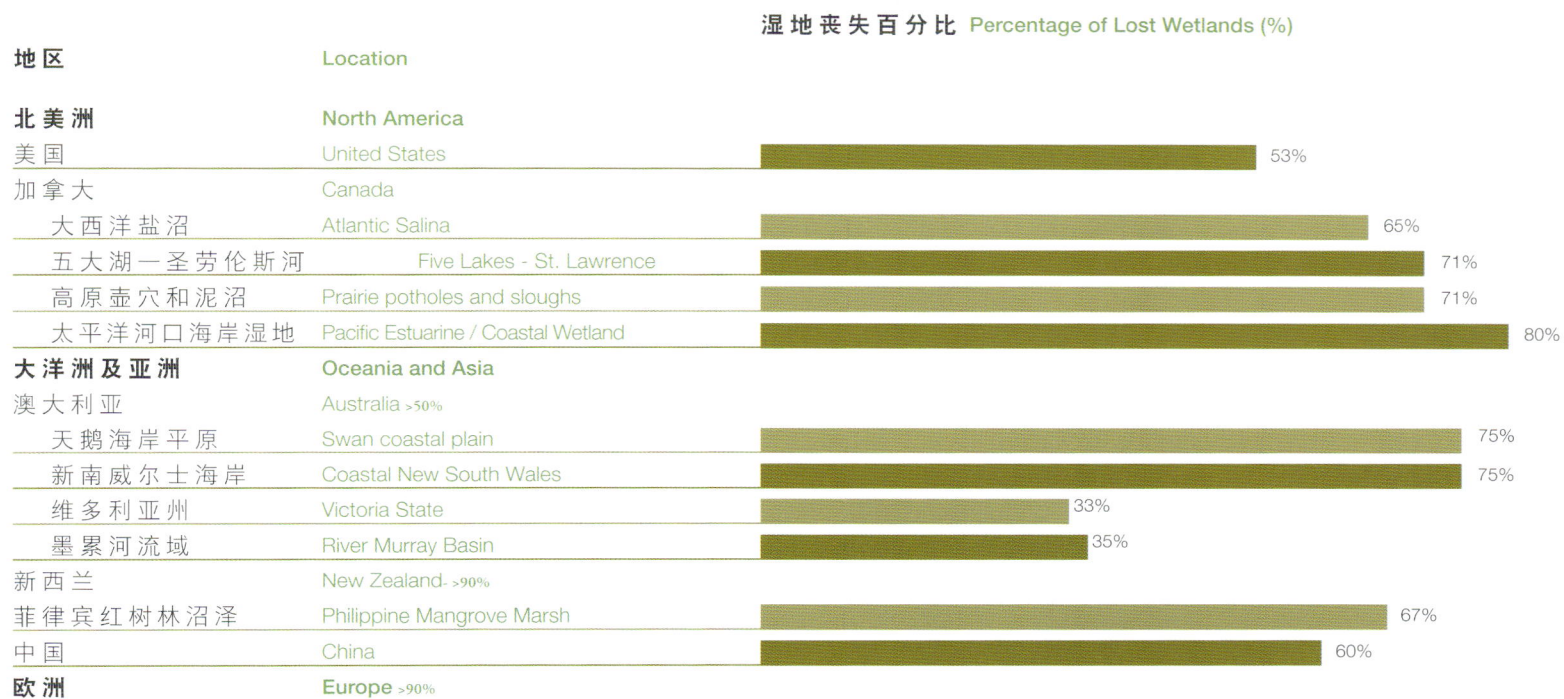

湿地丧失百分比 Percentage of Lost Wetlands (%)

地区	Location	
北美洲	**North America**	
美国	United States	53%
加拿大	Canada	
大西洋盐沼	Atlantic Salina	65%
五大湖—圣劳伦斯河	Five Lakes - St. Lawrence	71%
高原壶穴和泥沼	Prairie potholes and sloughs	71%
太平洋河口海岸湿地	Pacific Estuarine / Coastal Wetland	80%
大洋洲及亚洲	**Oceania and Asia**	
澳大利亚	Australia >50%	
天鹅海岸平原	Swan coastal plain	75%
新南威尔士海岸	Coastal New South Wales	75%
维多利亚州	Victoria State	33%
墨累河流域	River Murray Basin	35%
新西兰	New Zealand >90%	
菲律宾红树林沼泽	Philippine Mangrove Marsh	67%
中国	China	60%
欧洲	**Europe** >90%	

树林生态环境的丧失直接导致了台风灾害的增加，而在原先红树林地带建起的海边鱼塘、虾塘已经成为主要污染源之一。

虽然湿地在生态、社会和经济的可持续发展方面具有不可或缺的作用，但是，中国湿地恢复的专业技术、研究和实践仍处于起步阶段。尽管如此，在过去这20多年间，中国还是实施了多个不同生态环境类型的湿地恢复项目，包括沼泽、湖泊、河流和滨海湿地，这些项目也取得了一定的成效，使中国的湿地恢复逐渐走上正轨。但是有些湿地由于缺乏管理，当前恢复的湿地会再次退回到原来的生态水准，甚至退化到意想不到的更糟糕状态，因此在湿地恢复领域中国仍有待进一步的发展。下述案例研究概述了中国当前的湿地恢复项目及所应用的技术。

1.5.1 沼泽湿地

中国沼泽湿地的退化和丧失主要是由于泥炭开发和农用地开垦。东北黑龙江、松花江和乌苏里江构成的三江平原是中国最大的平原沼泽集中区，共有113万hm^{2}淡水沼泽（湿地国际，2002）。但是，解放以后，三江平原进行了大规模的土地开发，大面积的沼泽被挖沟排水变成耕地，粮食产量大幅度攀升，却丧失了大片沼泽湿地，继而失去了湿地的生态功能和生物多样性。

1998年，黑龙江省遭受到特大洪水的袭击后，率先发布中国第一个停止开垦湿地的政府禁令。同时，为降低人为干扰、争取沼泽的自然恢复，黑龙江省还制定了建立湿地自然保护区的紧急计划。

在三江平原的其他地区也开始广泛展开沼泽湿地的恢复工作，黑龙江安邦河下游的安邦河湿地恢复工程即是其中一例。该湿地最初面积达2万hm^2，但是由于20世纪70年代的开垦耕地政策，到90年代中期，该湿地已经减少到200hm^2。湿地资源的损失导致该地区生态失衡、降水量减少、洪涝干旱灾害频发，并出现土壤局部沙化和盐渍化以及动植物多样性下降。为解决这些问题，当地政府从1991年开始引水灌溉湿地，恢复湿地植被；1995年，开始尝试退耕

Similar extravagant losses have been recorded for mangrove habitats due to their conversion into aquaculture ponds, deforestation and shoreline development. China's mangroves have dwindled from 4,200,100 ha in the 1950s to just 1,487,700 ha in the 2004, meaning that 72% of all natural mangroves have been lost (Wetlands International, 2002). Mangroves play a crucial role in mitigating storm hazards and acting as habitats for near-shore fish. Over the years, lost mangrove habitats have resulted in the direct increase in the hazards associated with typhoons. Coastal fish and shrimp ponds constructed in areas once occupied by mangroves have also become a major source of pollution.

Despite the crucial role wetlands play in sustaining ecological, social and economic development, China unfortunately lags behind much of the world in technical expertise and in the research and implementation of wetland restoration. Nevertheless, a series of restoration programs have been conducted over the last 20 years for various kinds of wetland habitats including swamps, lakes, rivers and coastal areas. While the programs have put China on a corrective path and yielded positive results, there is still a ways to go. Due largely to ineffective administration, wetlands restored under existing programs are prone to degradation or a regression into previous or unexpected ecological states. The case studies presented in the following sections provide an overview of current projects and the technologies being applied to wetland restoration in China.

1.5.1 Restoration of Marshes

The loss and degradation of China's marshes are largely the result of peat mining and the conversion of wetlands to farmland. The flood plains of three rivers (the Heilong, Songhua, and Wusuli) in northeast China are home to the largest concentration of marsh wetlands in China, totalling 1.13 million ha of freshwater marsh (Wetlands International, 2002). In the latter half of the 20th century, these plains underwent extensive land development to drain large areas of marsh for conversion into agricultural lands. Agriculture boosted grain yields substantially, but at the cost of destroying vast tracts of marsh wetlands, which subsequently abolished the wetlands' ecological functions and biodiversity.

After being hit by unprecedented flooding in 1998, Heilongjiang Province became the first in China to issue an official ban on the cultivation of wetland areas and an emergency plan was concurrently formulated to establish wetland nature reserves with the aim of reducing disturbances and to facilitate the natural rehabilitation of marshes.

还湿。到2002年，已恢复的湿地面积达1220hm²。

1.5.2 湖泊湿地

湖泊是中国遭受破坏最严重的湿地类型。在过去这几十年中，中国损失了130万hm²湖泊湿地(湿地国际，2002)。随着对中国几大湖区大规模的恢复和重建项目的展开，已经出现令人欣慰的转机。

与其他湿地研究相比，中国湖泊湿地恢复的历史相对较长。中国科学院水生生物研究所20世纪70年代就开始了研究工作，首次利用水域生态系统藻菌共生的氧化塘生态工程技术，使污染严重的湖北鸭儿湖地区的水陆环境得到很大改善。

位于长江中游的洞庭湖一度是中国最大的淡水湖泊。然而，约有50%的湖区被开垦为农田，而且泥沙淤积，加之人为活动的破坏，造成湿地生态系统的严重退化，调蓄洪水的功能也大大衰退。1998年长江特大洪水发生后，洞庭湖地区实施退垸还湖。尽管湖区刚刚部分恢复原来的蓄水量，但成效显著，湿地面积减小的趋势也得到遏制。

乌梁素海地处内蒙古自治区巴颜淖尔市，是黄河流域最大的湖泊、中国十大淡水湖之一。自上世纪80年代末到90年代中期，有机物的污染致使湖泊水体富营养化，湖泊变黑变臭，水质一度恶化到劣五类。当地政府采取全面恢复措施治理湖水，疏浚湿地，收割芦苇，控制芦苇的过度生长。自2003年开始，有关部门开始从黄河大量引水向乌梁素海实施调水补水工程，帮助缓解长期以来的污染问题。2003年全年共自黄河引水1,350万m³，2004年补水达8,900万m³，2005年春又补水2,700万m³。

山东省南四湖的大型人工湿地也是近年来一个重要的湖泊湿地恢复项目。该项目2005年开始施工，规划面积200hm²，其中1/3用于人工种植芦竹、芦苇和莲藕，余下的2/3用于恢复当地水生植物，如湖苇、香蒲和藻类等。

More active marsh restoration projects have been undertaken in other parts of the three plains area of these rivers. Wetlands downstream from the Anbanghe River in Heilongjiang province had decreased from 20,000 ha to a mere 200 ha by the mid 1990s, due largely to conversion to farmland in the 1970's. The losses resulted in ecological imbalances, precipitation loss, flood and drought hazards, partial desertification, salinization, and a decline in biodiversity. To address these issues, in 1991 the county government introduced water to the wetland sites to bolster the growth of wetland vegetation. In 1995, it began tests on converting farmland into wetlands. A total of 1,220 ha of farmland had been converted by 2002.

1.5.2 Lake Restoration

Of all wetland types in China, lakes have suffered the most damage. In the past few decades, the country has lost 1.3 million ha of lake area (Wetlands International, 2002). However, encouraging signs can be found in the large-scale restoration and rehabilitation projects underway for most major lakes in China. In contrast to other channels of wetland research in the country, lake restoration has a relatively long history in China, with landmark research conducted by the Aquatic Biology Research Institute of the Chinese Academy of Sciences in the 1970s. The Academy pioneered the use of a symbiotic algae-fungus system to improve water and land conditions of the heavily polluted Ya'er Lake in Hubei Province. Three similar, large-scale restoration efforts are described below.

Dongting Lake, located in the middle of the Yangtze River, was once China's largest lake. However, approximately 50% of the lake area has been reclaimed for agriculture, while the lake itself has been degraded due to sedimentation and human activities. As a consequence, the lake's flood retention capacity was severely constricted. After massive flooding in 1998, measures were taken to convert farmland back to wetlands. As a result, Dongting Lake has made a noticeable yet partial recovery towards its original water capacity, and has successfully arrested further erosion of the area's wetlands.

Wuliangsuhai Lake, the largest along the Yellow River and one of China's largest, is located in Bayanzhuoer, northern China in the Inner Mongolia Autonomous Region. From the late 1980s to the mid-1990s, the lake experienced eutrophication as a consequence of organic pollution. Lake waters turned black, and water quality was once measured at the worst grade V-levels. The local government conducted comprehensive restoration projects to improve lake conditions, mainly by controlling excessive reed growth through dredging, harvesting machinery and manual cutting.

1.5.3 河流湿地

河流水利工程包括筑坝、筑堤、裁弯取直、河道整治、开挖管道和引水调水等。中国的水利工程建设历史悠久，可追溯到数千年以前，而且很多都取得了成功。但是近五年来，越来越多、越来越大的水利工程对湿地功能造成了负面影响，成为河流湿地的一大威胁。同时，河流生境面临着另一个威胁（见图1-10）：河流急剧地遭到污染。近期的统计资料非常令人担忧：中国江河中有70%被污染，流经城市的河流中则有90%受到污染。

恢复河流及河滨生境是生态恢复中一个新兴领域。与其他类型的湿地相比，河流湿地的恢复工作更具难度，因为许多河流都跨越了不同行政区、省份以至国家，因此，恢复工作需要各政府部门之间进行有效协调。由于河流恢复是一门新学科，因此在评估河流湿地恢复项目时，可以说中国还没有普遍认可的标准。这就使很多河流生态修复专案的投资和执行缺乏必要的知识和评估工具，难以监测、评估恢复项目的成功与否。结果，大型河流生态修复工程仅停留在改善水质和景观建设项目上，没有建立起一个更全面的环境恢复框架。

尽管如此，中国已经在评估生态工程的规划设计原则、河流生态恢复效果、模型开发以及河流廊道栖息地的建设等方面开展了小范围的研究工作，但是这些工作一般仅用于防洪、排水和景观建设等河流整治工程。一种广泛运用于河流湿地恢复的方法是利用水生植物改善水质，将河道中的混凝土护岸换为更加环保的材料，从而提高生态价值。水生植物的生长环境可通过河流结构来创造，如稳定塘、生物浮床和橡胶坝等（水利部国际合作与科技司，2005）。正如本手册第三部分所述，水生植物具有超强的吸收和去除营养物和其他污染物的能力，因而对改善水质有着很大的影响力。同样，一些研究开发出来的建筑材料也可用于河流整治，如生态混凝土、鱼巢砖、净水石笼以及各种土工网复合技术产品（水利部国际合作与科技司，2005）。其中，有些产品已经广

Water from the Yellow River and excess irrigation water have additionally been introduced into the lake since 2003, helping alleviate chronic pollution problems. A total of 13,500,000 m3 of water was introduced from the Yellow River in 2003. Water supplies to the lake reached 89,000,000m3 in 2004, and another 27,000,000m3 in the Spring of 2005.

A final example is the creation of a large-scale, constructed wetland in Nansi Lake, in Shandong Province. Launched in 2005, the project has a planned area of 200 ha, with a third designated for artificial growth of Giant reeds, Common reeds and Lotus plants, with the remaining two-thirds designated for the rehabilitation of local aquatic plants such as reeds, cattail and algae.

1.5.3 River Restoration

The hydraulic engineering of rivers (which includes the construction of dams and dykes, the realigning of waterways, creation of canals and water transfer schemes) has a long and often successful history in China, dating back thousands of years. Massive hydraulic projects that adversely affect the functioning of wetlands have increased in number and scale over the last half-decade. Another threat to river habitats is the alarming increase in pollution rates. Recent statistics show that worrisome pollution levels have been found in 70% of all rivers in China, and in 90% of urban rivers.

The restoration of compromised rivers and their associated riparian habitats is a relatively new field in ecological restoration. River restoration can be more problematic than the other wetland types, as river habitats usually cover multiple administrative, provincial or even national borders. Restoration therefore requires effective collaboration between various government departments. As a new discipline, especially in China, there are almost no universally agreed upon standards to evaluate the restoration projects of river wetlands. The result has been scant knowledge and availability of evaluation tools for the implementation and investment in restoration projects, with related ambiguities in monitoring and assessing a restoration's success.

Large-scale river restoration in China has consequently been limited to water quality improvement and landscape enhancement projects. There has been little momentum to approach restoration from a more comprehensive environmental framework. Nevertheless, there has been small-scale research into planning and design principles for ecological enhancement projects, evaluation methods and models of riverine restoration and the enhancement of riverine corridor habitat.

▲ 图 1-10：我国东部许多河流已经失去了大部分的河岸湿地，此图为长江三角洲一古镇，2003（摄影：何文珊）
Figure 1-10: Some rivers in eastern China have lost nearly all their riparian wetlands (Photograph by Wenshan He)

泛应用于水道治理工程，如北京北护城河综合整治工程（北京市水务局，2005）以及积水潭附近的河底和河岸改造等（人民日报，2005年12月29日）。

可以看出，河流湿地恢复的形式正在多样化，但水利工程、栖息地保护和景观建设三个方面还未能够更加有机地结合起来（水利部国际合作与科技司，2005）。一些破坏生态和环境的做法仍在普遍运用，如河流管道化、用混凝土衬砌等，这些低维护成本的防洪工程会破坏河流原有的湿地生态系统，阻断河流与地下水之间的相互补给。

This research has had a limited application in river improvement projects for flood prevention, dredging and landscape enhancement projects. One of the more widespread approaches to river restoration in China includes the use of aquatic plants to improve water quality, and the replacement of concrete-lining in river channels with more environmentally-friendly materials to improve ecological value. The conditions for aquatic plant growth can be created through in-stream structures such as stabilizing ponds, floating bio-beds, and rubber dams. (Department of International Cooperation & Science, the Ministry of Water Resources, 2005).

Aquatic plants have an extraordinary capacity to absorb and remove nutrients and other pollutants, and therefore have a significant impact on improving water quality. Likewise, engineering materials developed for river improvement include 'grasscrete', fish-lair bricks, water-purifying stone cages, and various geotextile products. (Department of International Cooperation & Science, Ministry of Water Resources, 2005). Some of these products have already been widely used in waterway improvement projects. For example, the comprehensive, environmental project on the northern fringe of Beijing city moat , the cement and stone paving of the riverbed and shores in the section near Jishuitan Lake was removed to make way for more environmentally-friendly engineering materials (*People's Daily*, December 29, 2005).

西班牙巴塞罗那市
业主：Hines/Diagonal Mar
Parc Diagonal Mar , Barcelona, Spain (EDAW)
Client: Hines/Diagonal Mar

获得2004年美国景观建筑师协会国家荣誉奖项
2004 National Honor Award, General Design Category, American Society
of Landscape Architects

EDAW | .COM

1.5.4 滨海湿地

中国的海岸线长达18000多公里，是中国人口最密集、城市化最集中的地区，分布着中国70%以上的大中城市，居住的人口约1.25亿。湿地资源为沿海地区的经济繁荣与发展做出了极为重要的贡献，但是其价值却未得到应有的认可。多年来，人们盲目围垦和改造滨海湿地，加剧了自然灾害造成的损害。例如，风暴潮每年都使沿海地区损失近百亿元人民币。因此，湿地恢复已经成为中国重要沿海地区的当务之急。

中国滨海湿地的恢复主要集中在红树林生态系统的恢复上（张乔民，2001），并且做得比较成功。科学家已经能够将红树林生态系统中各种动植物对环境的要求与种植区域特征一一对应；广东省2005年的红树林恢复面积达8000hm²，使该省红树林总面积增长至1.8万hm²。此外，张江、茂名、阳江、江门、珠海和汕头也成功地进行了红树林恢复工程（见图1-11）。 但是，这些工程中也遇到了一些问题，例如在项目区种植的海桑（Sonneratia spp）等外来红树林现在不断和本地红树林争夺空间。

在中国北方滨海湿地，2005年山东省东营开始黄河三角洲大汶流湿地恢复工程，湿地面积6000hm²，项目由大汶河管理站负责（水利部网站，2006年2月10日）。

1.5.5 人工湿地

中国是一个人工湿地类型和数量都很丰富的国家。主要的人工湿地类型有水稻田、鱼塘、水库等。随着国家对退耕还林、退田还湖、退塘还海的日益重视，与之相关的湿地恢复已经成为中国湿地恢复的主要内容，并且已经取得了一定的成效。人工湿地恢复的主要目标是改善水质、提高生物多样性，同时提供其他环境功能（见图1-12）。有关项目详见本书的案例分析部分。

从上述分析可以看出，中国的湿地恢复还刚刚起步，湿地保护与恢复工作与湿地退化的历史和现状极其不协调，理论和实践都还不够重视。在定位、规划和实施程式、管理维护湿地的有效方法上

Despite the emergence of more diverse applications, a tighter integration between irrigation projects, habitat conservation and landscape enhancement has not emerged (Department of International Cooperation & Science, Ministry of Water Resources, 2005). Ecologically destructive practices, such as the canalization of rivers and lining them with concrete are still the commonly practiced, low-maintenance approaches to flood control. This devastates a river's ecosystem, severing the connection between river and ground water resources.

1.5.4 Restoration of Coastal Wetlands

China's coastline spans over 18,000 km in length, and is the most heavily populated and intensely urbanized area in China. Over 70% of the country's large- and medium-sized cities and a population of 125 million lie along China's coastline. Wetland resources have played a paramount but often unheralded role in the economic boom and development of this coastal zone. In recent years, inappropriate cultivation and the reclamation of coastal wetlands have exacerbated the damage brought by natural disasters in coastal zones. In one instance, storm surges are now responsible for nearly RMB 10 billion worth of damage each year. Wetland restoration has become an urgent priority in China's important coastal areas.

The restoration of China's coastal wetlands has focused largely on degraded mangrove habitats (Zhang Qiaomin, 2001). These efforts have been relatively successful. Scientists have been able to match the environmental requirements of different mangrove species with the characteristics of the planting area. In Guangdong province, 8,000 ha of mangroves were restored in 2005, increasing the total mangrove area in the province to 18,000 ha. Other successful, on-going mangrove restoration projects can be found in Zhangjiang, Maoming, Yangjiang, Jiangmen, Zhuhai, and Shantou. Some problems have been encountered, however, where exotic mangrove species (e.g., *Sonneratia spp.*) have been planted in restoration programs, and are now out-competing native mangrove species.

Restoration of coastal wetlands in northern China is rather less extensive. An exception is the large-scale restoration of the Yellow River Delta wetlands in Dongying, Shandong. The province commenced restoration of over 6,000 ha of wetlands in the delta basin in 2005, under the supervision of Dawenliu River Administration Outpost (website, Ministry of Water Resources, February 10, 2006).

Confronted with the stark facts of a historical and on-going degradation, the protection and restoration of wetlands in China has been a largely insufficient narrative, lacking the appropriate attention both to theory and practice. An effective

▼ 图1-11:中国南部的红树林恢复（摄影:易道）
Figure 1-11: Mangrove restoration in southern China (photograph by EDAW)

▼ 图1-12:水生植物既是景观，又能净化水质（图片来源：www. photos.com）
Figure 1-12: Aquatic plants play roles in both landscape design and water treatment (Source: www.photos.com)

均需充分借鉴世界上其他国家同类项目的经验教训。此外，中国还需要制定重大政策和法律法规，以确保成功地开创可持续发展的湿地恢复事业。

wetlands restoration strategy now requires sound scientific principles as a basis to leverage experience gained on similar projects around the World and particularly in the United States.

1.5.5 Artificial wetlands

Artificially created or modified wetlands such as rice fields, aquaculture ponds and reservoirs are commonplace across the whole of China. Increasingly, these degraded wetland types are being targeted for restoration and enhancement to improve water quality, increase biodiversity, and provide other environmental services. Some successful examples of the restoration of artificial wetlands are discussed in Chapter 4 of the book, which describes a number of restoration case studies.

Wetland restoration is still a nascent field in China, and there is much to learn from the scientific community on positioning, planning, and implementation procedures, as well as effective approaches to the administration and maintenance of the wetland sites. Moreover, China requires the important policy, legal and regulatory instruments and corollaries that will aid it in the successful creation of a sustained wetland restoration enterprise in China.

Paya Indah 动物园和野生动物园
马来西亚吉隆坡
业主：马来西亚政府
Paya Indah Zoo and Wildlife Park
Kuala Limpur, Malaysia
Client: Government of Malaysia

获得2002年景观建筑协会昆士兰州景观规划优异奖荣誉奖项
2002 AILA Qld State Awards Merit Award for Planning in Landscape Architecture

Jungle Walks

Maintenance /
Site Operations Depot

Vehicle Access
from Toll Plaza

Vehicle
Maintena

Lift & Ram

Site Transport
Terminal

Car Park

Asian Tropical Theme

Time Travel
Themed Ride

Zoo Entry

Interpretation /
Technology Centre

Cafe

Retail

Australasian Tropical Theme

Canal Rides

Coach Park

Boat Harbour

Restaurants

Butterfly House

Overflow
Car Park

Maintenance
Route

Waste Water Treatment

Auditorium

Boat Station

Information

South American Tropical Theme

Petting Zoo

African Tropical Theme

Aviary
& Restaurant

Land Trails

Central Grasslands

Pedestrian
& Cycle Paths

Link to
Paya Indah

Free Ranging Savannah Animals

Forest

Vehicle Access
from Toll Plaza

Maintenance /
Site Operations Depot

Train Halt

The Dredge Attraction

Boat Route

Lift & Ram

Transport
Terminal

Vehicle
Maintena

Australasian Savannah

Car Park

Train Halt

Eurasian Savannah

Retail

Zoo Entry

Auditorium

Petting Zoo

Health
Club

American Savannah

Boat Route

Waste Water
Treatment

Cafe

Staff
Housing
Village

Zoo Resort

Rope Bridge

Kampong Air

Existing Surau

Zoo Entry

Restaurant & Jetty

Boat Harbour

Zulu Village

Arts & Crafts Street

Future Aquarium

Existing Water Tower

Entry to
Safari Plains

Train Halt

Car Park

Children's Adventure Play

Coach Park

Picnic Grounds

Plains Safari

Canoe Trails

Negara Hill Lookout

Land Train Link to
Paya Indah

Maintenance
& Store

Negara Lake

Horse
Riding Trails

Namibian Desert Village

EDAW | AECOM

Waste Water Treatment

Boat Route

第 二 章
Chapter 2

湿地保护战略与行动计划
Wetland Conservation Strategies and Regulations

2.1 《湿地公约》中的湿地恢复内容

《湿地公约》是为了保护湿地而签署的全球性政府间的保护公约，全称为《关于特别是作为水禽栖息地的国际重要湿地公约》，其宗旨是通过各个国家的行动和国际合作来保护与合理利用湿地（见图2-1）。自从《湿地公约》于1971年签署后，其一直致力于对现有湿地的保护，并向各缔约国推广湿地保护与合理利用的理念。

在签署初期，受当时人们对湿地生态系统服务价值认识上的局限，《湿地公约》只是一个以保护水禽栖息地为主要内容的国际公约，并未在湿地恢复领域开展工作。在全球湿地持续退化的大背景下，湿地恢复并非当时的《湿地公约》的重点，《湿地公约》也没能承担起退化湿地恢复和重建的职责。

《湿地公约》缔约国大会（以下简称COP）是传播湿地保护知识、制订《湿地公约》工作计划和战略方针、制订缔约国开展湿地保护工作的原则和指导方针的重要会议，每三年召开一次。随着加入《湿地公约》的缔约国数量不断增加，包括相关的非政府组织和机构（NGOs）参与，每届缔约国大会讨论后的建议书和决议都可以反映出世界各国湿地保护工作的水准和工作重心。《湿地公约》便据此制订工作计划和指导原则，并在每次缔约国大会的讨论中不断完善。

从COP1（1980）到COP9（2005）的会议建议书和决议可以看出；湿地恢复越来越受到缔约国的重视，湿地恢复的内容逐次增加。在COP4（1990）和COP6（1996）上，分别出现了以"湿地恢复"为主题的议题，更将讨论结果写入会议的建议书中。在COP7（1999）上，《将湿地恢复纳入到湿地保护与合理利用的国家计划中》被列入决议（决议VII.17）。COP8（2002）的决议则纳入了《湿地恢复的原则和指导方针》（决议VIII.16）。后来，缔约国采取一系列具体的行动来加强对湿地恢复，包括组织编写与湿地恢复相关的研究资料，在公约局的培训项目中增加了"湿地恢复与重建"的培训模块，制作湿地恢复的资料网站，即时增加湿地恢复的新知与

▼ 图2-1：湿地公约的标志
Figure 2-1: Ramsar logo

CONVENTION ON WETLANDS
CONVENTION SUR LES ZONES HUMIDES
CONVENCIÓN SOBRE LOS HUMEDALES
(Ramsar, Iran, 1971)

2.1 The Ramsar Convention on Wetlands

The Ramsar Convention on wetlands is an intergovernmental treaty for the conservation of the world's wetlands. Officially known as The Convention on Wetlands of International Importance Especially as Waterfowl Habitat, the aim of the Convention is the conservation and wise use of wetlands through national action and international cooperation (Ramsar website, www.ramsar.org, 2006). Since the treaty was adopted in 1971, the Ramsar Convention has been committed to this goal through all Contracting Parties.

In the early years of the Convention, Contracting Parties focused mainly on the preservation and effective management of existing wetlands considered valuable for waterfowl. As the world's wetlands continued to erode at an alarming rate, the Convention paid increasing attention to wetland restoration and during the past quarter century, the Convention has markedly increased its role in the stewardship of wetlands, adapting and communicating its perceptions and findings.

The Conference of Parties (COP) of the Ramsar Convention is an important meeting held once every three years to disseminate knowledge on wetlands conservation, formulate Convention action plans and strategies, and provide principles and guidelines on wetland conservation for Contracting Parties. With an increasing number of Contracting Parties (152 as of July 2006) and the extensive participation of relevant NGOs, the recommendations and resolutions presented at each subsequent COP reflects the heightened priority of and emphasis on global wetland restoration efforts. Similarly, The Ramsar Convention, in conjunction with its work plans and guidelines, has seen a steady improvement not only in Conference discussions, but in the success of its conservation efforts.

The recommendations and resolutions from COP-1 (1980) to COP-9 (2005) clearly show a steady improvement, both in the agenda of the Contracting Parties and the greater dissemination of Ramsar documentation. During COP-4 (1990) and COP-6 (1996), the topic of "wetland restoration" appeared on the discussion list and the results of the discussion were recorded in the ensuing recommendations. COP-7 (1999) passed Resolution VII.17 declaring that "Restoration" is "an element of national planning for wetland conservation and wise use". COP-8 (2002) passed Resolution VIII.16, or "Principles and guidelines for wetland restoration".

A series of actions were taken in COP-8 towards improving wetland restoration efforts, which included compiling new research and restoration methodologies, and adding a new module, "Wetland Restoration and Rehabilitation" to the

实践资讯；同时，加强对湿地恢复与重建的宣传的力度。在2003年的《湿地公约》科技会员会（The Convention's Scientific and Technical Review Panel，以下简称 STRP）第11次会议上，将湿地恢复与重建列入了STRP的二级优先项目。因此，《湿地公约》对湿地恢复和重建的发展过程可分为萌芽阶段、发展阶段、完善与提高阶段。

2.1.1 萌芽阶段 (COP1—COP3)

从《湿地公约》的签署到COP1，签署的文档以及会议制订的建议书中没有提到任何关于湿地恢复的内容。根据COP2提交的《湿地公约执行框架》建议书附件，湿地管理部分仅由个别缔约国提到："在可能的情况下重建退化湿地"（建议书2.3）。在COP3的建议书《缔约国管理机构的责任》中，提出"敦促缔约国相关管理机构重建完全退化的湿地"（建议书3.4），在建议书《国际重要湿地生态特征变化》中（建议书3.9），提出"敦促所有缔约国切实有效地阻止湿地退化，并在可能的情况下恢复退化湿地的价值"。

2.1.2 发展阶段 (COP4—COP6)

在这个阶段，湿地恢复越来越受到各缔约国的重视，在COP4的建议书《湿地恢复》中提到有些缔约国"湿地恢复方案已经在实施中"、"深信维持和保护现存的湿地要比退化后再修复更可取更经济"、"着重强调退化湿地在被破坏前就该进行恢复"、"保证退化湿地的恢复方案不能削弱对现有湿地的保护"，并建议缔约国"调查开展湿地恢复工程的可能性"、"管理机构应将湿地恢复作为任务"、"寻找机会实施湿地恢复"（建议书4.1）。在建议书《合理利用概念执行的指导方针》的附件中提到"要特别关注对效益和价值下降湿地的恢复"，但并没有把湿地恢复列入优先行动中（建议书4.10）。

在COP5决议（钏路，1993）《钏路声明和公约执行框架》的附件1中提到在国际重要湿地的保护和管理中要"恢复退化的湿地并补偿丧失的湿地"（决议V.1），附件2中建议缔约国在湿地保护方面加

Ramsar Bureau training program. Other outcomes of COP-8 were the creation of the Convention's website to disseminate regular knowledge updates on restoration theory and practice, and an intensified public education effort on wetland restoration and rehabilitation. At the 11th Meeting of the Scientific and Technical Review Panel (STRP) of the Ramsar Convention, wetland restoration and rehabilitation were placed into the second-tier top-priority level. In terms of understanding wetland restoration policy, the evolution of the Ramsar Convention can be divided chronologically into an incipient stage, a development stage and its present fruition stage.

2.1.1 The Incipient Stage (COP-1 – COP-3)

From the initial adoption of the Ramsar Convention in 1971 to COP-1 (Cagliari, 1980), the topic of wetland restoration appeared in neither the official documentation nor conference recommendations. At COP-2 (Groningen, 1984), a handful of Contracting Parties mentioned "rehabilitat[ing] the degenerated wetlands if possible" in the Wetland Management Section of the Annex of Recommendation 2.3, as "action points for priority attention". In COP-3 (Regina, 1987), Recommendation 3.4 notes the "Responsibility of development agencies towards wetlands" and urges the relevant management agencies of the Contracting Parties to rehabilitate wetlands that have suffered total degradation. Recommendation 3.9 calls for a "change in [the] ecological character of Ramsar sites", further urging all Contracting Parties to take measures that prevent the degeneration of wetlands and, where possible, for restoring the value of the already degraded wetlands.

2.1.2 The Development Stage (COP-4 – COP-6)

During this stage, Contracting Parties began to see the importance of wetland restoration. At COP-4 (Montreux, 1990), Recommendation 4.1 on Wetland Restoration states that "opportunities for wetland restoration be sought and put into operation", and that "all Parties examine the possibility of establishing appropriate wetland restoration projects". It additionally provides a value assertion by stating that "the maintenance and conservation of existing wetlands is always preferable and more economical than their subsequent restoration, while "restoration schemes must not weaken efforts to conserve existing natural systems". This important recommendation also urged Contracting Parties to "study the feasibility of restoration schemes" and to "establish management agencies tasked at wetland restoration", and to "seek opportunities to implement the restoration schemes". The Annex of Recommendation 4.10, "Guidelines for the implementation of the wise use concept", suggested paying special attention to the restoration of wetlands with failing efficiency and value. Despite a greater emphasis on process,

强国际合作，并要"开展湿地恢复工程"。在 COP5(1993) 的决议《湿地的合理利用》的附件中把湿地恢复技术研究作为《湿地及其价值知识》的内容之一，并在湿地管理培训的课程中考虑加入湿地恢复的内容。在此次会议期间，个别缔约国和国际组织谈到了一些已经开展的湿地恢复工作（建议书 5.1），并回顾了 COP3(建议书 3.9) 中的敦促缔约国开展湿地恢复部分（建议书 5.2）。

COP6(布里斯本，1996) 也把"湿地恢复"主题作为大会讨论的议题之一。从会议的决议来看，各缔约国已经认识到湿地恢复和重建，能够产生正面的人为干扰来改变湿地生态特征（决议 6.1 附件）。此次会议的建议书《保护泥炭地》中提出"鼓励缔约国对退化的泥炭地进行恢复"（建议书 6.1）。接着，在建议书《湿地恢复》中认识到"世界各国普遍面临着湿地退化和丧失的威胁，特别是过去 50 年，有些国家已经有 70% 的湿地丧失"、"湿地退化现象在发达国家十分显著"、"欧洲已经开始着手保持、恢复和提高对生物多样性至关重要的湿地"、"湿地的自然恢复对于解决水管理问题、改善地下水和地表水的质量以及减少洪灾的损失是非常重要的方法"，更呼吁"将湿地恢复纳入到国家自然保护和水土管理政策中去"、"敦促 STRP 与公约局、有关缔约国及伙伴组织合作详细制订湿地恢复和湿地监测的指南"、"基于缔约国的湿地资讯表，制订一个需要进行恢复的重要湿地名录"、"进一步敦促缔约国给予湿地恢复更高的优先，并采取恢复措施"、"允许缔约国将湿地恢复作为将来 COP7 国家报告的一个部分"（建议书 6.15）。在建议书《有关缔约国的新增国际重要湿地》（建议书 6.17）中，个别缔约国陈述了其正在进行中的改善湿地生态特征的恢复方法。

2.1.3 完善与提高阶段 (COP7— COP9)

(1)COP7(圣荷西，1999)
COP7 的召开确定了"湿地恢复"在《湿地公约》中的地位，并要求将其列入到各缔约国的国家计划中。在 COP7 的决议《国家湿地政策发展和执行的指导方针》（决议 VII.6）中，鼓励缔约国"要认识到将湿地恢复写入国家湿地政策中

the Recommendation does not designate any priority level for wetland restoration.

At COP-5 (Kushiro, 1993), Annex 1 of Resolution 5.1, "The Kushiro Statement and the framework for the implementation of the Convention", states that the conservation and management of wetlands of international importance requires "the restoration of degenerated wetlands and the compensation for lost wetlands". Annex 2 of the same resolution urges Contracting Parties to strengthen their international cooperation and to "establish restoration projects". The Annex of Resolution 5.6, "Wise use of wetlands", lists technical research of wetland restoration as a critical part of "Knowledge of Wetlands and Their Values" and deliberates on the merits of adding wetland restoration into wetland management training courses. During COP-5, a few Contracting Parties and international organizations discussed works in progress on wetland restoration (Recommendation 5.1, 1993), and reviewed the commitments of Recommendation 3.9, urging fellow parties to start working on wetland restoration (Recommendation 5.2, 1993).

COP-6 (Brisbane, 1996) was something of a landmark conference. Wetland restoration became a separate topic of discussion and through the forum's resolutions Contracting Parties asserted that wetland restoration and rehabilitation was greatly aided by positive human interference to improving a wetland's ecological character (Annex of Resolution VI.1). Adding to the wider definition of wetlands, Recommendation 6.1, the "Conservation of Peatlands", encourages Contracting Parties to restore degenerated peatlands. Recommendation 6.15, the "Restoration of Wetlands", provided the starkest language up to that date on the urgency of rehabilitation efforts, stating that "the world faces a threat of wetland degeneration and loss, and some countries have lost 70% of their wetlands over the past 50 years", and that "wetland degeneration is a remarkable problem in developed countries". It further stated that "measures have been taken in Europe to conserve, restore and improve the wetlands that are of vital importance to biodiversity", while "the natural restoration of wetlands is an extremely useful method in resolving water management problems, improving underground and surface water quality and reducing flood damage".The Recommendation also suggested that "wetland restoration be included in the national policies of natural conservation and the management of water and soil", specifically stating that "the SRTP of the Ramsar Bureau, the Contracting Parties and the Convention's partner organizations cooperate to formulate detailed guidelines on wetland restoration and monitoring". Recommendation 6.1 also calls for creating a list of important wetlands that require immediate restoration, based on current information available from the Contracting Parties, and urged

的好处，以确保湿地恢复在管理计划与政府支出中给予足够的优先考虑，并且能够促进当地退化湿地的重建"。

同时在决议VII.6的附件中提出"必须尝试对湿地进行重建、恢复与再造"，"在起草国家湿地政策时要考虑防止湿地进一步丧失，并鼓励恢复湿地以保持其完整性"，加强学习与"湿地恢复、湿地重建以及减轻或补偿湿地的丧失"相关的知识。

在决议《建立并巩固社区和本地居民参与湿地管理的指导方针》（决议VII.8）的附件中，将"湿地的恢复与重建取代湿地的退化"作为湿地合理利用管理的标准之一。在《湿地公约的发展计划》（1999—2002）中，将"湿地的恢复与重建作为环境管理工具"纳入到国家或者地方的环境政策中去。

在COP7中，由于"退化湿地的恢复"已经在世界范围内广受重视，并被许多国家或地区的公约采纳，一些与会学者详细阐述了湿地恢复产生的背景，以及将湿地恢复纳入到湿地保护与合理利用国家计划中的必要性，并给出了具体的方法和指导方针，这些观点最后被纳入到COP7的决议《将湿地恢复纳入到湿地保护与合理利用的国家计划中》（决议VII.17）。

COP7的决议VII.24提出"对丧失的湿地栖息地和其他功能的补偿"，决议VII.27通过了《公约局工作计划》（2000—2002）（决议VII.27）《公约局工作计划》的"执行目标2.6"，为鉴定需要恢复和重建的湿地，并执行必要的恢复方法，旨在提供退化湿地的恢复方法。目标是增加合适的理论研究和恢复实例。在全球范围内协助STRP在湿地恢复方面的工作，建立一个基于网路资源的恢复技术与实例研究手册（《公约局工作计划》，2001）具体行动计划为：

行动一，利用国家和地方的湿地调查或者监测项目来确定需要恢复和重建的湿地。COP7的技术会议II的评估表明，仅有为数不多的缔约国开展了需要进行恢复和重建的湿地的调查工作，同时制定了到COP8时要至少有50个缔约国完成此项工作。

the parties to "accord higher priorities for wetland restoration", by taking active rehabilitation measures and "allowing Contracting Parties to include wetland restoration as part of their COP-7 national report". Recommendation 6.17, "Ramsar Sites in the territories of specific Contracting Parties", explains restoration methods for improving wetlands' ecological characteristics, using examples from restoration projects underway at that time.

2.1.3 The Fruition Stage (COP-7 – COP-9)

(1) COP-7 (San José, 1999)

COP-7 confirmed the increased status of wetland restoration in the Ramsar Convention by integrating wetland restoration into the national reports of Contracting Parties. Resolution VII.6, "Guidelines for developing and implementing National Wetland Policies", encourages Contracting Parties to "recognize the benefits of integrating wetland restoration in national wetland policies, in order to ensure the priority of wetland restoration in administration plan and government expenditure and facilitate the rehabilitation of local wetlands".

Similarly, the Annex to Resolution VII.6 states clearly that attempts "must be made to rehabilitate, restore and rebuild the wetlands", while concurrent "national wetland policies must be formulated to prevent further loss of wetlands and encourage restoration efforts to maintain the integrity of the wetlands". The resolution further states that public education should be conducted in tandem with restoration efforts.

The Annex to Resolution VII.8, "Guidelines for establishing and strengthening local communities' and indigenous people's participation in the management of wetlands", calls for the presence of wetland restoration and rehabilitation instead of using degradation as the defining standard in the wise use of wetlands. In the Development Plan of the Ramsar Convention (1999 - 2002), the restoration and rehabilitation of wetlands is integrated into national or local environmental policies as a mechanism for environmental management.

A Danish delegate noted at COP-7 that the restoration of degraded wetlands had become a major global concern that was already being enforced in a number of national or regional treaties. His report explained the historical background of wetland restoration, and the need to incorporate restoration into national planning guidelines for conservation and wise use. These viewpoints, representing something of a consensus, were later included in Resolution VII.17, "Restoration as an element of national planning for wetland conservation and wise use".

At COP-7, Resolution VII.24 proposed "compensation for the

行动二，提供丧失和退化湿地恢复和重建的方法。目标是争取在接下来的三年里给"湿地公约的合理利用资源中心"增加合适的恢复研究实例和恢复方法的资讯。

行动三，制订退化湿地恢复与重建的方案，特别是针对主要河流的交汇区或者是具有高度自然保护价值的区域内的湿地恢复与重建。从缔约国的报告可以看出一些国家已经开展了湿地恢复与重建工作，尽管有个别地方开展了重点工程项目，但目前似乎普遍进行的都是小规模尺度的试验计划。公约将持续促进湿地的恢复与重建，特别是对能够使人们身心愉悦，有利于人体健康和滨海环境生产力的湿地进行恢复与重建。目标是到COP8，所有的缔约国都完成优先恢复湿地的确定工作，而且至少要有100个以上的缔约国具体开展湿地的恢复与重建工程。

行动四，在COP7组织召开一次关于湿地恢复与重建的技术会议，并在当地、省级或者集水区尺度上确定10块最好的湿地恢复的案例。另外，有些湿地恢复的实例研究被收集到其他的项目中，例如《巩固和加强地方社区和本土居民共同参与湿地管理的指导方针》（决议VII.8）包括湿地恢复与重建的原理。

(2)COP8(瓦伦西亚，2002)
COP8出台了《湿地恢复的原理与指导方针》（决议VIII.16）并将湿地恢复纳入到《湿地公约》的工作计划以及战略计划中，号召各缔约国在减少湿地退化的同时，也要开展湿地恢复工作。在决议《湿地的综合性问题与滨海湿地区域管理的原理和指导方针》（决议VIII.4）中提出要重视滨海湿地在海陆过程中所扮演的角色，并"考虑恢复退化的滨海湿地，恢复其在海陆过程中所起到的积极作用"，由于滨海湿地极易退化，而且退化后要花非常昂贵的代价才能恢复，有些甚至不可能得到恢复，因此要"让缔约国考虑与滨海湿地退化、丧失以及修复的问题"。在决议《泥炭地全球行动的指导方针》（决议VIII.17）中表明，泥炭地的合理利用管理就包括对其恢复和重建。

《湿地公约战略计划》（2003—2008）（决议VIII.25）第一部分的内容为湿地保护与合

lost wetland habitat and other functionalities". The Convention Work Plan (2000 - 2002) was also passed in Resolution VII.27, while Implementation Target 2.6 of the Work Plan offered goals for identifying wetlands in need of restoration and rehabilitation, implementing necessary restoration methods, and providing for restoration tools for degraded wetlands. The aim was to introduce new theory and case studies to aid the STRP in wetland rehabilitation on a truly global scale, and to create a common research manual on restoration technologies with case studies, based on materials to be made available online (Convention Work Plan, 2001). The specific actions were as follows:

Action No. 1: Identifying wetlands in need of restoration and rehabilitation based on information from national and local surveys or monitoring. The evaluation of STRP during COP-7 showed that only a few Contracting Parties undertook a national inventory for wetlands in need of restoration and rehabilitation. The Panel expected the survey to have been completed in at least 50 Contracting Parties by COP-8.

Action No. 2: Providing the methods for the restoration and rehabilitation of lost and degraded wetlands. This action aimed to contribute information on case studies and methodology for wetland restoration to the Wise Use Resource Center over the following three years.

Action No. 3: To formulate plans for the restoration and rehabilitation of degraded wetlands, especially those associated with major river systems or areas with a high conservation value. The reports of Contracting Parties indicate that some nations had already started working on wetland restoration and rehabilitation on generally limited-scale test projects, with major projects underway in a few instances. The Ramsar Convention would continue to encourage the restoration and rehabilitation of wetlands, especially those beneficial for human health and enjoyment, and those that improve the productivity of coastal areas. This action aimed to ensure that priority sites identified by COP-8 for restoration have been completed for all Contracting Parties, and that wetland restoration and rehabilitation projects were underway in over 100 Contracting Parties.

Action No. 4: Organizing a technical meeting on wetland restoration and rehabilitation according to mandates for COP-7, and identifying the 10 best cases of wetland restoration at local, provincial or water basin levels. Some case studies on wetland restoration were to be used in other projects. For example, Resolution VII.8, "Guidelines for establishing and strengthening local communities' and indigenous people's participation in the management of wetlands", included the principles of wetland restoration and rehabilitation.

Wing Eld Springs 社区总体规划
美国内华达州Sparks 市
业主：Loeb企业
Wing Eld Springs Master Planned Community
Sparks, Nevada
Client: Loeb Enterprises

获得35届年度金奖
内华达州建筑协会 "最佳新规划社区"
Merit Award, 35th Annual Gold Nugget Awards
"Best New Planned Community," Builders Association
of Northern Nevada

EDAW | AECOM

理利用的进展、将来的挑战以及常规目标。其中的"常规目标"为合理利用所有的湿地。第27条提出了湿地公约会通过执行目标和行动来有计划地完成常规目标，这些目标集中在7个方面，其中就包括退化湿地及其价值和功能的恢复与重建，同时也认识到无论是从生态上、经济上还是文化上，保持现有的湿地远远要比破坏后再重建更可取、更经济。执行目标4"湿地的恢复与重建"内容为确定需要优先恢复与重建的湿地，并通过实施必要的方法来进行恢复，要保证这些湿地在恢复和重建后能获得长期的环境效益、社会效益和经济效益。其具体行动包括以下几个方面：

行动一，在可能的情况下，对于遭受破坏和退化的湿地区域，制定湿地恢复或重建的方案，特别是那些主要河流的交汇处或者具有高度自然保护价值的区域。

行动二，组织编写与湿地恢复相关的研究资料和湿地恢复的新方法，并广泛传播这些资讯资料（决议IV.13）。公约局鼓励缔约国提供新的实例研究和新的理论，支持STRP做关于这一主题的工作，并在公约局的网站上增加相关资讯。

行动三，在实施《湿地恢复的原则和指导方针》时，对于拟恢复的湿地，要充分认识到这些湿地在文化遗产方面的重要意义，确保在恢复过程中保留这些，要考虑《在国际重要湿地进行有效管理时重视湿地文化价值的指导原则》。

行动四，将湿地恢复的培训作为国家培训需求评估的一部分，增加湿地恢复专业技术方面的培训机会，创立相关的培训模组作为《湿地公约》培训项目的一部分。

行动五，向社区内对湿地恢复感兴趣的利益共同体宣传《湿地恢复的原则和指导方针》（COP8决议III.16），使当地社区和本地居民参与湿地恢复和维护工作。

行动六，利用《湿地公约》湿地恢复网站上的资讯和资源，并补充关于湿地恢复工程和经验的资讯，使其应用范围更广，特别是提供示范工程来详细说明《湿地恢复的原则和指导方针》（决议

(2) COP-8 (Valencia, 2002)

By COP-8, wetlands were a central topic at the Ramsar Convention. COP-8 adopted Resolution VIII.16, "Principles and guidelines for wetland restoration", which officially incorporated wetland restoration into the work plan and strategy of the Convention. The Convention called for Contracting Parties to continue their analysis and work on wetland restoration while simultaneously asking them to reduce the degradation of existing wetlands. Resolution VIII.4, "Wetland issues in Integrated Coastal Zone Management (ICZM)", is important in that it notes the paramount role of coastal zones in geological evolution and recommends the restoration of degraded coastal zones to regain their positive effects in geological evolution. Since coastal zones are extremely vulnerable ecosystems whose restoration could be costly undertakings, the Resolution reminded Contracting Parties to consider the unique problems of degradation, loss and the subsequent rehabilitation of coastal areas. Resolution VIII.17, "Guidelines for Global Action on Peatlands" extended wise use management and restoration principles to peatlands.

The Ramsar Strategic Plan (2003–2008) explains the progress, future challenges and standardized goals of conservation and wise use of wetlands. Article 27 specifically states that the Ramsar Convention will design implementation targets and action plans to achieve the normal goals, which fall into seven categories, including the restoration and rehabilitation of degraded wetlands as well as their value and functionalities. This article also recognizes that maintaining existing wetlands is more economical and thus preferable to rebuilding damaged ones. Part II lists 21 Implementation Targets. Specifically, Implementation Target 4, "Restoration and rehabilitation of wetlands", requires identifying priority wetlands where restoration or rehabilitation would be beneficial and yield long-term environmental, social or economic benefits, and for implementing necessary measures to recover these sites. The specific actions for Implementation Target 4 are as follows:

Action No. 1: Formulating plans for the restoration and rehabilitation of degraded wetlands, especially those associated with major river systems or areas with a high conservation value. All Contracting Parties with lost or degraded wetlands must make a priority list of wetlands slated for restoration and that wetland restoration must be completed or underway in at least 100 Contracting Parties.

Action No. 2: Compiling new research and methodology on wetland restoration and making such information widely available (Resolution 4.13). The Ramsar Bureau encourages Contracting Parties to provide new case studies and theories

VIII.16）的应用问题，根据2003—2005年的全球目标，在《湿地公约》湿地恢复网站上增加湿地恢复的理论和实例研究，由公约局维护湿地恢复的网站，并增加进一步相关的信息（见图2-2，2-3）。

行动七，进一步发展湿地恢复的方法和指导方针，包括湿地恢复的专业词汇表以及针对小型水坝和湿地恢复的指导方针（决议VIII.16）。

在"执行目标20"中，增加了包含湿地恢复与重建的培训模组。

(3)STRP第11次会议
2003年4月，STRP第11次会议在二级优先项目议题6.3中提出了湿地恢复与重建的内容，会议通过了2003—2005年度的《STRP工作计划》，将湿地恢复与重建列为次要优先领域，并由相关的组织牵头承担这一领域，具体包括：组织编写与湿地恢复相关的研究资料和湿地恢复的新方法，并广泛传播这些资讯资料（决议4.13）；增加湿地恢复专业技术方面的培训机会，创立相关的培训模组作为《湿地公约》培训项目的一部分；将介绍有关湿地恢复的工程和经验的资讯放到《湿地公约》相关网站上，特别是提供一些湿地恢复示范工程的资讯，这些示范工程要能很好地阐述COP8通过的《湿地恢复的原则和指导方针》（COP8决议III.16）；进一步发展湿地恢复的方法和指导方针，包括湿地恢复的专业辞汇表以及针对小型水坝和湿地恢复的指导方针。

这几项内容主要由IUCN生态系统管理委员会（IUCN-commission on Ecosystem Management）和湿地国际作为牵头单位组织实施，并与湿地科学家协会（SWS）从事湿地恢复工作方面的并有兴趣加入STRP委员会工作的代表进行磋商。

(4)COP9(坎帕拉，2005)
COP9的决议（决议IX.1）的附件E"湿地调查、评估和监测的综合框架"中将湿地的恢复与重建作为湿地管理的一部分。决议IX.2的第二部分为与《湿地合理利用与生态特征变化概念框架》相关的问题，在由非STRP承担的行动中包括：要求加强滨海湿地的恢复；在采取减少

on wetland restoration, support the STRP in its commitment in this field, and to disclose relevant information on the Ramsar websites.

Action No. 3: Recognizing the cultural and archaeological significance of wetlands being considered for restoration, so as to ensure that this significance is maintained during the implementation of Principles and Guidelines for Wetland Restoration, or "Guiding principles for taking into account the cultural values of wetlands for the effective management of sites".

Action No. 4: Identifying training needs in wetland restoration to become part of the national training needs assessments, identifying training opportunities and expertise in wetland restoration and creating relevant training modules as part of the Ramsar Wetland Training Initiative.

Action No. 5: Disseminating principles and guidelines for wetland restoration to community stakeholders who have an interest in wetland restoration, and involving local communities and indigenous peoples in restoring and maintaining wetlands. Cooperating with local communities may help with the compilation of wetland restoration plans.

Action No. 6: Utilizing the information and resources of Ramsar's restoration website and contributing information on wetland restoration projects to the website. This will make demonstration projects available that illustrate the application of "Principles and Guidelines for Wetland Restoration" (Resolution VIII.16). Theories and case studies must be added to Ramsar's restoration website based on the global target of 2003 – 2005. The Ramsar Bureau is in charge of running the website and updating relevant information.

Action No. 7: Further development of tools and guidance on wetland restoration, including a glossary of wetland restoration terminology and guidance on small dams and wetland restoration (Resolution VIII.16). Implementation Target 20 includes the training modules for wetland restoration and rehabilitation.

(3) The 11th Meeting of the STRP
The 11th Meeting of the STRP included wetland restoration and rehabilitation in Agenda 6.3 of priority projects, and passed the STRP Work Plan 2003 – 2005 in April of 2003. Wetland restoration and rehabilitation is listed as a secondary top-priority with relevant organizations in charge of specific tasks including: compiling and publicizing new research and methodology on wetland restoration (Resolution 4.13); identifying training opportunities and expertise in wetland restoration and creating relevant training modules as part of

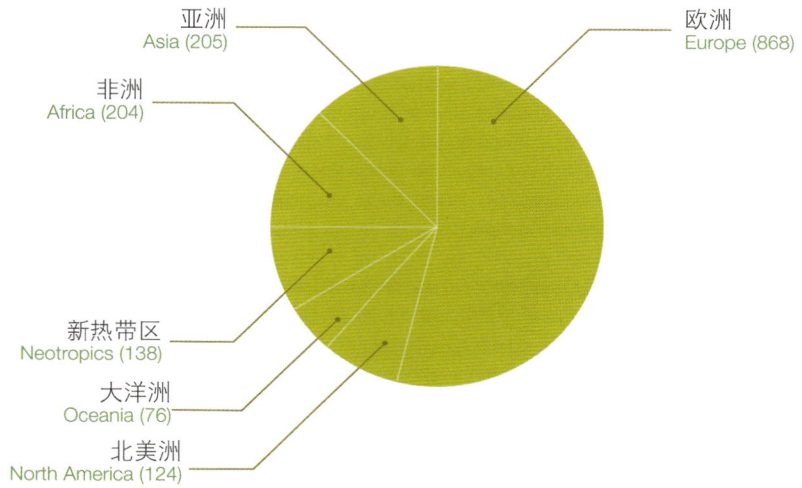

亚洲
Asia (205)

欧洲
Europe (868)

非洲
Africa (204)

新热带区
Neotropics (138)

大洋洲
Oceania (76)

北美洲
North America (124)

▲ 图2-2: 国际重要湿地的全球各大洲的分布数量
（资料来源：湿地国际网站 www.wetlands.org. 2006）
Figure2- 2: Global distribution and number of Ramsar sites. The total number
of Ramser sites is 1,615 as of August 2006.
(Source: Wetlands International website: www.wetlands.org. 2006)

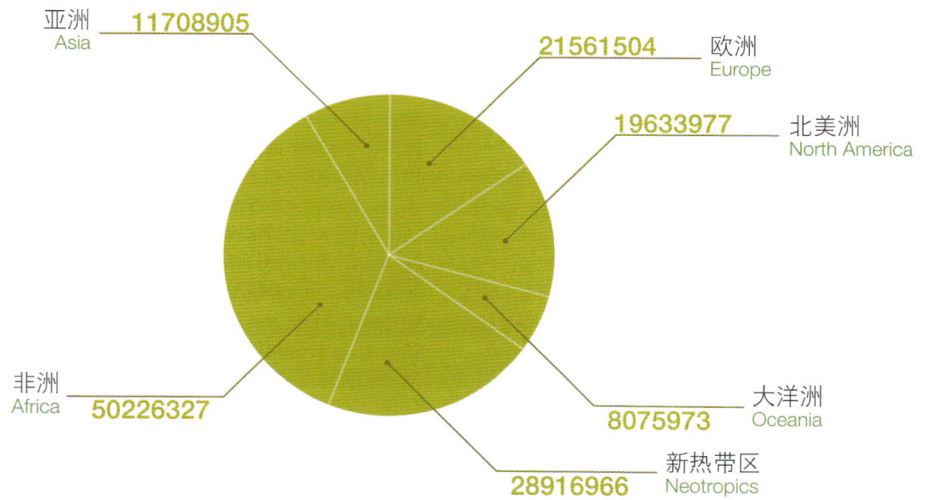

亚洲
Asia
11708905

欧洲
Europe
21561504

北美洲
North America
19633977

非洲
Africa
50226327

大洋洲
Oceania
8075973

新热带区
Neotropics
28916966

▲ 图2-3:国际重要湿地在全球各大洲的分布面积共计为140123652hm²
（资料来源：湿地国际网站 www.wetlands.org. 2006）
Figure 2-3: Global Ramsar site areas total 140123652 hm²
(Source: Wetlands International website: www.wetlands.org. 2006)

贫困的行动部分，要求通过可持续利用和湿地恢复来减少自然灾害对人类的影响，同时在制订的主要的方针政策中考虑采用湿地恢复的方法。第三部分为与《湿地调查、评估与监测的综合框架》相关的问题，在不是由STRP承担的行动中提到，在开展全国范围内的湿地调查时，要依据决议VII.16和VIII.6采取的方针政策以及结合湿地生态效益和生态特征的前后变化来确定适合进行恢复和重建的湿地。在第四部分的流域管理中提到了要通过保护和恢复集水区来调节径流和水质。第六部分的内容为与湿地管理相关的问题，STRP要承担的行动包括为缔约国制订一套自然灾害发生后湿地恢复的指南，这些指南应该在吸收现有的指南基础上形成，在新的指南中要充分体现湿地在自然灾害发生后在生态系统恢复中的作用，着重强调湿地生态系统在减轻自然灾害影响中扮演的角色；编辑并宣传新的退化或丧失湿地恢复和重建的研究与方法。对于其他非STRP组织在此方面要做的工作包括：编辑并宣传新的退化或丧失湿地恢复和重建的研究与方法；建立退化湿地恢复与重建的方案，特别是主要河流的交汇区或者是具有高度自然保护价值的区域；将湿地恢复与减贫联系起来，通过合作提供工作、技能以及机会并集中恢复社区所依靠的湿地生态系统服务及效益的地区；帮助恢复受自然灾害影响的生态系统；制订生态系统管理与恢复，以及抵御自然灾害的应急计划。

COP9还强烈呼吁有红树林湿地的缔约国修改国家战略中危害红树林生态系统的内容，并敦促缔约国加强对珊瑚礁、海草床以及其他相关生态系统的恢复应该在国家战略计划中执行。在其附件中提到，管理那些进行渔业生产的湿地，应该探究被毁坏的湿地的恢复问题。

在决议IX.8，战略1.5为确定需要优先恢复与重建的湿地，并通过实施必要的方法来进行恢复，要保证这些湿地在恢复和重建后能获得长期的环境效益、社会效益和经济效益。预计到COP10，所有的缔约国都完成优先恢复湿地的确定工作，并至少已经有100个缔约国已经开展了湿地的恢复与重建工程；继续在《湿地公约》湿地恢复的网页添加新的实例研究及恢复方法。

the Ramsar Wetland Training Initiative; updating Ramsar websites with information about restoration projects and experience, particularly information on projects that can illustrate the application of Principles and Guidelines for Wetland Restoration passed in COP-8 (Resolution VIII.16); and further developing the tools and guidelines for wetland restoration, including a glossary of wetland restoration terminology and guidance on small dams and wetland restoration.

The tasks were to be implemented under the guidance of IUCN Commission on Ecosystem Management and Wetlands International in consultation with delegates from the Society of Wetland Scientists (SWS).

(4) COP-9 (Kampala, 2005)
Resolutions made at COP-9 integrate wetland restoration and rehabilitation into wetland management and include specific articles on restoration and rehabilitation. Part II of Resolution IX.2 "Issues relating to the Conceptual Framework for the wise use of wetlands and the maintenance of their ecological character", outlines actions to be taken by parties other than the STRP, which include: working together urgently to promote and actively support the recovery of coastal wetlands; taking action that contribute to poverty reduction by adopting measures to protect humans against natural disasters through the sustainable use and restoration of wetlands; increasing priority given to wetland restoration in all relevant mainstream policy sectors.

In Part III, "Issues relating to the Integrated Framework for wetland inventory, assessment and monitoring", actions to be taken by parties other than the STRP include undertaking national inventories of and identifying wetlands where restoration or rehabilitation would be appropriate due to their present and/or former ecosystem benefits/services. In Part IV, "Issues relating to the integrated framework for Ramsar's water-related guidance", actions to be undertaken by parties other than the STRP include conserving and restoring water basin areas to regulate water flow and quality. In Part VI "Issues relating to wetland management", extensive actions by the STRP will include: developing restoration guidelines for Contracting Parties that may be implemented following natural disasters, including drawing on existing and using new guidance relating to the role of wetlands in implementing responses to ecosystem rehabilitation and vulnerability; updating existing guidelines that emphasize the role of wetland ecosystems in mitigating the effects of natural disasters; and compiling and disseminating information on new research and methodologies for the restoration and rehabilitation of lost or degraded wetlands.

在决议 IX.9 中，认为保护和恢复泥炭地以及洪泛平原的集水区能够阻止自然洪灾的发生。

在决议《国际重要湿地名录中的湿地的状态》（决议 IX.15）中，提到许多国家已经开始着手恢复生态特征发生变化的国际重要湿地。

Actions to be taken by parties other than the STRP include: compiling and disseminating information on new research and methodologies for the restoration and rehabilitation of lost or degraded wetlands; establishing wetland restoration/rehabilitation programs, where feasible, at destroyed or degraded wetlands, especially those associated with major river systems or areas of high nature conservation value; linking wetland restoration to poverty reduction by incorporating provisions for work, skills and opportunities into restoration projects and by focusing on the restoration of ecosystem benefits and services upon which communities depend; helping restore ecosystems affected by natural disasters; and establishing emergency plans to manage and restore ecosystems against natural disasters.

Further resolutions urge Contracting Parties with mangrove ecosystems in their territories to modify national policies and strategies that could have harmful effects on these ecosystems. In addition, Contracting Parties with coral reef, sea grass beds and other associated ecosystems in their territories should implement national programs for the protection of these ecosystems. The Annex of this Resolution also suggests that management of wetlands for fishery production take into account the restoration of destroyed wetlands.

Resolution IX.8 creates Strategy 1.5 to "identify priority wetlands where restoration or rehabilitation would be beneficial and yield long-term environmental, social or economic benefits, and implement the necessary measures to recover these sites". It is expected that by COP-10 all Contracting Parties with lost or degraded wetlands will have identified priority sites for restoration and that restoration projects will be underway or completed in at least 100 Contracting Parties. New case studies and methodologies will continue to be added to the Ramsar's restoration website.

Resolution IX.9 acknowledges that the protection and rehabilitation of peatlands and other catchment or floodplain wetlands contribute to natural flood prevention. Resolution IX.14 urges Contracting Parties to take measures to protect humans against natural disasters through the sustainable use and restoration of wetlands, and links wetland restoration to poverty reduction by incorporating the provision of work, skills and opportunities into restoration projects and by further focusing on the restoration of ecosystem benefits and services upon which communities depend. Lastly, Resolution IX.15 notes that many Contracting Parties have started work on restoring wetlands of international importance, after monitoring the changing ecological character of these sites.

2.2 美国《清洁水法案》

2.2.1 简介

《清洁水法案》的全称为《联邦水污染控制法案的1972年修正案》，该法案是美国国会通过的影响最为深远的环境法律之一（见图2-4）。《清洁水法案》为管理美国水域中的污染物排放制定了法律架构，对于改善美国水质很有裨益。在该法案制定前，美国可通行航道中只有1/3可供饮用或游泳；到了1998年，美国2/3的地面水都可安全游泳（Daniels, 2003）。

《清洁水法案》目的在于"恢复和保持美国水域的化学、物理和生物完整性"，因此，有关改善水质和清除污染物排放的管理规定就成了该法案的一个重要组成部分。同时，该法案又是湿地保护的重要依据（湿地被视为美国水道之一）。该法案的第404条特别对水道的疏浚和填平做出了规定，以确保必须经过申请许可、进行评估后方可进行湿地开发。

《清洁水法案》最初在1972年通过，后在1977年、1981年、1987年、1990年和2002年增添了修正案，以反映美国不断变化的环境状况。总体来说，该法案通过增添修正案而变得越来越严格，一个很好的例子就是填平湿地进行开发所需的许可证要求：起先只需申请很宽松的"全国许可证"就可填平最多40468.6m²（10英亩）的湿地，到了1996年，用这一许可证只能填平最多12140.58m²（3英亩）的湿地，到了2000年则只能填平最多2023.43m²（0.5英亩）的湿地；这就意味着现在如要填平超过2023.43m²（0.5英亩）湿地的话，就必须申请一个更严格的"特殊许可证"（见图2-5）。

美国联邦机构——环境保护署（Environmental Protection Agency，以下简称EPA）负责执行《清洁水法案》，各州政府（包括水资源委员会、自然资源部或环境局）则负责水质日常管理，美国联邦工程（The ArmyCorps of Engineers，以下简称ACOE）受委派对湿地疏浚和填平的许可证申请进行评估。

The United States Clean Water Act, officially known as the 1972 Amendments to the Federal Water Pollution Control Act, is one of the most far-reaching environmental pieces of legislation to be passed by the US Congress. The Clean Water Act (the Act) establishes the legislative structure for regulating pollutant discharges into the water systems of the US, and has been instrumental in significantly improving the water quality of the country's waterways. Before the Act was passed, only one-third of the nation's navigable waterways were safe for drinking or swimming. By 1998, a quarter century after the passage of the Act, two-thirds of the nation's surface waters have become safe for swimming (Daniels, 2003).

The stated objective of the Act is to "restore and maintain the chemical, physical, and biological integrity of the nation's waters". Thus, regulations aimed at improving water quality and eliminating pollutant discharge constitute a core component of the Act. At the same time, the Act is a critical tool for protecting wetlands, which are recognized as part of the country's waterways. Section 404 of the Act specifically regulates the dredging and filling of waterways, ensuring that wetlands are not developed without first undergoing review, making it a required part of the permit application process.

First passed in 1972, the Clean Water Act was subsequently amended in 1977, 1981, 1987, and again in 1990 and 2002 to reflect the changing needs of the environment and of the country. In general, the standards required by the Act have become more stringent over time. An example is the permit requirements for developments involving the filling of wetlands. Initially, only an application for the more lenient Nationwide Permit was needed fill up to 10 acres of wetlands. The threshold since then has been lowered to three acres in 1996 and 0.5 acre in 2000, which now means that an application for the more stringent Individual Permit is required for fillings that involves more than 0.5 acre of wetlands.

The Clean Water Act is administered by the Environmental Protection Agency (EPA, a federal agency), while day-to-day regulation of water quality is mainly carried out by state governments, including water resources boards, departments

▼ 图2-4:《清洁水法案》是美国国会通过的影响最为深远的环境法律之一（图片来源：www.photos.com）
Figure 2-4: The United States Congress passed the Clean Water Act in 1972
(Source: www.photos.com)

▼ 图2-5: 该湿地靠近美国加州圣克鲁兹，受《清洁水法案》保护 (摄影：Rowan Roderick-Jones)
Figure 2-5: This freshwater wetland near Santa Cruz, California, is protected under the Clean Water Act (Photograph by Rowan Roderick-Jones)

湿地无净损失

1989年，美国总统布什颁布了"湿地无净损失"政策，虽然这一政策有利于湿地的长远保护，但仍有许多不足之处，比如，该政策允许在"湿地损失不可避免或无关紧要"的情况下填平湿地。固然，"湿地无净损失"并非"湿地无损失"，建设新湿地或重建已退化的湿地确能补偿已被破坏的现存自然湿地；不过，很多研究和实践表明，重建和恢复自然湿地的结构和功能绝非易事。因此，在该政策颁布10年之后，"在美国48个州内仍出现了湿地数量和质量的下降"。1998年，美国总统克林顿颁布了更加大胆的政策，准备从2005年起每年增加40468.6hm²（10万英亩）湿地。这一政策能否实现预期的效果尚待观察（Bush, 2002）。

2.2.2《清洁水法案》的湿地保护措施

《清洁水法案》中主要有两个部分提到了湿地：第404条"疏浚或填埋材料许可证"和"水质标准、污染控制"部分。

《清洁水法案》第404条规定，土地所有权人或开发者未从ACOE取得许可证的，不得疏浚或填平水道，以此保护湿地；必须根据EPA制定的规范来进行这一评估过程（Daniels, 2003）。

划定湿地范围

《清洁水法案》对美国水域的疏浚和填平做出了限制，这一"水域"概念包括了领海、沿海及内陆的水域、湖泊、河道和溪流（及附近湿地）、可通行航道的支流（及附近湿地）和洲际水域及其支流（包括附近湿地）。

海洋、湖泊、河流和溪流的边界很容易确定，然而，湿地的边界就较难限定了（见图2-6）。为此，ACOE在1987年推出了《划定湿地范围》手册，提供用于确定湿地范围的具体规范。这本手册将湿地定义为"被地表水或地面水淹没或饱含水分的土地，其受淹或含水的频率和持续时间足以令适应含水土壤条件的植被成长蔓延"。虽然划定湿地范围的具体方法并不在本书范畴之内，但可从上述定义中得知判断湿地的三个基本要素：水文状况、湿地植被和水成土（见图2-7）。

No Net Loss of Wetlands

In 1989, President George H.W. Bush initiated a policy of "no net loss of wetlands". While a positive and ambitious goal, the policy contains many exceptions. For example, it allows wetlands to be filled if the loss is "unavoidable" or "insignificant". Furthermore, "no net loss" is not the same as "no loss", and indeed, creating of new wetlands and restoration of degraded wetlands are allowed to compensate for destroying existing natural wetlands. Of course, it has been documented by numerous studies that recreating the complete biological and hydrologic functions comparative of a natural wetlands is no easy task. Therefore, a decade after the declaration of the no net loss policy, there is "still a decline in wetland quantity and quality in the lower 48 states". In 1998, President Clinton modified this by calling for an increase of 100,000 acres of wetlands each year, beginning in 2005. It remains to be seen how this policy will be achieved (Bush, 2002).

of natural resources, or environmental departments (Daniels, 2003). The Army Corps of Engineers (ACOE) is delegated with the responsibility of conducting permit review processes regarding the dredging and filling of waterways.

2.2.2 Clean Water Act Wetland Protection Efforts

Wetlands are primarily addressed in two different areas of the Act: Section 404 Permits for Dredged or Fill Material, and in the Section on water quality standards and pollution controls. Section 404 of the Act protects wetland habitats by prohibiting the dredging and filling of waterways, unless the landowner or developer applies for a permit from the ACOE. This involves a review process using guidelines set by the EPA (Daniels, 2003).

Wetland Delineation

The Act restricts dredging and filling "in the waters of the US". This is an intentionally broad definition that includes the "territorial seas", "coastal and inland waters, lakes, rivers, and streams that are navigable waters of the United States, including their adjacent wetlands", "tributaries to navigable waters of the United States, including adjacent wetlands" and "interstate waters and their tributaries, including adjacent wetlands".

While the boundaries of oceans, lakes, rivers and streams can be easily defined, wetland edges are considerably more difficult to delineate. In 1987, the ACOE produced a manual called the "Wetlands Delineation Manual" to provide detailed guidelines for marking the extent and borders of wetlands. In the manual, wetlands are defined as "areas that are inundated or saturated by surface or groundwater at a frequency and duration sufficient to support a prevalence of vegetation typically adapted for life in saturated soil conditions". Although the actual methodology of

▼ 图2-6: 淡水沼泽地亦视作为湿地，定义为"美国水道"，所以规定予以保护，使免于填平与疏通（图片来源：www. photos.com）
Figure 2-6: Wetlands such as this fresh water marsh are included in the definition of "waters of the US" and are therefore protected from dredging and filling (Source: www.photos.com)

▼ 图2-7: 美国《清洁水法案》定义湿地土壤的一种标准，水道中氧化的根部和灰色杂质的部分为典型的水成土 (摄影: Rowan Roderick-Jones)
Figure 2-7: The presence of wetland soils is one criterion that defines wetlands under the Clean Water Act. Note the presence of oxidized root channels and the grey "mottling" typical of hydric soil. (Photograph by Rowan Roderick-Jones)

对湿地范围进行精确的划定，可以避免不必要的混淆，并且免于在申请许可证过程中和随后执行许可证条款过程中引发法律诉讼。ACOE的《划定湿地范围》手册提供了一套所有申请者都能够遵循的标准化规范，对于在划定过程中保持客观性非常重要。

应当指出的是，ACOE并非对美国所有的湿地都具有管辖权。美国最高法院在2001年作出判决，认定其许可权仅限于包含在可通行水域内的湿地，而与河流或溪流不接壤的独立湿地则并不受其管辖，其管辖权归各州。据估计，美国1/5的湿地现在不受联邦管辖（Daniels, 2003）。

申领许可证的过程
ACOE根据湿地开发计划的影响范围和程度，对"全国许可证"和"特殊许可证"的申领作出评估。

"全国许可证"较为宽松，可以允许对环境略有不利影响的开发计划。如果受到开发计划影响的湿地和开阔水域少于2023.43m²（0.5英亩），而且该开发计划属于预先允许的开发范围之一，那么就可以使用"全国许可证"。这一许可证现有共约40大类的允许开发范围，如：

- 美国海岸警卫队批准并按照其标准安装的航运辅助设施；

- 用于防治水土流失的固岸工程；

- 非主要道路交叉口的填埋。

一个开发计划如果需要填平2023.43m²(0.5英亩)以上的湿地，就必须取得更严格的"特殊许可证"。对该开发计划的评估包括：项目是否需要水源（即能否在湿地以外修建该项目）？项目是否符合公众利益？最后，项目是否遵守以下联邦法律：

- 《全国环境政策法案》，该法案要求开发计划考虑是否还有其他可供选择的地点；

- 《海岸地区管理法案》，确保开发计划适合该州对海岸地区的规划；

wetlands delineation is beyond the scope of this handbook, the definition signifies that, for an area to be qualified as a wetland, three criteria must be present: wetland hydrology, hydric vegetation, and wetland soils.

The precise delineation of wetland boundaries is necessary to avoid confusion and potential lawsuits resulting from the permit application process, as well as the subsequent enforcement of permit terms. The manual produced by the ACOE is essential in injecting objectivity into the delineation process, and for providing standardized guidelines.

It should be noted that the jurisdiction of the ACOE does not strictly cover all wetlands in the country. In 2001, the Supreme Court ruled that jurisdiction of the ACOE should only apply to wetlands that are part of "navigable waters", which do not include isolated wetlands not connected to a tributary, stream or river, reverting the right to regulate these areas back to the states. It is estimated that as much as one-fifth of the nation's wetlands are no longer under federal control (Daniels, 2003).

Permitting Process
Depending on the area affected and the impact level of the proposed activity, the ACOE reviews the development using the Nationwide Permit or Individual Permit process.

The Nationwide Permit is the more lenient of the two. It authorizes certain activities that cause only minimal adverse effects on the environment. If the total area of wetlands and open waters affected by the activity is 0.5 acres or less, and if the proposed activity meets one of the categories of pre-authorized activities, a permit is granted for the project (Copeland, 1999). There are roughly 40 categories of authorized activities under the Nationwide Permit. Examples include:

- Placement of aids to navigation approved by, and installed according to US Coast Guard requirements;

- Bank stabilization activities necessary to prevent erosion; and

- Fills for minor road crossings. (Copeland, 1999)

Developments that involve filling over 0.5 acres of wetlands require approval through the more stringent Individual Permit process. The review includes judging whether the proposed project is water dependent, that is, if it can be sited away from the wetland, whether the activity advances the public interest, and whether the proposal complies with a variety of federal laws, including:

- National Environmental Policy Act (NEPA), which requires

- 《海洋保护、研究和禁用法案（向海洋倾倒废弃物法案）》，确保开发计划并未侵入海岸和海洋区域中的禁止开发区；

- 《濒危物种法案》，确保开发计划不会对濒危物种的栖息地造成破坏。

"全国许可证"的办理平均需要14天（见图2-8），而"特殊许可证"则平均需104天。ACOE86%工作量（约55000个案例）即为"全国许可证"的许可范围；每五年，还要对"全国许可证"许可范围进行重新评估。"特殊许可证"申请几乎都能得到批准，当然其中约一半在批准时都须带有附加条件（Daniels, 2003）。

以上的申请许可过程确保了只有政府部门进行了环境影响评估后才能进行湿地开发计划。不过，为了效率起见，一些预计不会对环境造成太大影响的开发计划就交由"全国许可证"申请过程，以减小管理负担，而那些预计会对环境造成不良影响的开发计划则必须执行"特殊许可证"申请过程（Copeland, 1999）。

联邦湿地许可过程

1. 申请前的会议：ACOE地区办公室和土地所有人之间的会议，讨论项目的概念规划；ACOE提出初步建议。

2. 提交申请：土地所有人向ACOE地区办公室提交申请，项目管理者审查该申请；向公众发布通知，在15至30天内向公众征询意见；可以召开公众听证会。

3. 确定"全国许可证"还是"特殊许可证"：项目管理者决定该项目是否符合"全国许可证"的40大类允许开发范围，如符合的话，项目管理者对是否颁发"全国许可证"提出建议；如颁发"全国许可证"，则应包括建设、最佳管理实践和补偿措施；如果项目将有重大不良影响，则应考虑"特殊许可证"，并进行更详细的评估。

4. EPA的否决：EPA有权否决ACOE已批

consideration of possible alternative sites;

- Coastal Zone Management Act (CZMA) for ensuring that the proposed use is compliant with state plans for the coastal area;
- Marine Protection, Research, and Sanctuaries Act (Ocean Dumping Act) for ensuring that the proposed project does not fall within coastal and ocean regions designated as off limits to development by the marine sanctuaries program; and

- Endangered Species Act, for ensuring that the proposed project does not destroy the habitat of any declared "threatened" or "endangered" species (Daniels, 2003).

On average, Nationwide Permits require 14 days to process, compared to 104 days for Individual Permits. Approximately 55,000 cases or 86% of the ACOE's workload fall under this category (Copeland). All Nationwide Permit categories are reviewed for reauthorization every five years. As for Individual Permits, nearly all applications are approved, although conditions are attached to roughly half of these (Daniels, 2003).

The permitting process described above ensures that development on wetlands is not allowed before its impact is reviewed by a government agency. However, to balance reviews with the need for efficiency, the regulatory burden is minimized by channeling certain activities that are deemed not to cause significant environmental impact through the Nationwide Permit process, leaving only proposals that are certain to cause more significant impact to the Individual Permit process (Copeland, 1999).

The Federal Wetlands Permitting Process

1. Preapplication meeting. A meeting between ACOE district office and landowner, where the conceptual design of the project is discussed. The ACOE provides preliminary suggestions.

2. Application submission. Landowner submits application to the ACOE District Office. The project manager reviews the application. Public notice is published with a public comment period of 15-30 days. A public hearing may also be held.

3. Nationwide Permit or Individual Permit? The project manager decides whether the activity meets one of the 40 Nationwide Permit categories. If so, then the project manager recommends whether to grant the permit. A permit may include specific conditions about construction, Best Management Practices and mitigation measures. Alternatively, if the project will have major impacts, an

▼ 图2-8: 在湿地修建灯塔会很快获得"全国许可证"批准（图片来源：www.photos.com）
Figure 2-8: Lighthouses, if built on a wetland, can be quickly approved by the Nationwide Permit (Source: www.photos.com)

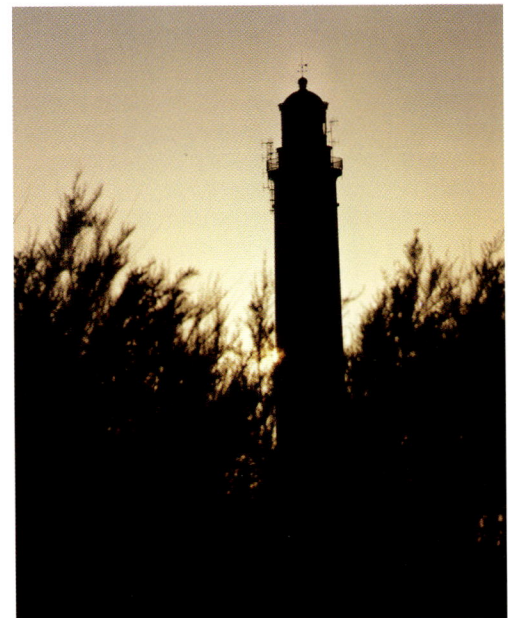

5. 监督和执行：ACOE地区办公室负责对"全国许可证"和"特殊许可证"定内容的监督和执行（Daniels，2003）（见图2-9）。

必须指出的是，有很大一部分人类活动不受《清洁水法案》第404条的限制（见图2-10），因此，这些活动可以自由进行湿地填平或疏浚，无需评估或申请许可，这些活动包括：

- 普通农耕、造林和牧场经营活动；

- 对目前处于使用状态的设施（如堤、坝、堰、堤道和交通建筑)进行维护；

- 农林业道路的建造和维护等。

这是《清洁水法案》的一个重大漏洞，使其不能有效地保护湿地免受许多常见人类活动的破坏。

湿地外补偿措施

作为申请许可证过程的一部分，申请人也可对开发计划中将要填平的湿地提出湿地外替代补偿措施（Daniels，2003）。开发者可通过以下方式对湿地进行补偿：

在别处恢复湿地

开发者可以提议在别处恢复已退化的湿地，面积可与原有相同或更大；该重建湿地最好位于原有流域。

建设新湿地

虽然建设新湿地能在理论上"代替"在开发中损失的湿地，但是在实际操作中，新湿地置换改变了原有栖息地的土壤、水文和植物，成功机率较低，因而不宜提倡。而且，美国国家科学院在2001年的一份报告中指出，人工湿地并不能重新创建类似自然湿地的功能（Daniels，2003）。将曾被疏浚或填平的湿地重新恢复原貌是更为明智的选择。

求助于"补偿银行"

开发者还可以通过向"补偿银行"购买"补偿点数"来保护湿地。"补偿银行"可由政府部门、土地信托或其他非营利机构设立（Daniels，2003），可凭藉其已创建或重建的湿地来出售"补偿点数"，或用开发者的支付款来继续创建或重建湿地。这样，"补偿银行"就可以让开发者不必直接参与创建湿地

Individual Permit may be required, which involves a much more detailed review.

4. EPA veto. The EPA has power to veto a permit granted by the ACOE. However, this very rarely happens.

5. Monitoring and enforcement. The ACOE district office is responsible for monitoring and enforcing the terms of the Nationwide and Individual permits (Daniel 2003).

It should be noted, however, that a wide range of activities have been exempted by the Clean Water Act from the control of Section 404 Permits. These activities could include the filling or dredging of wetlands without the need for review or permit application. Activities include (Section 404[f]):

- Normal farming, forestry, and ranching activities;

- Maintenance of currently serviceable structures, including dikes, dams, levees, causeways and transportation structures; and

- Construction and maintenance of farm roads or forest roads, etc.

This loophole has considerably reduced the effectiveness of the Act to protect wetlands from many common activities.

Offsite Mitigation Measures

As a part of the permitting process, the applicant can also suggest offsite mitigation measures to replace wetlands that are filled as a result of project activity (Daniels, 2003). There are several common ways a developer can "mitigate" wetlands loss.

Restoring wetlands

The developer can propose to restore an area of degraded wetlands, most likely of a larger size, elsewhere. The restored wetland would preferably be located in the same watershed.

Constructing new wetlands

While constructing new wetlands can in theory "replace" wetlands lost to development, in reality new wetlands displace and alter the soil, hydrology, and plant life of existing habitats. It is not recommended due to its high failure rate. Moreover, a 2001 report by the National Academy of Sciences found that artificial wetlands are unable to closely replicate the functions of a natural wetland (Daniels, 2003). Restoring wetlands that have been previously drained or filled is a more effective solution.

Contributing to a mitigation bank

Developers can also preserve wetlands by buying mitigation credits from special credit banks, which can be set up by a

▼ 图2-9: ACOE并非对美国所有的湿地都具有管辖权：独立的湿地如加州的罗蒙纳（摄影：EDAW）
Figure 2-9: The Jurisdiction of the Army Corps of Engineers does not cover isolated wetlands such as these vernal pools in Ramona, California (Photograph by EDAW)

▼ 图2-10: 美国大多数农耕活动不受第404条的限制（图片来源：www. photos.com）
Figure 2-10: Most farming activities are exempted from Section 404 Permits (Source: www.photos.com)

湿地恢复手册 原则·技术与案例分析

（Daniels，2003）。"补偿银行"的另一大好处是，能够战略性重建或保护大量成片湿地，而开发者只能各自为政地重建分散、不连贯的小块湿地。

监督和执行

审查了开发者的申请，并对其作了批准或是拒绝的决定后，更加重要的任务是对开发者和土地所有权人进行监督，以确保其切实遵守许可证中约定的内容，避免未批准的活动，并实行其已允诺的补偿措施。不过，美国国家科学院在2001年的一项研究表明，ACOE并未跟踪或验证补偿措施是否已得到实行（Daniels，2003）。如果没有切实有效的监督和执行，那么即便申请过程中提出的目标再好，湿地退化的恶果仍会一发而不可收拾。

水质标准和污染控制

《清洁水法案》提出的水质标准和污染控制也间接保护了湿地。

通过控制点源污染和维持最低水质标准来改善水道水质，这是《清洁水法案》的一个首要目标。该法案不仅严格限制湿地中的污染物排放，还要求在临近湿地的水道中改善水质，以减小对湿地的外部损害，维护湿地的总体健康和自然恢复力。同时，湿地因为对水的过滤和净化作用，能帮助满足水道的最低水质要求，因而得到有效保护。《清洁水法案》中有关水质标准和污染控制的一些程式简列如下：

国家清除污染物排放系统（National Pollutant Discharge Elimination System，以下简称NPDES）

《清洁水法案》规定，如未得到NPDES许可证授权，就不能从任何点源向可通行水域排放污染物（见图2-11）。污染管理部门可以通过限制和逐步降低所颁发的许可证数量，来减少水道污染物排放和改善水质。不过，该许可证并不能管辖所有的污染物排放；而且，即便用该许可证对污染物排放加以控制，也并不能确保清除污染或将水质改善至可饮用或可游泳的标准（Daniels，2003）。

废水处理设施的建设和规划

《清洁水法案》第201条指出，EPA已向

government agency, a land trust, or other non-profit organizations. (Daniels, 2003). The mitigation bank can either sell mitigation credits for the wetlands they have already created or restored, or use the payment from a developer to create or restore additional wetlands. An added advantage of mitigation banks is that they are more able to strategically restore and protect larger parcels of wetlands than individual developers, who can only propose restorations of separate, unrelated sites.

Monitoring and Enforcement

The responsibility of the ACOE does not end with the application review, or the granting or denial of the permit. Equally important is the monitoring of developers and landowners to ensure that they actually comply with the conditions of the permit, that no unauthorized activities are committed, and that mitigation proposals promised are carried through. Unfortunately, a National Academy of Sciences study in 2001 noted that the ACOE did not routinely track or verify whether mitigation had actually been completed (Daniels, 2003). Without adequate monitoring and enforcement, for all the good intentions of the permit reviewing processes, degradation of wetlands may still continue.

Water Quality Standards and Pollution Controls

Wetlands are also indirectly protected by the water quality standards and pollution controls required under the Act.

The water quality improvement of waterways through the control of point-source pollution and the maintenance of minimum water quality standards is a primary goal of the Act. Not only is pollutant discharge into wetlands strictly regulated, improving water quality in waterways surrounding wetlands reduces external pressure on wetlands systems, helping to maintain their overall health and resiliency. Furthermore, in order to meet minimum water quality standards for waterways, the value of wetlands in filtering and purifying water provides incentives for their protection. The following is a brief overview of the Clean Water Act program elements that help reduce pollution and maintain water quality standards.

National Pollutant Discharge Elimination System (NPDES)

Unless authorized by a NPDES permit, the Clean Water Act prohibits the discharge of any pollutants into navigable waters from a point source (Daniels, 2003). By controlling and gradually reducing the number of permits given out, the controlling authority can significantly reduce pollution discharge into waterways, and in theory, improve water quality. However, not all pollutants discharged into a waterway are covered by the permit. Moreover, even if pollutant discharge is controlled by permitting, it does not guarantee that pollution is eliminated or sufficiently reduced in a way that improves water quality to potable standards (Daniels, 2003).

▼ 图2-11：《清洁水法案》规定，如未得到NPDES的许可证授权，就不能从任何点源向可通行水域排放污染物（图片来源：www.photos.com）

Figure 2-11: The Clean Water Act prohibits the discharge of any pollutants into navigable waters from a point source unless authorized by a NPDES/SPDES permit (Source: www.photos.com)

各州和地区拨出300亿美元的废水处理专款，用于废水处理厂的建设和升级，这些专款为改善美国水质起了很大作用（Daniels, 2003）。

非点源污染控制

自从《清洁水法案》颁布以来，由于NPDES和相应的州级清除污染物排放系统（State Pollutant Discharge Elimination System，以下简称SPDES）的制定，以及废水处理厂的不断建设和升级，点源污染已得到了有效的控制（Daniels, 2003）。现在，非点源污染（如农场土地和城市道路）已成了主要的水道污染源。要继续改善美国水质，必须解决这一问题。有鉴于此，《清洁水法案》的1987年修正案添加了针对非点源污染的第319条，该条要求各州制定、执行非点源污染控制方案，并为非点源污染控制方案提供贷款和政府专款。此外，EPA还规定，所有对4046.86m²（1英亩）以上土地进行开辟、平整或挖掘的建设工程，都必须取得NPDES或SPDES的排涝许可证（见图2-12）；不过，《清洁水法案》并未对大多数农业活动提出具体要求。

每日最大总承载量计划

《清洁水法案》第303(d)条规定，各州必须列出本州内受到损害的水道，执行每日最大总承载量计划，优先处理受损水道，将其改善至可饮用或可游泳的标准。对于列出的每一条受损水道，各州必须测定其在满足联邦和州的可饮用或可游泳水质标准下的最大可承受污染排放量，包括测算、限制来自各个点源或非点源的各种污染物（Daniels, 2003）。

确定了需要优先改善水质的水道之后，就可在两方面保护湿地。其一，如上所述，水质的改善有助于减轻湿地的生态压力，让湿地动植物免受毒性伤害，并使湿地有能力自然处理污染物。其二，湿地提供了很有价值的环境功能，如过滤沉积物和污染物、有效降低BOD、防止水土流失等。为此，各州环境保护部门应保护自然湿地，或建设用于处理污染的湿地，以吸收、清除污染，达到每日最大总承载量标准。

▼ 图2-12：美国环境保护署规定，所有对4046.86m²（1英亩）以上土地进行开辟、平整或挖掘的建设工程，都必须取得NPDES或SPDES的排涝许可证（图片来源：www. photos.com）
Figure 2-12: The EPA requires an NPDES/SPDES stormwater permit on construction sites that involve clearing, grading, and excavating one or more acres of land. (Source: www.photo.com)

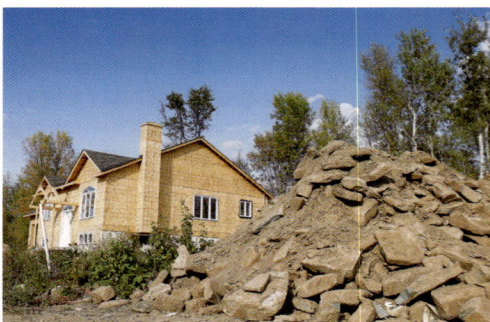

Wastewater Treatment Construction and Planning

Under Section 201 of the Act, the EPA has provided more than US$ 30 billion in wastewater treatment grants to states and localities for the construction and upgrading of wastewater treatment plants. These grants have been instrumental in improving the nation's water quality (Daniels, 2003).

Nonpoint Source Pollution Control

As a result of the NPDES program (and the corresponding state programs, SPDES) and the upgrading and construction of wastewater treatment plants, point source pollution has been significantly decreased since the passage of the Act (Daniels, 2003). Non-point sources such as farm fields and city streets are now the major contributors to waterway pollution. Addressing this latter issue is the next necessary focus for further improvements to the nation's water quality. In light of the above, the 1987 amendments to the Clean Water Act included Section 319 to address non-point source pollution. This includes requiring states to develop and implement plans both to control non-point pollution and provide loans and grants for these purposes. In addition, the EPA requires an NPDES/SPDES stormwater permit on construction sites that involve clearing, grading, and excavating one or more acres of land, and provisions for storm sewer systems, though the Act once again does not specifically address most agricultural practices (Daniels, 2003).

Total Maximum Daily Load (TMDL) Plans

Under section 303(d) of the Clean Water Act, states are required to create a list of impaired waterways and to implement TMDL plans to prioritize a clean-up of those waterways to standards that are fit either for swimming or drinking. For each impaired waterway on the list, the state is required to determine the maximum amount of pollution that can be discharged while still meeting state and federal water quality standards suitable for drinking or swimming. This includes calculating and placing pollution limits on individual point and non-point sources for each pollutant (Daniels, 2003).

The establishment of a priority list for identifying the lowest quality waterways for improvement helps wetland protection in two ways. First, as previously mentioned, water quality improvement reduces stress on wetland systems, reduces the potential for animals and plants in the wetlands to be poisoned, and reduces the likelihood that wetlands will be taxed beyond their ability to effectively treat pollutants. Additionally, wetlands provide valuable environmental services, including flood control, filtering sediments and pollutants, removing significant amounts of biological oxygen demand, and contributing to erosion control. All are incentives for state environmental protection agencies to either protect natural wetlands, or to construct treatment wetlands that assimilate and help remove pollutants in order to meet TMDL standards.

2.2.3 结论

《清洁水法案》在保护湿地上取得了多方面的显著成效，具体如下：

- 未经联邦部门许可，不得填平或疏浚湿地。第404条可以说是《清洁水法案》中对于保护湿地最关键的一条，对湿地造成影响的开发计划必须接受评估才能获得批准。

- 用两种许可证分担湿地开发计划的批准，这样就在湿地保护与管理效率之间作出了适当的平衡。"全国许可证"可批准一些对环境略有不利影响的开发计划，而预计产生严重环境影响的开发计划则需执行更加详细的"特殊许可证"评估过程。

- 批准过程促使开发者支持"补偿银行"。"补偿银行"能够在关键地点作战略性湿地重建，保护大量成片湿地，而开发者能购买"补偿点数"，这样就不必自行进行湿地补偿。

- 严格限制向湿地排放污染物。《清洁水法案》规定，如未得到NPDES或SPDES的许可证授权，就不能从任何点源向可通行水域（包括湿地）排放污染物。

- 美国水道水质已有显著改善，减轻了湿地系统的生态压力。水质的改善使得湿地有能力自然处理污染物。

不过，多年来，《清洁水法案》在执行过程中也暴露了一些重大的缺陷：

- 对湿地栖息地多采取间接保护。《清洁水法案》主要是针对改善水质的法律，其中除了第404条以外，对湿地保护仅有间接涉及。

- 批准过程过于被动。只有当土地所有权人或开发者即将对湿地进行开发时，才启动批准程式；虽然申请有可能遭到管理部门拒绝，但大多数仍然获得通过，而且许可证中所附加的条件仅限于对湿地所受的损

2.2.3 Conclusions

In many different respects, the Clean Water Act has been broadly successful in protecting wetlands. Notable achievements of the Act can be summed up as follows:

- Wetlands cannot be filled or dredged without first being reviewed by a federal agency. Developments that affect wetlands need to first undergo a reviewing process before a permit is obtained.

- The need for wetland protection is suitably balanced with the need to minimize regulatory burdens through channeling the permit process. Nationwide Permits authorize certain activities that cause only minimal adverse effects on the environment, allowing these proposals to bypass more detailed reviews required by the Individual Permit process.

- Discharging pollutants into wetlands is strictly regulated. The Act prohibits the discharge of any pollutants into navigable waters – which include wetlands – from a point source unless a NPDES/ SPDES is obtained.

- The permitting process provides incentive to support mitigation banking. Mitigation banks can strategically restore wetlands in critical locations and protect larger parcels of wetlands, from which developers can purchase credits.

- Water quality of US waterways has shown marked improvement, thus reducing stress on wetland systems. Water quality improvement reduces the likelihood that a wetland is taxed beyond its ability to cope with pollutants.

However, the Act has a few major shortcomings:

- Protection of wetland habitats is mainly indirect. As a piece of legislation that is primarily designed to improve water quality, with the exception of Section 404 permits, wetlands are only indirectly addressed and protected.

- The permitting process is reactive. The process is only activated when a wetland is about to be developed by a landowner or developer. While permits can be denied, the majority of them are approved, while conditions attached to approved permits often serve as items for damage control. Legislation is needed to encourage pro-active restoration of

害加以控制。新建的补偿湿地在功能上无法与被破坏的原有湿地相比，因此，为了鼓励对退化湿地进行积极的重建，必须对《清洁水法案》作出必要的增补。

- 事后监督和执行不力。对于开发者是否按申请过程中所作的承诺新建或重建湿地，EPA和ACOE常缺乏必要的事后监管；如果没有切实有效的监督和执行，那么即便申请过程中提出的目标再好，湿地退化的恶果仍会一发而不可收拾。

- 漏洞较多。《清洁水法案》最大的不足之处莫过于其中的许多"例外情况"，如大多数农业活动不受第404条限制，这会导致因疏失而造成湿地损害。此外，ACOE对于独立的湿地没有管辖权，独立湿地因此归各州政府管辖，造成了保护程度不一的情况。

2.2.4 对中国的启示

目前美国没有专门针对湿地栖息地保护的法律，在一系列不同程度涉及到湿地的法律中，《清洁水法案》为湿地栖息地提供了最有效的保护。然而，《清洁水法案》仍不是专门针对湿地保护的法律。由于湿地栖息地对于自然环境具有巨大价值，中国急需一个全面针对湿地保护的专门法律。这一法律必须清楚地界定其管辖范围。为了避免混淆，必须对"湿地"概念有明确的定义；在美国，由于"湿地"定义不一，造成了标准多变、模糊不清，还引起了土地所有权人与联邦/州部门之间的大量摩擦（Daniels,2003）。为了在界定湿地范围上保持客观性，必须制定标准化规范文件（就像ACOE的《划定湿地范围》手册），这样就能减少以后各方摩擦。

最后，尽管申请许可证是对湿地开发加以限制的关键法律工具，对许可证条款的监督和执行也同样重要，因为只有这样才能防止湿地最终受到破坏。监督不应只是政府的责任，非赢利的民间环保组织和赢利性的法律事务所，如自然资源保护委员会（Natural Resources Defense Council）和环境保护协会（Environmental Defense），都可以参与《清洁水法案》的

degraded wetlands and the protection of other valuable wetlands. Mitigated wetlands may not be of the same quality as wetlands that are destroyed.

- Follow up monitoring and enforcement is inconsistent. The EPA or ACOE often do not follow up on whether developers created or restored wetlands as required as a part of the permit review process. Without adequate monitoring and enforcement, degradation of wetlands can continue unchecked.

- Many loopholes. Perhaps most damaging are the many exceptions that create considerable loopholes in the legislation. For example, most agricultural activities are exempt from Section 404 Permits, which can result in the wholesale destruction of wetlands with little oversight. In addition, the ACOE has lost jurisdiction over isolated wetlands, which are reverted back to protection by states, which approach wetland protection with the varying degrees of commitment.

2.2.4 Implications for China

There is currently no federal legislation that specifically protects wetland habitats in the US. From a series of legislation that addresses wetlands, the Clean Water Act offers the most comprehensive protection for wetland habitats. Given the importance of wetland habitats to the environment, a specific and comprehensive legislation designed for wetlands protection is nonetheless warranted. In China, such legislation should clearly define its jurisdiction, preferably covering all wetlands. Care must be taken in defining "wetlands" to avoid confusion. Conflicting interpretations could lead to and have already created, shifting standards, uncertainty, and considerable friction between landowners and local and federal agencies (Daniels, 2003). A document such as the "Wetlands Delineation Manual" is important in injecting objectivity in defining the extent of wetland areas.

Moreover, while the permit process is a critical legislative tool used to control development in wetlands, equally important is the monitoring and enforcing of permit terms, without which wetlands can still be destroyed unchecked. However, monitoring need not be the sole responsibility of government agencies. Private and non-profit environmental groups and for profit legal firms also act as agents that help monitor and enforce the Act. This includes organizations such as the Natural Resources Defense Council and Environmental Defense (Daniels, 2003). The establishment of such civil organizations in China would not only put pressure on violators, but

监督和执行（Daniels, 2003）。如果能在中国建立类似的组织，不但会对违规者产生压力，还会促使政府更好地对相关法律的执行情况进行监督。

《清洁水法案》在加强湿地保护的具体政策方面提供了具有建设意义的蓝本，可供中国借鉴（见图2-13）。但是，正如上文所说，该法在加强湿地保护方面尚待继续改进。中国在制定和实施相关的法规时，应加强对现有湿地的积极保护，在制定湿地损失的判别依据和惩罚条款的同时，也应该制定评估各类湿地生态效益和恢复效果的程式。我国目前正在制定排污许可证制度，有关的经验也可用于"湿地补偿银行"制度，以减缓当前中国湿地快速丧失和退化。可喜的是，我国相关湿地管理部门已经着手开始了这方面的工作，颁布了可对湿地提供间接保护的法律、法规和法令。

would sanction governments to properly monitor relevant laws.

Legislation promoting the conservation of wetlands in the US could be used as a guideline for future legislation in China. The Clean Water Act shows constructive examples of specific policies that improve wetlands protection, but also, as the above suggests, the Act also shows instances where it could be improved to better protect wetland areas. Implementation of wetlands legislation in China would promote the consideration and active protection of existing wetland areas, and would further acknowledge the benefits of their restoration. Legislation along those lines is needed to reduce the current and alarming rate of wetlands loss in China. Laws in China should also aim to promote an action plan to conserve sensitive areas from future development. The following section outlines the laws, regulations, and directives that are currently in place in China that indirectly provide protection to wetlands.

◄ 图2-13：一个靠近浙江的湿地，被当作农地排水后的情形(摄影：Rowan Roderick-Jones)
Figure 2-13: This former wetland near Cixi, in Zhejiang Province, was drained for agriculture. Regulations that specifically address wetlands, such as the Clean Water Act, and the "no net loss" policy adopted by the United States, will help to ensure that wetland loss is controlled and mitigated in China. (Photograph by Rowan Roderick-Jones)

2.3 中国与湿地恢复有关的法规和导则

2.3.1 与湿地相关的国家法律法规

我国目前尚没有一部关于湿地保护和恢复的综合性专门法律法规，可适用于规范湿地保护与恢复的条款，多分散于其他有关的法律法规中。《中华人民共和国宪法》（1978）、《中华人民共和国刑法》（1979)和《中华人民共和国民法通则》（1987)明确了环境资源在国家法律中的重要地位，涉及了与湿地相关的资源类型，例如《宪法》中只是提到了"荒地、滩涂"，但"湿地"一词并未在这三大法律中出现。

到目前为止，在中国颁布的一系列有关自然资源及生态环境保护的法律法规有《中华人民共和国自然保护区条例》（1994)、《海洋自然保护区管理办法》（1995)、《林业事业费管理办法》（1997)、《中华人民共和国海洋环境保护法》（1999)、《林业工作站管理办法》（2000)和《中华人民共和国农业法》（2002)、《城市规划强制性内容暂行规定》（2002)等明确出现"湿地"一词，并将湿地纳入其管理范围。

另外，国家出台的涉及湿地的法律法规还有：

- 《水产资源繁殖保护条例》（1979)

- 《风景名胜区管理暂行条例》（1985)

- 《中华人民共和国河道管理条例》（1988)

- 《中华人民共和国野生动物保护法》（1988)

- 《中华人民共和国河道管理条例》（1988)

- 《中华人民共和国森林法》（1988)

- 《中华人民共和国环境保护法》（1989)

- 《中华人民共和国海洋石油勘探开发环境保护管理条例》（1990)

- 《中华人民共和国防止船舶污染海域管

2.3 Chinese Law, Regulations and Directives on Wetland Restoration

2.3.1 National laws governing the regulation of wetlands

Similar to the US, there is no explicit law or regulation in China for the conservation and restoration of wetlands. The articles applicable for the conservation and restoration of wetlands can be found in a wide range of different laws and regulations. The Constitution of the People's Republic of China (1978), The Criminal Law of the People's Republic of China (1979) and the Civil Code of the People's Republic of China (1987) all to some extent confirm the importance of environmental resources, including resources that relate specifically to wetlands. For example, "uninhabited lands" and "tidal marshes" are mentioned in the Constitution. However, the term "wetland" never itself appears in the three major laws associated with wetlands.

However, under current Chinese laws and regulations concerning natural resources and the conservation of ecosystems, the term "wetland" is explicitly mentioned. It is mentioned in the following: Regulations of Natural Reserves of the People's Republic of China (1994); the Marine Reserve Administrative Rules (1995); Forestry Funding Administrative Rules (1997); Marine Reserve Conservation Law of the People's Republic of China (1999); Forestry Outpost Administrative Rules (2000); Agricultural Law of the People's Republic of China (2002); and the Provisional Regulations on Compulsory Measures in City Planning (2002). Naturally, wetland conservation and restoration falls under the jurisdiction of all these laws and regulations.

Additionally, there are numerous laws and regulations closely associated with wetland conservation and restoration efforts, despite a complete lack of the term "wetland" in any one. They are:

- The Regulations on the Reproduction Conservation of Aquatic Resources (1979);
- Provisional Administrative Regulations of Sightseeing Areas (1985);
- Waterway Administrative Regulations of the People's Republic of China (1988);
- Wildlife Conservation Law of the People's Republic of China (1988);
- Forestry Law of the People's Republic of China (1988);
- Environmental Conservation Law of the People's Republic of China (1989);
- Administrative Regulations on the Environmental

理条例》（1990）

- 《中华人民共和国水土保持法》（1991）

- 《中华人民共和国陆生野生动物保护
  条例》（1992）

- 《中华人民共和国水生野生动物保护
  条例》（1993）

- 《中华人民共和国基本农田保护条
  例》（1994）

- 《中华人民共和国水污染防治法》
  （1996）

- 《中华人民共和国防洪法》（1997）

- 《中华人民共和国土地管理法》（1998）

- 《中华人民共和国渔业法》（2000）

- 《中华人民共和国草原法》（2002）

- 《中华人民共和国水法》（2002）

虽然上述法规没有出现"湿地"一词，但是其内容均与湿地保护与恢复密切相关。

从全国对自然资源保护法律法规中可以看出，对环境保护以及退化自然资源的恢复工作坚持"全面规划、合理布局、预防为主、防治结合、综合治理和污染者付费、利用者补偿、开发者保护、破坏者恢复"的原则。

目前，《中华人民共和国湿地保护条例》的起草工作基本完成，条例草案将"恢复保障湿地生态功能"作为立法目的之一，草案中规定"国家退化湿地所在地省、自治区、直辖市政府编制、实施拯救计划的职责"，要进行"对影响重要湿地生态功能的建设用地的事后监管"，建设单位和个人开展"湿地恢复和建设"或"依法缴纳生态补偿费的义务"，以调动单位和个人依法保护、恢复、重建和合理利用湿地的积极性。

Conservation in Oil Exploration & Exploitation of the People's Republic of China (1990);

- Administrative Regulations on Ships' Sea Pollution of the People's Republic of China (1990);

- Water & Soil Conservation Law of the People's Republic of China (1991);

- Implementation Regulations on Land-based Wildlife Conservation of the People's Republic of China (1992);

- Implementation Regulations on Water-based Wildlife Conservation of the People's Republic of China (1993);

- Principal Farming Fields Conservation Regulations of the People's Republic of China (1994);-

- Water Pollution Prevention Law of the People's Republic of China (1996);

- Flood Prevention Law of the People's Republic of China (1997),

- Land Administration Law of the People's Republic of China (1998);

- Fishery Law of the People's Republic of China (2000);

- Prairie Law of the People's Republic of China (2002);

- The Water Law of the People's Republic of China (2002).

A set of principles are evident in the above laws and regulations with regards to environmental conservation and the restoration of degraded natural resources. There are strategic planning guidelines at the national level that serve as a reasonable implementation blueprints at local levels. Prevention measures should precede comprehensive restoration efforts, and developing parties are responsible for wetland conservation, while parties that commit environmental infractions are responsible for the restoration.

For the time being, the drafting of Wetland Conservation Regulations of the People's Republic of China has been authorized by China's State Council, and the draft has adopted "restoring and guaranteeing wetland ecological functionalities" as a guiding principle and objective. As stated in the draft, provincial level governments are in charge of the formulation and implementation of restoration plans of their degraded wetlands, and are obligated to conduct a subsequent supervision of construction areas that could damage the ecological functions of important wetlands. On the other hand, developers or the owners of construction properties shall provide the necessary means for the restoration and rehabilitation of wetlands, or pay ecological compensation fees. This has been a proven incentive for all parties to commit themselves to the practical conservation, restoration, rehabilitation and wise use of wetlands.

2.3.2 与湿地保护相关的地方性法律法规

与湿地保护和恢复相关的地方性法律法规

我国各级地方政府为加强对湿地及其生物多样性的保护，恢复和保障湿地生态系统的基本功能，促进湿地资源的可持续利用，全面保护湿地，根据国家有关法律、法规，结合本地区实际情况，制定并颁布了一些与湿地保护有关的地方性法规、实施办法以及实施细则。

综合性保护湿地的地方法规

我国首部地方性湿地保护条例《黑龙江省湿地保护条例》于2003年8月1日正式实施，《甘肃省湿地保护条例》、《湖南省湿地保护条例》、《广东省湿地保护条例》和《陕西省湿地保护条例》也于2004年、2005年和2006年相继发布实施，这些地方性的湿地保护条例为国家湿地保护立法和其他地区的地方立法提供了经验和参考。在这些地方法规中，规定了湿地恢复的行为主体、监管机构以及补偿等事宜。

《黑龙江省湿地保护条例》（2003）的第8条规定"应当鼓励和支持退耕还湿等湿地恢复，并对在湿地保护和恢复工作中做出突出贡献的单位和个人给予表彰奖励"。第36条规定，在湿地内"擅自进行开发建设活动的"、"排放湿地水资源的"以及"挖沟、筑坝、开垦湿地的"，必须停止违法行为，恢复湿地原状或采取其他补救措施，并依情节轻重处以相应数额的罚款。

《甘肃省湿地保护条例》（2004）第8条规定应当采取一定的措施，保护和恢复湿地功能，对在保护区内"因缺水导致湿地功能退化的，应当建立湿地补水机制，定期或者根据恢复湿地功能需要有计划地补水"，并在第20条规定了依情节轻重处以相应数额的罚款。第21条规定在非保护区内"擅自进行采挖、爆破、倾倒废弃物等活动的"、"擅自开发利用湿地或占用湿地的"、"在天然湿地内修建设施的"、"引进有害生物物种的"，应"由县级以上湿地行政主管部门或其委托的湿地管理机构责令其停止违法行为，限期恢复原状"。

2.3.2 Local laws and regulations

Local laws and regulations on wetland conservation and restoration

Based on national laws and regulations, local governments at different levels have formulated and issued a series of laws, regulations, policy procedures and implementation rules for the conservation of wetlands and their biodiversity. These laws aim for the restoration and maintenance of the principle functionalities of wetland ecosystems, the sustainable use of wetland resources, and the comprehensive conservation of wetlands in China.

Comprehensive local conservation laws and regulations

The Wetland Conservation Regulations of Heilongjiang Province, which went into effect on August 1, 2003, became the first set of local regulations on wetland conservation. Shortly thereafter, the Wetland Conservation Regulations of Gansu Province and Hunan Province went into effect in 2004 and 2005, respectively. These pioneering laws have become reference points and templates for legislation at both national and local levels. These ground-breaking regulations dictate the responsible parties, supervising organizations and compensation amounts for wetland conservation.

Article 8 of Heilongjiang's Regulations (2004) encourages restoration efforts such as converting farmlands back into wetlands and recommends awarding parties that have made outstanding contributions to wetland conservation and restoration. Article 36 of the regulations proscribe unauthorized activities in wetlands, such as construction, the illegal draining of water resources, trench-digging, dam-building and farm cultivation. Parties that infringe on these laws are not only subject to fines in accordance with their offence, but are obligated to restore wetlands to their original status or provide their equivalent in compensation.

The Wetland Conservation Regulations of Gansu Province (2004) call for specific measures (Article 8) to conserve and restore wetland functionalities in conservation areas. For example, for wetland degradation resulting from water shortages, a water supply system should be installed to replenish water levels to the wetland in question, as needed.

Article 20 of the Gansu regulations provides guidelines for fines that correspond to the level of an offense. There are further regulations that outlaw unauthorized activities in non-conservation areas of wetlands, such as mining, the

《湖南省湿地保护条例》（2005）的第11条规定"县级以上人民政府应当采取措施，对退化的湿地进行恢复改造"，并"鼓励和支持自愿从事湿地恢复改造的活动"。第12条还规定了"对可控水位的重要沼泽类型湿地确定合理的水位"，"当水位出现异常时"，"应当采取恢复合理水位的相应措施"。

针对某块湿地的专门保护条例

江西省和上海市分别出台了针对其辖区内的鄱阳湖湿地和九段沙湿地的保护条例，对两块湿地恢复也做了详细的规定。

《江西省鄱阳湖湿地保护条例》（2003）的第28条中规定了鄱阳湖湿地区域人民政府在湿地恢复中应起到的作用，认为鄱阳湖湿地区域人民政府应当加强水土保持和湿地植被保护与恢复工作，严禁毁草开垦。并且经鄱阳湖湿地区域人民政府划定的植被恢复区，应当实行封洲禁牧。第35条规定了湿地资源利用补偿的原则，提出了开发利用鄱阳湖湿地资源应当坚持谁开发谁保护、谁破坏谁恢复、谁利用谁补偿的原则。

《上海市九段沙湿地自然保护区管理办法》（2003）中第20条规定"对保护区生态环境造成严重破坏的，责令限期恢复原状或者赔偿损失"，并可处相应数额的罚款。

针对某种类型湿地的保护条例

《福建省海洋环境保护条例》（2002）中明确了破坏湿地承担恢复责任的主体，规定"造成海岛地形、岸滩及海岛周围海域生态环境破坏的，开发者应当承担整治和恢复的责任"。

《海南红树林保护规定》（2004）的第6条规定"禁止任何单位和个人占用或者征用自然保护区内的红树林地"（见图2-14），"必须占用或者征用红树林地的"，要"由用地单位依照国家有关规定缴纳森林植被恢复费用"；第12条规定"破坏红树林自然保护区生态环境的"，要"限期恢复原状或者采取其他补救措施"。

《江苏省湖泊保护条例》（2004）的第

use of explosives, waste disposal, developing or occupying the wetlands, and building artificial facilities in natural wetlands. The regulations are under the supervision of wetland management agencies organized or entrusted by the local administration just above the county level government. Parties that violate these provisions should restore any wetland in question to its original status over a prescribed period.

The Wetland Conservation Regulations of Gansu Province (2005) also instruct local administrations above the county level to take measures to restore degraded wetlands and encourage voluntary activities for wetland restoration. Article No. 12 demands that standard water levels be set for important swamp-type wetlands whose water levels are controllable, ensuring that measures can be taken in the case of fluctuations or irregularities to restore water levels back to normal.

Special conservation regulations for specific wetland sites

Special conservation regulations for specific wetland sites have been issued in Jiangxi Province and in the Shanghai Municipality, providing detailed stipulations on the conservation of Boyanghu Lake and Jiuduansha wetlands, respectively.

Conservation Regulations for the Boyanghu Lake Wetlands in Jiangxi Province (2003) define a critical role for the government in the restoration of the lake's wetlands (Article 28): the local government covers water and soil conservation, the conservation and restoration of wetland vegetation, and prevents their cultivation into farmlands. The government additionally has the power to decide the appropriate ratio between restoration for wetland vegetation and pastures. Article 35 provides principles on the compensation for the wetland resources utilization.

Shanghai's Natural Reserve Administrative Rules for the Jiuduansha Wetlands (2003) stipulates that parties causing severe damage to the natural reserve ecosystem shall restore the site in question to its original status or compensate for the damages caused, and are further subject to fines based on the gravity of the offence (Article 20).

Conservation regulations for specific types of wetlands

The Conservation Regulations for the Marine Environments of Fujian Province (2002) stipulates that parties responsible damaging the wetlands and/or causing ecological damage to an island's topography, seashore or sea area around

17条规定"已经围垦或者圈圩养殖的，批准湖泊保护规划的人民政府应当按照防洪规划的要求和恢复湖泊生态条件的需要，制定实施退田（渔）还湖、退圩还湖方案的计划"，并且规定了实施"还湖计划"的资金筹措方案。第二十二条中规定"围湖造地或者在湖泊保护范围内圈圩的，由县级以上水行政主管部门责令其停止违法行为、恢复原状"，并规定了具体的处罚金额。对于"不恢复原状的，由县级以上水行政主管部门指定有关单位代为恢复原状，所需费用由责任人承担"。

由此可见，许多省份都根据各自的湿地类型制定了地方性的法律法规，在全国性湿地法律出台之前，这些法律法规起了积极的作用。但是，和美国的《清洁水法案》一样，这些法律法规在监管方面的力度不够，另外，一些罚款和补偿方面的条款较为模糊，导致了具体操作上的漏洞。

其他相关的地方性法律法规
为加强水资源的管理，保护水资源，发挥水资源的综合效益，实现水资源可持续利用，我国各地方政府根据《中华人民共和国水法》、《中华人民共和国环境保护法》、《中华人民共和国水污染防治法》等法律、法规，结合当地实际情况，制定了地方性的保护水资源的法律法规，这些法律法规是保护和恢复湿地的有益补充。

▼ 图2-14: 海南省万宁县红树林
（摄影：Rowan Roderick-Jones）
Figure 2-14: Mangroves near Wanning, Hainan
(Photograph by Rowan Roderick-Jones)

islands in its development projects are responsible for the restoration of the site in question.

Mangrove Conservation Regulations for Hainan Province (2004) dictate that occupying and/or acquiring mangrove sites inside natural reserves are prohibited, unless a compensation fee for vegetation restoration is paid according to relevant national regulations (Article 6). The rules further instruct that any party causing damage to a mangrove reserve's ecosystem shall restore the site in question to its original status, or take equivalent restoration measures (Article 12).

The Lake Conservation Regulations of Jiangsu Province (2004) provision for the local government to convert lake areas already encircled with dykes for proximate farmlands and fisheries back into lake areas with flood-prevention requirements that serve a lake's ecological needs (Article 17). The article further provides funding packages for these conversions.

A further provision abrogates the practice of encircling lake areas with dykes in order to make farmlands and fisheries, a practice immediately effective upon implementation and under the supervision of water management agencies above the county level. Responsible parties in violation of these regulations will restore the site in question to its original status and is again subject to fines that are based on the level of transgression (Article 22). Moreover, if the responsible party fails to restore the site in question to its original status, a deputy organization assigned by water management agencies above the county level will restore the site at the expense of the responsible party.

Other relevant local regulations
Based on the national Water Law, Environmental Conservation Law, and the Water Pollution Prevention and Control Law of the People's Republic of China, local governments at different levels have formulated and issued local regulations on water resources conservation that conform to the guidelines set by the above laws.

The regulations intend to strengthen the management of water resources, improve water quality and ensure safe water use. By encouraging wise development, utilization, economization and the conservation of water resources, these local regulations aim at the efficient and sustainable use of water resources. As a result, local regulations have been a beneficial complement to their national counterparts in wetland conservation and restoration.

2.3.3 国家规划与导则中的湿地恢复

中国湿地保护行动计划

在《中国湿地保护行动计划》中制订的中国湿地保护与合理利用行动目标中，提出了"实施封山植树、退耕还林，平垸行洪、退田还湖，以工代赈、移民建镇，加固干堤、疏浚河湖的湿地综合治理措施，建立退化湿地的恢复与合理利用的示范"。

到2020年 "力争使退化湿地得到不同程度恢复治理"，"一批重要的湿地资源得到恢复"。并提出了"退化湿地生态系统恢复与重建技术研究、示范（项目8）"、"长江中下游湿地的恢复和重建（项目21）"等以湿地恢复为主的保护行动。

中国湿地保护战略研究中的湿地恢复

湿地恢复越来越受到国家的重视，在制定中国湿地保护战略中，湿地恢复占有重要位置。在我国湿地保护的总目标中提到通过湿地及其生物多样性的就地保护与污染控制、土地利用方式调整等管理，形成自然湿地保护体系，全面维护湿地生态系统的生态特性和基本功能，使湿地面积萎缩和功能退化的趋势得到扭转。实施重点生态区域退化湿地的恢复和治理，有计划地恢复自然湿地及其生态功能。

在我国湿地保护的战略重点中也要求大力推进退化湿地的生态恢复。以逐步恢复和重建退化湿地生态系统、促进受威胁的湿地物种恢复为出发点，开展先进的科学研究、将修复技术与退化湿地处理相结合的方式应用于湿地恢复中。

通过建立湿地生态恢复示范并注重政策、法制等社会性对策，利用社会的力量和作用推进湿地生态修复，鼓励社区、利益部门和民众的主动参与，示范湿地重建对社会带来的益处。此外，除了直接针对湿地恢复和重建的法律外，还需要其他法律的辅助——因为湿地退化的原因很多，有自然的，也有人为的，在恢复和重建退化湿地时，应该尽可能地根除人类活动对湿地的负面影响。其他如针对物种管理、水质改善、水利工程方面的法律应该协助起到积极的作用"（见图2-15，2-16）。

2.3.3 Wetland restoration in national planning and directives

Action Plan for China's Wetland Conservation

The Action Plan for China's Wetland Conservation, published in 2000, aims for the conservation and wise use of Chinese wetlands. It includes objectives for cordoning off mountains for re-forestation and the conversion of farmlands back into forests, the removal of lake enclosure dykes for flood prevention, and the conversion of dyke-enclosed farmlands back into lake-land areas.

It was far reaching enough to help displaced farmers resettle in adjacent townships and to provide them with new jobs as compensation. The plan also calls for reinforcing major dams and dredging lakes and rivers, establishing demonstration projects for wetland restoration and wise use and ensuring that by 2020, restoration projects are underway in all degraded wetlands in China and completed at a number of important wetland sites. The Action Plan recommends specific conservation actions aimed at wetland restoration, such as Project 8, which calls for the research and demonstration of technologies for the restoration and rehabilitation of degraded wetland ecosystems. Similarly, Project 21 provides a restoration action plan for the Yangtze River Basin.

Wetland Restoration as Part of a National Strategy for Wetland Conservation

In recent years, wetland restoration has become an exigent national concern and conservation strategies have become a high priority. The broad goals of Chinese conservation efforts converge around plans for conservation systems of natural wetlands that protect biodiversity, and provisions for pollution control and land use adjustments. The aim is to maintain the overall ecological character and principal functions of wetland systems. At the same time, success will be measured in a strategy's ability to slow or halt the trend of shrinking wetland areas and degraded functionalities.

The extensive adoption of the ecological restoration of wetlands starts with the rehabilitation of degraded ecosystems and the restoration of endangered species. It calls for applying the latest scientific research, technologies and methods towards that end.

Critical public outreach includes restoration demonstrations to encourage communities and societies to participate in protection efforts by showing the human benefit of wetlands rehabilitation. Multiple causes, both natural and human, result in the degradation of wetlands, which further results in their diminished ecological capacities. All of this affects and threatens our very existence. A key premise of any restoration

▼ 图2-15：深圳大鹏半岛的鱼池曾经是红树林湿地（摄影：Rowan Roderick-Jones）
Figure 2-15: These fish ponds on Shenzhen's Dapeng Peninsula have been reclaimed over time from former mangrove wetlands (Photograph by Rowan Roderick-Jones

▼ 图2-16：天津团泊湖边的农业活动能对当地经济起到一定的贡献作用，但是当地的湿地却持续受到人为干预（摄影：Rowan Roderick-Jones）
Figure 2-16: Reed harvesting at Tuanbo Lake near Tianjin benefits the local economy but constitutes a significant human disturbance to the natural ecosystem (Photograph by Rowan Roderick-Jones)

国务院办公厅关于加强湿地保护管理的通知

2004年，在"国务院办公厅关于加强湿地保护管理的通知"中提到"对开垦占用或改变湿地用途的，应责令停止违法行为，采取各种补救措施，努力恢复湿地的自然特性和生态特征，并严格按照有关法律、法规予以处罚"。在相关部门制定土地利用总体规划、海洋功能区域规划时，要"确保自然湿地能够得到有效保护和恢复"。同时，"要把加强湿地保护，恢复湿地功能，作为改善生态状况和全面建设小康社会的一件大事，予以高度重视，并切实抓紧抓好"。

全国湿地保护工程规划

国家有关部门共同编制的《全国湿地保护工程规划》（2002—2030年，以下简称"规划"）得到了国务院批准。该规划明确说明了到2030年，我国湿地保护工作的指导原则、任务目标、建设布局和重点工程，对指导开展中长期湿地保护工作具有重要意义。在《规划》的总体目标中提出，要通过加强对水资源的合理调配和管理、对退化湿地的全面恢复和治理，使丧失的湿地面积得到较大恢复，使湿地生态系统进入良性状态。到2030年，使全国湿地保护区达到713个，国际重要湿地达到80个，使90%以上天然湿地得到有效保护。完成湿地恢复工程140.4万hm²，在全国范围内建成53个国家湿地保护与合理利用示范区。

《规划》将我国湿地按地域划分的东北湿地区、黄河中下游湿地区、长江中下游湿地区、滨海湿地区、东南和南部湿地区、云贵高原湿地区、西北干旱湿地区以及青藏高寒湿地区等八个区域，均将湿地恢复作为建设重点。

- 在东北湿地区，重点通过湿地保护与恢复及生态农业等方面的示范工程，建立湿地保护和合理利用示范区，提供东北地区湿地生态系统恢复和合理利用模式；

- 在黄河中下游湿地区，重点加强黄河干流水资源的管理及中游地区的湿地保护，利用南水北调工程尝试性地开展湿地恢复的示范；

- 在长江中下游湿地区，重点通过退

and rehabilitation project for a degraded wetland is the elimination of negative interference from human activities.

Directives from the State Council Office on Strengthening Wetland Conservation Administration

The State Council's Directives on Strengthening Wetland Conservation Administration, issued in 2004, decreed the immediate cessation of converting wetlands into farmlands and the further alteration of wetland functionalities. Parties that infringe upon this ruling shall take measures to restore the natural and ecological character of the wetland prior to being fined. Government offices will further ensure that natural wetlands receive adequate conservation and restoration action plans on general land use. This applies to maritime functional zones as well. The directive is clear that governments at different levels should make a wetland agenda a high priority.

The National Planning of Wetland Conservation Projects

The National Planning of Wetland Conservation Projects (2004 – 2030), jointly compiled by national government offices and approved by the State Council, states the guiding principles, objectives, construction layouts and key projects for wetland conservation in China by 2030. The document is an important roadmap for mid- to long-term commitments on wetland conservation.

The main components of the National Planning strategy include strengthening the wise allocation and administration of water resources and enhancing the overall restoration and rehabilitation of degraded wetlands. The ultimate goal is to recover a substantial amount of wetland areas from degradation and to induce benign cycles for wetland ecosystems. By the year 2030, it is expected that there will be 713 wetland natural reserves and 80 wetland sites of international importance in China, while over 90% of all natural wetlands in China will be placed under effective conservation protection. A total of 1,404,000 ha of wetland restoration will have been completed, and 53 national demonstration zones of wetland conservation and wise use plans will have been established.

The National Planning blueprint divides China's wetlands into eight regions: Northeast China, Yellow River (Middle and Downstream), the Yangtse River (Middle and Downstream), Coastal Zone, Southeast and South China, the Yunnan-Guizhou Plateau, the Northwest China Dry Zone, and the Qinghai-Tibet Plateau. Wetland restoration is a key element for ecological reconstruction in all the eight regions.

- Northeast China: Demonstration zones on wetland conservation and wise use will be established for wetland

田还林、还湖、还泽、还滩、还草及水土保持等措施，使长江中下游湖泊湿地的面积逐渐恢复，改善湿地生态环境状况；

- 在滨海湿地区，重点以生态工程为技术依托，对退化海岸湿地生态系统进行综合整治、恢复与重建；

- 在东南和南部湿地区，重点加强水源地保护和流域综合治理，在河流源头区域及重要湿地区域开展植被保护和恢复措施；

- 在云贵高原湿地区，重点加强流域综合管理，保护水资源和生物多样性，进行生态恢复示范，对高原富营养化湖泊进行综合治理，恢复和改善湿地生态环境；

- 在西北干旱湿地区，重点加强天然湿地的保护区建设和水资源的管理与协调，采取保护和恢复措施缓解西部干旱荒漠地区由于人为和自然因素导致的湿地环境恶化、湿地面积萎缩甚至消失的趋势；

- 在青藏高寒湿地区，重点加强保护区建设及植被恢复等措施，保护世界独一无二的青藏高原湿地。

《规划》安排的5个方面的重点建设工程就包括湿地恢复优先工程，主要包括两个方面的内容：一是加强水资源的调配与管理，确定全国、流域和省区水资源配置方案及水资源宏观控制指标体系和水量分配指标。在重要湿地区和重要河流流域开展水资源调配与管理工程，适当增加关键区域生态用水比例，逐步恢复原有的湿地生境。二是开展湿地恢复和综合整治工程，包括在生物多样性丰富的低产农田区实施退耕还湖（泽、滩）工程，在退化和被改造的滩涂区实施恢复与重建工程，在土地沙漠化趋势严重的湿地区实施工程退牧育林还草、封沙育林育草、休牧（轮牧）育林育草工程，已退化沼泽草地进行改良，恢复天然植被和水禽栖息地，在沿海退化红树林地区进行红树林生态恢复工程的试验示范等。

and ecological agriculture to showcase implementation modes of wetland restoration and wise use across the region.

- Yellow River (Middle and Downstream): The region is focusing on the administration of water resources of the primary Yellow River and for wetland conservation (at mid-stream for the latter). Demonstration projects will be underway, tentatively based on the massive "South-to-North" water diversion project.

- Yangtze River (Middle and Downstream): Inappropriately cultivated farmlands will be converted back to forests, lakes, swamps, shores and grasslands, while water and soil conservation efforts will be a high priority. The lake area in the region is expected to expand and wetland ecosystems to show marked improvements.

- Coastal Zone: The focus is on ecosystem improvement projects that serve as a technical basis for the comprehensive restoration and rehabilitation of shoreline wetlands.

- Southeast and South China: The region is focusing on the conservation of river sources and a comprehensive rehabilitation of river basins. Conservation and restoration of vegetation projects will focus on river sources and in important wetlands.

- Yunnan-Guizhou Plateau: Its tasks are the comprehensive administration of river basins, water resources and of the region's notable biodiversity. Other objectives include the establishment of demonstration zones for ecosystem restoration, the overall rehabilitation of over-eutrophic lakes in the Plateau, and the restoration and improvement of wetland ecosystems.

- Northwest China Dry Zone: The region will focus on establishing natural wetland reserves and the administration and coordination of water resources. Restoration measures and improvement actions will be taken to address the chronic degradation and the dwindling of wetland environments due to human activities and natural elements.

- Qinghai-Tibet Plateau: Efforts are underway to establish natural reserves and restore vegetation in order to conserve unique wetlands located on the Plateau.

The priority projects designated for wetland restoration, or five key projects stipulated in the National Planning directive, share two aspects. The first is enhancing the

全国湿地保护工程实施规划（2005—2010年）

国务院近期批准的《全国湿地保护工程实施规划(2005—2010年)》提出，5年内计划投资90亿元，优先启动湿地保护、湿地恢复、可持续利用示范和能力建设等4项重点建设工程，通过加大湿地自然保护区建设和管理，期望至2010年能够使我国50%的自然湿地、70%的重要湿地得到有效保护，基本形成自然湿地保护网络体系。

在湿地恢复的建设工程中，重点对国家级自然保护区和国家重要湿地区域内的退化湿地进行恢复，计划恢复各类湿地58.8万hm²。在吉林向海、黑龙江扎龙、黑龙江洪河、山东黄河河口、黄河禹门口－潼关段河滩、河北白洋淀和衡水湖、新疆塔里木河下游、内蒙古黑河下游居延海、江苏太湖、云南滇池、山东南四湖等急需补水的重要湿地实施生态补水示范工程。

实施湿地污染控制工程，包括选择污染严重、生态价值大的江苏阳澄湖和滆湖、新疆环博斯腾湖、内蒙古乌梁素海开展富营养化湖泊湿地生物控制示范。

实施湿地生态恢复和综合整治工程，包括退耕(养)还泽(滩)示范工程11万hm²，湿地植被恢复工程31.6万hm²，栖息地恢复工程24.3万hm²，红树林恢复1.8万hm²。

▼ 图2-17: 简单却巧妙的建筑可降低对湿地的干扰。香港米埔湿地内的观鸟屋可以让游客在观鸟的同时免于干扰自然环境
(摄影：Rowan Roderick-Jones)
Figure 2-17: Human disturbances in restored wetlands can be reduced by creating ways for humans to interact subtly with the environment. This bird-blind at the Mai Po wetlands in Hong Kong allows wildlife to remain undisturbed under the visitor's eye. (Photograph by Rowan Roderick-Jones)

allocation and administration of water resources by devising plans for water resource allocation at national levels, catchment levels and provincial levels. This includes establishing a macro-control index system and allocation quotas for water resources, the allocation and administration of water resources for important wetlands and major river areas. Ecologically important areas will receive a larger water allocation quota, with the aim of gradually restoring wetland habitats to their original status.

The second key aspect will be extensive measures for comprehensive wetland restoration and rehabilitation. This includes: converting low-yield, high-biodiversity farmlands into lakes, swamps and shores; conducting restoration and rehabilitation in degraded and converted tidal marshes; brining forestation into heavily sanded wetlands and removing or alternating pasturage for rehabilitation; or improving the ecosystem of degraded swamp grasslands. It could further include restoring natural vegetation and waterfowl habitats, and establishing demonstration zones for ecosystem restoration in degraded seashore mangroves.

Implementation Plan of the National Wetland Conservation Project (2005 – 2010)

The Implementation Plan of the National Wetland Conservation Project (2005 – 2010), recently approved by the State Council, recommends that RMB 9 billion (US$ 1.12 billion) be invested in four priority wetland restoration projects, for rehabilitation, sustainable use demonstration sites, and for wetlands reconstruction over the next five years. The emphasis on the construction and administration of natural wetland reserves will engender a network of natural wetland conservation efforts, expected to be established by 2010, with benchmarks that place 50% of all natural wetlands and 70% of important wetlands into effective conservation.

The focus of the above is on the restoration of degraded wetlands in national natural reserves and wetland areas of national importance. Up to 588,000 ha of various wetland types are expected to receive restoration. Demonstration projects on improving ecological water supplies will be underway in important wetlands with severe water shortages, covering vast areas such as Xianghai, Jilin Province; Zhalong, Heilongjiang Province; Honghe, Heilongjiang Province; the Yellow River estuary in Shandong Province; the Yumenkou–Tongguan section of the Yellow River; Baiyangdian Lake and Hengshuihu Lake in Hebei Province; downstream of Talimuhe River in the Xinjiang Autonomous Region; Huyanhai Lake, downstream

旧金山湾国家野生动物栖息地总体规划和设备规划
美国加州旧金山
业主：美国内政部
San Francisco Bay National Wildlife Refuge
Master Plan and Facilities Program
San Francisco, California
Client: US Department of the Interior

from the Heihe River, Inner Mongolia Autonomous Region; Taihu Lake, Jiangsu Province; Dianchi Lake, Yunnan Province; and Nansihu Lake, Shandong Province.

Demonstration projects for wetland pollution control, especially for the biological control of eutrophic lakes, will be undertaken in heavily-polluted wetlands that otherwise have a high biodiversity value. The latter includes areas such as Yangchenghu Lake and Zhenhu Lake, Jiangsu Province; Huanbositenghu Lake, Xinjiang Autonomous Region; and Wuliangsuhai Lake, Inner Mongolia Autonomous Region.

Upon completion of these ambitious rehabilitation and restoration projects, it is expected that 110,000 ha of farmlands and pastures will be converted back into swamps and shores as demonstration zones, while 316,000 ha of wetland vegetation, 243,000 ha of natural habitat and 18,000 ha of mangroves will be recovered.

第 三 章
Chapter　3

湿地恢复的设计过程
Wetland Restoration Design Process

湿地恢复是一项复杂的工程，需要综合不同专业的知识，包括生物学、生态学、水文学、水力学、地理学、经济学以及规划等领域。不同的湿地恢复项目有不同的目标，也因而可能涉及更多更专业的知识。比如，为了吸引某些鸟类在自然保护区内的湿地觅食，就需要水禽或涉禽专家在栖息地设计方面提供建议。同样，面对全球海平面上升的影响，岛屿的滨海湿地的恢复便可能需要全球变化的专业分析。因此，如何设计和进行湿地恢复是一个科学决策过程，包括了目标设立、过程设计、工程实施和生态监测等各个环节。只有经历一个较为完整的决策过程，湿地恢复才能有效地融合各领域的知识和顾及各方的要求，反映科学家对湿地发育和演替方向的判断、区域规划的要求、土地所有者的意图和能力、公众意识等。

与开发和利用湿地相比，湿地恢复只有很短的历史。美国是较早开始恢复湿地的国家，在20世纪80年代，《清洁水法》的执行推动了"湿地补偿"，即"湿地恢复"的前身。但是，当时采用植被覆盖面积作为湿地补偿评价因数的做法现今已被认为有失偏颇，因为湿地功能的复原程度才是衡量湿地生态恢复程度的有效指标。这个认识上的变化改变了湿地生态恢复设计的过程。本章将在借鉴美国湿地生态恢复的经验基础上，结合中国的湿地状况以及社会经济发展特点，为读者提供一个适用于中国国情的湿地生态恢复设计过程。

3.1 确定问题

湿地是一个动态系统，因此必须将湿地的正常变化同其所面临的生态、环境变化区分开来（见图3-1）。季节性波动是湿地的自然特色。内陆湿地具有洪季和枯季的变化，而受到海陆交互作用影响，滨海湿地的面积和形状亦会有所改变。但是，湿地的生态恢复旨在缓解和消除不良的生态和环境影响，将湿地恢复到可自我更新、自我维持、自我调节的动态系统。

目前中国湿地普遍面临的生态和环境问题有：

1. 生物多样性下降：可能的原因包括生境的丧失和破碎化、饵料减少、

Wetland restoration is a complex process that requires a comprehensive understanding of multiple disciplines such as biology, ecology, hydrology/hydraulics, geography, economics and planning. The different goals of restoration projects often require more specialized knowledge at varying areas of expertise. For example, for wetlands inside a nature reserve where attracting birds is a project goal, experts on water or wading fowl must be consulted for habitat landscaping. Or, when the restoration of marine wetlands is affected by a rise in global sea levels and may therefore require the expertise of a marine biologist. The design and implementation of wetland restoration warrants a scientific decision-making process comprised of interconnected steps that include goal-setting, procedural designing, project implementation and ecological monitoring. It is only through this kind of complete decision-making process that a restoration project may incorporate knowledge from different disciplines, weigh the interests of different parties, consider scientific conclusions in wetland development and succession, and consider regional planning requirements, land owner intentions and capabilities, as well as public opinion.

Compared to the considerable span of time that marked the degradation and destruction of wetlands, wetland restoration has had a relatively short history with regards to development and use. The United States, one of the forerunners in wetland restoration, passed the Clean Water Act in 1972, which required "wetland mitigation", an early version of wetland restoration, for wetland losses. However, the Act evaluated a restoration's success by the size of vegetation coverage in the wetland site, a practice that is today considered insufficient, where standards dictate that only the recovery of wetland functions can serve as an effective index of wetland ecological restoration. This current understanding has completely changed the process of wetland ecological restoration and this chapter aims to draw on the experience of wetland restoration practices in the US in order to provide insight into a restoration design process that is applicable to China, based on the specific character of a wetland's conditions and China's socio-economic development.

3.1 Problem Statement

Wetlands are dynamic systems, the natural functions of which can be altered by anthropogenic, ecological and environmental forces. Some fluctuations, such as those that occur during the course of several seasons, are a natural trait of wetlands. For example, inland wetlands can change dramatically between rainy and dry seasons, while marine wetlands may change in area and shape due to

过度捕猎、疾病爆发、外来物种入侵和气候变化等；

2. 水文状况持续改变而导致的侵蚀或干旱：可能的原因包括沉积物含量的变化、流域水利工程、气候变化和生物作用等；

3. 水质恶化以及富营养化：可能的原因包括点源污染、非点源污染（包括干湿沉降）、生物群落变化等。

land-sea interactions. Natural equilibrating processes allow wetlands to mitigate negative impacts resulting from ecological and environmental factors, and maintain a dynamic self-renewing, self-sustaining and self-regulating system. Figure 3-1 shows how human and ecosystem interactions are taken into consideration as part of a master plan for a new development community bordering a flood control lake with a proposed bird sanctuary in Tianjin, China.

Currently, wetlands in China face the following ecological and environmental problems:

1. Declining biodiversity: possibly resulting from loss and fragmentation of habitats, prey shortages, over-hunting, disease outbreak, foreign species invasion and climate change.

2. Erosion and droughts due to continual hydrological alterations: possibly resulting from changes in sediment concentrations, watershed irrigation constructions, climate changes and biological actions.

3. Declining water quality: possibly resulting from point source pollution, non-point source pollution (including wet and dry subsidence), and biological community change.

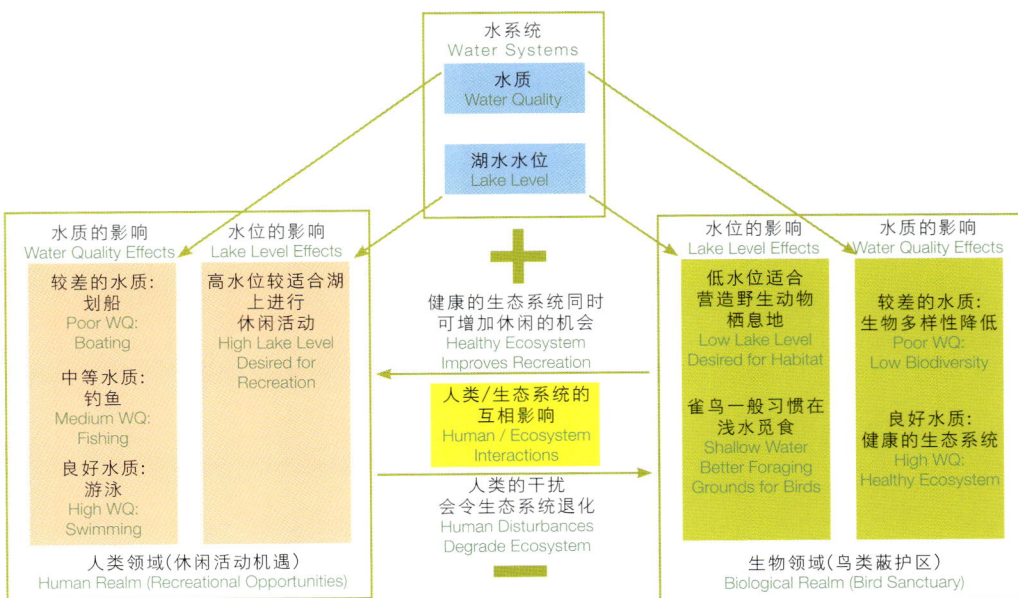

水系统
Water Systems

水质
Water Quality

湖水水位
Lake Level

水质的影响
Water Quality Effects

水位的影响
Lake Level Effects

较差的水质：
划船
Poor WQ:
Boating

中等水质：
钓鱼
Medium WQ:
Fishing

良好水质：
游泳
High WQ:
Swimming

高水位较适合湖上进行休闲活动
High Lake Level
Desired for
Recreation

健康的生态系统同时可增加休闲的机会
Healthy Ecosystem
Improves Recreation

人类/生态系统的互相影响
Human / Ecosystem
Interactions

人类的干扰会令生态系统退化
Human Disturbances
Degrade Ecosystem

人类领域（休闲活动机遇）
Human Realm (Recreational Opportunities)

水位的影响
Lake Level Effects

水质的影响
Water Quality Effects

低水位适合营造野生动物栖息地
Low Lake Level
Desired for Habitat

雀鸟一般习惯在浅水觅食
Shallow Water
Better Foraging
Grounds for Birds

较差的水质：
生物多样性降低
Poor WQ:
Low Biodiversity

良好水质：
健康的生态系统
High WQ:
Healthy Ecosystem

生物领域（鸟类蔽护区）
Biological Realm (Bird Sanctuary)

◄ 图 3-1：都市生态系统管理：天津团泊湖水系与人类和生物领域之间的相互影响（图片来源：易道）
Figure 3- 1: Urban Ecosystem Management Program: Interactions between water systems, and human and biological realms at Tuanbo Lake, Tianjin (Copyright EDAW)

除了一些已经显现的生态及环境问题之外，湿地的生态系统还有可能存在潜在的风险，随着时间的推移而逐渐显露。比如，有机污染物的排放浓度即使不高，但可能会通过食物链的累积作用而影响高等动物乃至人类自身。此外，另有一些潜在风险是人为因素，也会影响湿地恢复的效果，就如流域上游的水利工程如果拦截过多泥沙的话，便会导致下游河口地区不可逆转的滩涂蚀退（见图3-2）。这些风险应在进行湿地恢复工作之前加以明确，并通过湿地恢复来减少甚至去除潜在风险，或者通过调整恢复方式或位置来回避风险带来的影响。一般来说，湿地的潜在风险来自：

- 即将进行的土地利用规划；

- 流域上中游的变化；

- 全球变化（如海平面上升、全球变暖等）；

- 湿地长期的缓慢变化，如有机污染物的累积和传递等。

一块湿地同时会存在多个问题和面对许多风险，比如中国东部的许多湖泊都同时间面临了水质恶化、生物多样性下降、湖泊淤浅等问题，但是对于湿地恢复而言，首要工作是明确各主要问题的特征并确立一个或若干议题作为主要的针对性问题，以确立必须采用的措施和技术。确立主要问题和风险的依据有：湿地管理者的意愿（比如，水质恶化是社区建设和水域管理中要解决的主要问题）、生态和环境的退化程度、各问题之间的关联性及其它相互关系、湿地的潜在风险、以及公众意愿等。

Apart from apparent ecological and environmental problems, there are additional potential hazards that chronically plague wetland ecosystems. For example, organic pollutant discharge, even in low concentrations, may affect higher mammals and humans through accumulation in the food chain. Some potential hazards are entirely artificial and may negate the efforts of wetland restoration. For instance, excessive sedimentation from irrigation practices in upstream portions of a watershed may lead to irrevocable mudflat erosion in the downstream estuarine area. These hazards must be identified before restoration and countermeasures are established in the project design. Potential threats to wetland ecosystems generally include:

- Planned land uses;

- Upstream and midstream changes in the watershed;

- Global environmental changes (such as rising sea levels and global warming); and

- Long-term, gradual changes, such as the accumulation of organic pollutants.

A single wetland site can face multiple problems and hazards. For instance, many lakes in eastern China suffer simultaneously from water deterioration, a declining biodiversity, and stagnation. The principal task in wetland restoration is determining the character of these problems and identifying one or more key issues as top priorities, so that appropriate countermeasures may be taken and restoration technologies applied accordingly. The identification of key threats and issues is a function of the intention of those managing the wetland, the level of ecological and environmental degradation, and the relationship and interaction between these problems, and public opinion.

图 3 - 2: 安徽蚌埠龙子湖在未处理的污水
侵蚀的影响下，岸线和滨水湿地发生退
化 (摄影 : Rowan Roderick-Jones)
Figure3- 2: Degradation of shoreline and riparian
wetlands from raw sewage and erosion at Dragon
Lake in Bengbu, Anhui Province (Photograph by
Rowan Roderick-Jones)

3.2 确立湿地恢复的目标、意义

由于湿地种类繁多,生态服务功能丰富,并且所处的社会、经济背景不同,不同地方的湿地恢复必然有不同的目标。但是,差异之中,一些恢复方向却是共同的,可为日后恢复项目的目标确定提供指导(USEPA, 2000)。

1. 保护和保育水生生物资源:湿地是大部分水生生物的栖息地,因此,其生态恢复通常是保护和保育水生资源的最直接、最有效的方式。

2. 恢复生态完整性:一个退化的湿地生态系统在其结构、过程上都有缺憾,比如,食物链上重要环节的缺失等。一个完整的生态系统是一个具有弹性的、可自我维持的自然系统,能够承受一定程度的胁迫和干扰。在生态恢复中,必须考虑到湿地的一些主要过程,比如关键种群增长、营养物质循环(见图3-3)、演替、水文过程以及沉积物的冲淤动态等。恢复这些过程的完整性应该成为湿地恢复的主要目标。

3. 恢复湿地的自然水系:水文是湿地的最重要特征之一,决定了湿地的植被类型和其他生物群落。许多湿地资源的丧失都与水系的改变有关,比如河流渠化、任意调水、过度排水、水流切断等。如果水系得不到调整和恢复的话,湿地恢复只会事倍功半,甚至影响生物群落的恢复。如果水系的改变不能逆转,比如潮滩湿地的过度淤涨,就必须通过水文调查、生物群落调查和经济分析来确定湿地恢复是否可行。若可行的话,则必须谨慎设计水系的结构,以满足恢复湿地以及目标物种的栖息地要求。

4. 恢复湿地的自然功能:河流、湖泊、河口、沼泽等自然湿地的结构和功能是紧密相连的。当重建自然结构(包括生物群落和水文结构)时,其功能通常能得到相应的恢复。在最早期的湿地恢复项目中,管理者通常只以湿地植被覆盖面积作为湿地恢复成功的标准,而忽略了湿地的自然群落特征及其实际功能,因此,许多经过恢复的湿地在栖息地功能、水质净化功能和景观功能方面都未能得到实质性的改善(Lewis III, 2000)。

3.2 Goals and Objectives

Wetlands boast a range of ecological and social services, and they are located across widely varying socio-economic regions. As such, a wetland restoration project may have diverse goals. That said, there are a few generally applicable principles and common goals in wetland restoration that could serve as initial guidelines (USEPA, 2000).

1. Protect and nurture aquatic biological resources: Wetlands serve as a habitat for many aquatic species. The ecological restoration of wetlands is usually an effective way to protect and nurture aquatic biological resources.

2. Restore ecological integrity: A degraded wetland ecosystem has structural and functional deficiencies, such as a broken link in the food chain, while an integrated wetland ecosystem is a resilient, self-sustaining system capable of withstanding certain levels of stress and disturbance. Natural processes such as the growth of key biological communities, nutrient cycles, vegetative succession, sediment transport and fluid dynamics must be taken into consideration in wetland restoration. It is always a principle goal of wetland restoration to restore the integrity of these ecological functions.

3. Restore natural hydrology: As one of the most important features of wetlands, hydrology determines vegetation and other types of bio-communities. Losses of wetland resources can be attributed to alterations in hydrology, such as canalization of rivers, water allocation, excessive drainage and water flow disruption. If changes to wetland hydrology are irrevocable, such as from dredging a tidal flat, a feasibility assessment on wetland restoration must be conducted by means of hydrological surveys, bio-community surveys and economic analysis. If wetland restoration is still deemed feasible, the designer must carefully plan out a water system structure to meet the requirements of a wetland's vegetation and wildlife.

4. Restore the natural functions of the wetland: Natural riverine, lake, estuarine and marsh wetlands reveal the close interdependence between structure and function. The rehabilitation of natural structures (including bio-communities and hydrologic structure) usually leads to the restoration of relevant functions. In early restoration projects, project managers judged restoration success by vegetation coverage and neglected its functioning ecosystem. As a result, many restoration sites did not show substantial improvements in habitat, water purification or aesthetic quality (Lewis III, 2000).

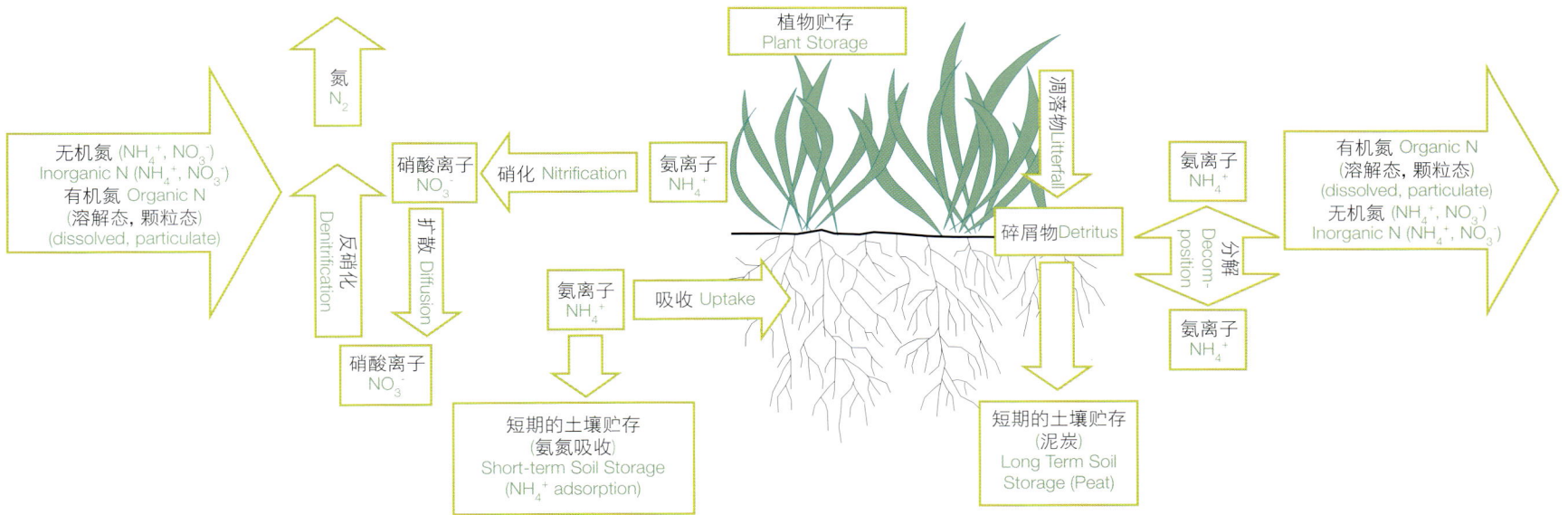

无机氮 (NH$_4^+$, NO$_3^-$)
Inorganic N (NH$_4^+$, NO$_3^-$)
有机氮 Organic N
(溶解态，颗粒态)
(dissolved, particulate)

氮
N$_2$

反硝化
Denitrification

硝化 Nitrification

硝酸离子
NO$_3^-$

扩散 Diffusion

硝酸离子
NO$_3^-$

氨离子
NH$_4^+$

吸收 Uptake

短期的土壤贮存
（氨氮吸收）
Short-term Soil Storage
(NH$_4^+$ adsorption)

植物贮存
Plant Storage

氨离子
NH$_4^+$

凋落物Litterfall

碎屑物Detritus

短期的土壤贮存
（泥炭）
Long Term Soil
Storage (Peat)

氨离子
NH$_4^+$

分解 Decom-position

氨离子
NH$_4^+$

有机氮 Organic N
(溶解态，颗粒态)
(dissolved, particulate)
无机氮 (NH$_4^+$, NO$_3^-$)
Inorganic N (NH$_4^+$, NO$_3^-$)

▲ 图 3 - 3: 自然湿地的氮循环过程 (绘制：易道)
Figure 3-3: The diagram showing the nitrogen cycle in a typical wetland illustrates natural Wetland processes and functions(Graphic:EDAW)

正是湿地功能在生态恢复目标中的重要性，导致湿地恢复具有因地制宜、因目标而异的复杂性。在实际工作中，湿地丰富的功能不可能被面面俱到地兼顾。为了提高湿地恢复的社会和经济效益，及利用湿地的自我组织、自我调节能力以增加复原速度，便必须确立需要优先恢复的功能，并将这些功能的复原程度作为检验湿地恢复是否成功的标准之一。在确立湿地恢复的功能性目标时，通常需要遵循如下原则：

- 确立湿地恢复目标时考虑流域现况及景观尺度。许多湿地资源生物和保护生物都是迁徙或洄游物种，一生中不同的阶段在不同的生境生活，所以其生活史中各阶段的生境都对其种群具有直接影响。此外，考虑景观尺度能够更好地确立恢复湿地的栖息地功能，尤其是在生境破碎化的地方，湿地恢复往往有助于补充已经缺失的廊道，或者在迁徙路线上增强驿站的功能。

- 针对导致湿地退化的原因设立生态恢复目标。大部分湿地恢复项目都只涉及流域或整个区域中退化最严重的一部分。被恢复的湿地范围以外的各种干扰仍然会对生态恢复产生各种不利的影响。如果导致湿地退化的因素仍然存在，并且没有任何预防或缓解措施的话，湿地恢复是很难成功的。比如，在滨海湿地生态恢复中，必须考虑湿地对风暴潮等自然灾害的响应和抵抗能力，而在中下游的湿地恢复可考虑设置高效的人工湿地来缓解和防止上游污染对湿地生物的影响。

- 湿地生态恢复的目标必须具有明确性、可操作性和可衡量性。没有明确目标的湿地恢复是难以成功的。湿地所处的社会、经济背景也应在确立目标时考虑在内。湿地恢复不仅在工程实施的时候需要经济投入，后续的监测和调整也是需要一定投入的。需要较高施工、维护费用的湿地恢复项目不适宜在经济较为落后的地区开展。湿地恢复的监测必须有明确的可衡量或测量的评价因素，以便能够及时发现问题并采取调整措施。

The importance of wetland functions in ecological restoration is a complex issue that should respond to different locations and goals. In practice, a restoration project may not be able to accommodate all functions of a wetland system. Therefore, it is best to prioritize key functions to be restored and take the recovery of these functions as part of the success criteria of the restoration project. With this method, the project manager may enhance the socio-economic benefits of wetland restoration, take advantage of self-regulating abilities of the wetland, and speed up the restoration process. Common issues to consider for any wetland project include:

- Plan at multiple geographic scales: Many wetland species that are protected or farmed for economic production are migratory, and their populations are thus affected by numerous ecosystems. Recognizing the scale of a landscape may aid in the recovery of habitat functions of the wetland site, especially in areas with fragmented habitats, where restoration projects can help to replenish lost wildlife corridors or add outposts for established migration routes.

- Set restoration goals based on sources of disturbance: Many wetland restoration projects involve only the most degraded areas in a watershed or region. As a result, wetlands under restoration are still susceptible to impacts from chronic disturbances outside a restoration site. Restoration projects are less likely to succeed if disturbances continue without monitoring or mitigation measures. For example, natural disasters and disturbances like storms and tides must be considered in the restoration of marine wetlands, while for inland wetlands, the impact of upstream pollution may be the primary agent of disturbance.

- Ensure the clarity, feasibility and measurability of restoration goals: Wetland restoration without a clear goal is unlikely to meet its criteria for success. The goal must also incorporate social and economic considerations based on the wetland's location. Restoration projects require continuous economic investment in both their construction and management phases. Restoration projects that have high construction and management costs are generally inappropriate in less developed areas. Clear, measurable success criteria must be established to evaluate a completed restoration project, so that problems can be detected and adjustments made immediately.

- 湿地恢复通常具有参考系统。虽然湿地不可能复原到与退化前一模一样的状态，但是，它仍然应该锁定一个参考系统，作为恢复的目标。比如，要恢复一片供水禽栖息的湿地，就需要参考在相同或相近纬度带下的类似自然栖息地，即所谓的效法自然。

- 谨慎选择物种。尽可能选用乡土物种和同一气候带的物种。由于湿地植物具有耐淹、耐盐或耐贫瘠的特性，多为世界分布类型，因此湿地植被具有隐域植被的特性。比如，芦苇从内陆到滨海，从淡水到咸水，均有分布（吴征镒，1991）。在湿地恢复时，必须分析所种植物种的生态位和共用同一生态位的其他物种的数量，并预测该物种的扩散速度。如果该物种在湿地中缺乏竞争对手和天敌，就必须考虑换用其他物种或采取控制性措施（见图3-4）。

Establish a reference site: Although a wetland site cannot be restored to its precise pre-degradation state, a reference site should be established as the ideal end result of the restoration. For instance, wetland restoration for a water fowl habitat may be compared to a similar habitat which currently exhibits the desired results of the restoration site, such as one having a diverse, abundant bird population. In this way, natural ecosystems become the model for restoration efforts.

Careful selection of species in a site: Native species from a single climatic zone are preferable for placement in a restoration site. Wetland vegetation has clear intrazonal characters, as different plants can withstand submersion, salt and infertile soils to varying degrees. Some reed species, for example, can be found both inland and along the coast, in freshwater and saltwater (Wu Zhengyi, 1991) habitats. When a particular species is intended for a restoration project, its ecological niche and the other species which share that niche should be well understood. The proliferation rate of the intended species should be estimated. Exotic species should be avoided when possible. Often, exotic species will have no competitors or natural enemies on the site, and may therefore proliferate uncontrollably and out-compete native species. Figure 3-4 shows the zonation of high marsh, low marsh, riparian and upland vegetation as they relate to water levels for a restoration project at Yangcheng Lake, in Kunshan, China.

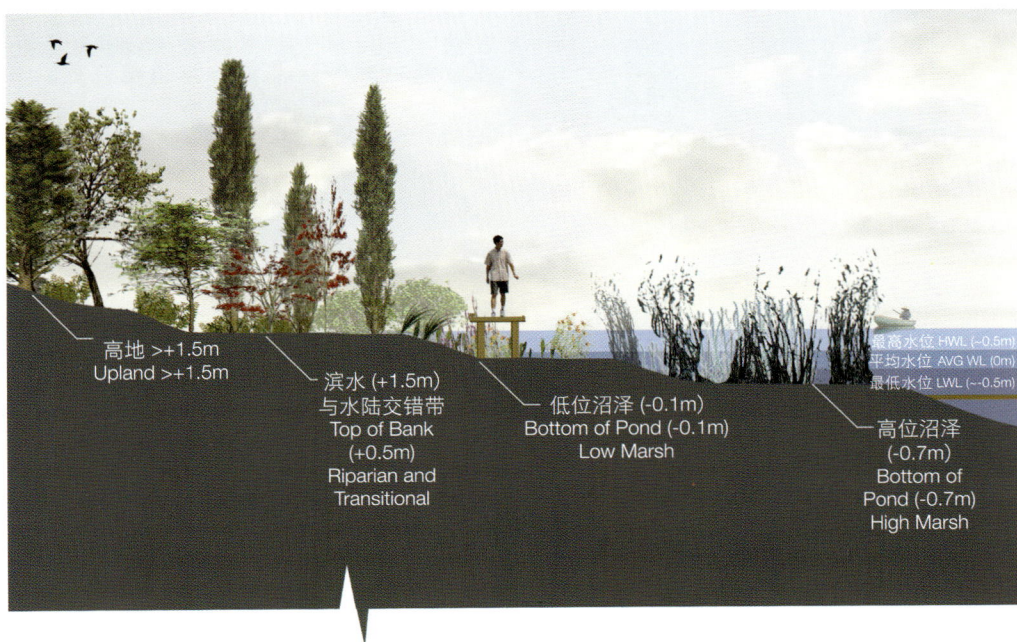

高地 >+1.5m
Upland >+1.5m

滨水 (+1.5m)
与水陆交错带
Top of Bank
(+0.5m)
Riparian and
Transitional

低位沼泽 (-0.1m)
Bottom of Pond (-0.1m)
Low Marsh

最高水位 HWL (~0.5m)
平均水位 AVG WL (0m)
最低水位 LWL (~-0.5m)

高位沼泽
(-0.7m)
Bottom of
Pond (-0.7m)
High Marsh

图 3-4：规划中的江苏省昆山阳澄湖滨岸湿地（图片来源：易道）
Figure 3-4: Wetland zonation along Yangcheng Lake, Kunshan, China (Source:EDAW)

增强湿地的自我维持功能。要确保所恢复湿地的长期稳定性，最好的方法就是增强其自我维持功能，以降低如长期的供水设施、植被管理或每次大潮汛过后的修补等人为辅助工作在湿地发育和演替中的作用。这是湿地生态完整性的体现，也能够降低湿地管理成本。

根据当地情况，采用被动或主动的湿地恢复方式（见图3-5）。被动的湿地恢复方式需要很长的时间，并且必须去除和控制那些导致退化的因素，如过度捕捞、干旱化和外来物种入侵等。这种恢复方式比较适合资金投入少且人为干扰可得到严格控制的区域。主动的恢复方式属于生态工程，直接改变湿地的结构和过程，如进行植被恢复，调整水系，甚至引入关键种。这种恢复方式需要一定的资金投入，并且必须开展适应性管理,因为湿地是一个复杂的系统，各种物理、化学、生物过程相互影响，致使生态恢复的实施结果具有明显的不确定性。通过监测人为干预带来的后果，能及时修正或防止意外事件的发生。

从中国的湿地恢复工作进展情况看，目前的湿地生态恢复较多针对如下目标：

- 加强其作为野生动物栖息地的功能，比如，各种湿地自然保护区和国际重要湿地；

- 提高湿地景观价值，比如，城市湿地公园；

- 提高物质循环功能，净化水体，比如，以水质净化为目标的人工湿地；

- 加强湿地的水文调节功能，比如，大部分的滨水湿地恢复区域和城市雨水花园；

- 提高湿地的资源生产，比如，湖泊与渔场的生态恢复。

Encourage a wetland's natural equilibrium: The best way to ensure the long-term stability of a restoration site is by enhancing the self-preserving, equilibrating functions of the wetland. This may significantly relieve the need for continual human assistance, such as long-term water supply facilities, vegetation management, and repairs after major floods. The self-preservation function is a reflection of a wetland ecosystem's integrity and may reduce site management costs.

- Utilize passive and active rehabilitation methods based on local conditions: Passive restoration requires substantial time and requires the removal and control of disturbances such as excessive hunting or fishing, soil drying and foreign species invasion. It is mainly used in areas where economic investment is low and human disturbance is easily controlled. Active restoration is achieved through ecological engineering, which requires the structural and procedural alteration of a wetland site, such as the restoration of vegetation, adjustments to the water system, and the introduction of key species. Active restoration is generally more costly than passive restoration. Because of the complex interactions among physical, chemical and biological processes in wetlands, predicting the results of restoration efforts can be difficult. Therefore, adaptive management is an important aspect of each method. Monitoring the results of any artificial (human) interference will allow managers to implement immediate corrective actions or prevent unintended results.

The goals of restoration projects in China generally fall into the following categories:

- Enhancing wildlife habitats, such as natural reserves and wetlands that provide habitat for protected species;

- Creating aesthetic value, such as urban wetland parks;

- Improving circulation and water quality, such as artificial wetlands for water purification;

- Enhancing hydrological regulatory functions, such as for most floodplain restoration sites and urban rainwater gardens; and

- Increasing the commercial value of wetlands, such as harvesting biomass, rice production, and the enhancement of fish ponds, among others.

◄ 图 3 - 5: 清除污染源是被动恢复的一种，
如下图中这些管道将水产养殖产生的污
水排入海南的一个咸水湖中，而创造水
文环境（如上图）则是一种主动恢复方
式。(摄影 : Rowan Roderick-Jones)
Figure 3-5: The removal of pollution sources using
pipelines to carry aquaculture discharge to a
lagoon in Hainan (below), is one type of passive
restoration, while creating appropriate hydrology
by regrading (above) is an active resoration
approach
(Photographs by Rowan Roderick-Jones)

3.3 湿地的现状评估

3.3.1 湿地恢复的可行性与可能性

湿地的生态恢复是针对湿地退化而言的。在进行生态恢复之前，必须明确湿地退化的原因、程度和未来趋势。对于一些人力可控制的退化因素，比如富营养化，应该在湿地恢复的同时加强对污染物排放的控制力度；对于一些非人力可控制的退化因素，比如热带风暴和海平面上升，应该在生态恢复设计中加强风险评估和缓冲减灾措施；因此，在湿地恢复之前开展现状评估具有重要意义。

在选择最合适的重建地点时，必须考虑一些关键因素，以保证所恢复湿地在功能和结构上的持久性。以下几个问题可用来初步验证湿地恢复或重建项目的可行性，并有助于实现重建价值的最大化：

1. 这处湿地是否有永久性的水源？

2. 该处的地形、土壤条件是否具有蓄水的能力？

3. 主要的人为干扰是什么？这些人为干扰对湿地的水环境和生物环境会有怎样的影响？

4. 湿地可能为当地社区带来什么好处（例如水质改善、洪水控制或生物多样性带来的特殊景观等）？

只有当这些问题都得到确切的答案后，才能着手开展湿地恢复设计，否则，就要考虑经济上持续投入的可能性。一般来说，除非政府在该地区规划上有特殊要求或长期持续的管理计划，一般我们不赞成在不适合恢复湿地的地方进行湿地构建。比如说，在一个干旱地区，如果蒸发量远远高于降雨量和径流量之和的话，人工构建的湿地很容易干涸，或者需要高额的代价来建造引水工程或改变土壤结构，从而导致整个区域或流域内水分配的失衡。虽然美国在内华达沙漠中建立了拉斯维加斯，但是，其代价是高昂的，并且为了城市内高效的水循环系统和废水处理设施，每年的维护成本要远远高于其他普通城市。尽管这个城市是一个成功的奇迹，但是这样的模式并不适合中国目前的国情。

3.3 Site Assessment

3.3.1 Site Selection

While the location of available land generally drives restoration opportunities, the environmental planner will often have significant control over the specific location of a wetland restoration project. When considering the most appropriate location for a restoration site, several key factors must be considered in order to ensure the long term integrity of the wetland's function and structure. Initial questions to ask, ones that assess feasibility and maximize the value of a wetland restoration project, include:

1. Is there a permanent source of water for the wetland?

2. Is the topography suitable or can it be manipulated to achieve ponding?

3. What are the primary human disturbances and how do they impact a wetland's water, and biological systems?

4. What are the services that the wetland might provide to the local community (such as water quality improvement, flood control, recreation, protecting habitat species, etc.)?

These fundamental questions should be answered prior to initiating the design process. In some cases, the planner may find that a restoration may not be feasible. For example, due to the lack of a permanent water source, a project may have to be re-evaluated for other potential measures. Unless the government has special requirements or specific management schemes in its regional planning, the creation of wetlands in places unfit for wetland restoration is generally inadvisable. One such instance would be in arid climates, where evaporation far exceeds total precipitation and surface runoff, creating conditions wherein artificially created wetlands would be prone to drying out. This may be mitigated at high cost through the construction of a water supply system or the alteration of the site's soil structure and topography. Las Vegas, in the United States boasts an efficient water supply and waste water treatment system, yet one with much higher maintenance costs than most American cities. The creation of a constructed wetland for treating wastewater in the arid climate of Las Vegas is an example of how permanent water sources can allow for wetland "creation" in climates that are normally unsuitable for natural wetlands.

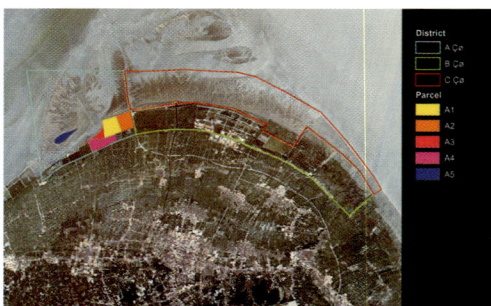

▼ 图 3-6: 全球环境基金在做宁波-慈溪湿地项目时曾考虑过许多地方，最后才将恢复可能性最大的一处定为该项目的开发地。择定恰当的地点时，一定要在经济、社会和环境诸方面之间进行协调，如农村人口迁移、商业生产、土地价值和当前的生态环境质量等 (提供：慈溪建设局)
Figure 3-6: Numerous site areas were considered for the Ningbo–Cixi Global Environment Facility before planning the project on a site with the highest restoration potential. Appropriate site selection had to balance economic, social, and environmental aspects such as rural population relocation, commercial production, land value and existing habitat quality. (Image courtesy of Cixi Construction Bureau)

3.3.2 现场特征

在自然状态下，湿地是位于水生和陆地生态系统之间相对狭窄的生态交错区。只有对景观特征有透彻了解以后，才能理解湿地的类型、功能以及湿地如何与邻近的生态系统发生关系的。一旦明确了湿地地形、湿地形态和自然作用的模式以及基本特征后，就能根据该地块的生态承载力做出高屋建瓴的规划设计。

举例来说，只有了解土地坡度是陡峭、平缓还是垂直之后，才能更好地理解地形如何影响排水、蓄水、侵蚀以及将来的土地发展潜力如何。土壤和植被类型的分布格局以及关键物种的缺失也值得关注。必须明确湿地的类型，例如是季节性的还是永久性的？如有洪泛平原的话，会是在哪里？发生洪水时，洪水的流向会在哪里，最大流量会是多少？（见图3-7）

规划者的视野一定要超越项目现场的边界，以便确认这里的湿地的生态系统是如何与临近或地区性的湿地、开放空间和水生生态系统相关联的。只有通过对雨季和旱季所进行的观察和资料收集，才能获得深入而有层次的答案。

通常，湿地现状评估的主要内容包括：

- 气候：温度、湿度、降水量、风速、风向、风期、首末次霜冻、雪、霜、雾、逆温、飓风、龙卷风、海啸、台风等；

- 地质：岩石、年代、形成、规划、剖面、特性、地震活动、岩崩、泥崩、基岩；

- 水文：淡水来源、潮汐、波浪、水井、水量、水质、地下水位；

- 地形：区域地形、地形特征、等高线、剖面、坡度、坡向、数字高程模型；

- 植被：组成、群落、物种、分布、物种数量、珍稀濒危物种、火灾历史、演替；

- 野生动物：栖息地、动物种类、数量、普查资料、珍稀濒危动物、科研和教育价值；

3.3.2 Initial Site Characterization

In the natural landscape, wetlands occupy a narrow niche between aquatic and upland ecosystems. Understanding wetland form and function as it relates to adjacent ecosystems requires an intimate understanding of a landscape's character. Identifying patterns and fundamental characteristics in topography, form and natural processes all give wetland planners a broad picture of the land's ecological potential.

For example, the observation of land slopes – whether they are steep, shallow, or hilly – will improve one's understanding of how landforms affect drainage, ponding, erosion, and future land development potential. One should observe gradients in soil and vegetation types or the lack of these characteristics. Another question is whether or not there are different types of wetlands on a site, such as seasonal or permanent wetlands, and where its floodplains are located, if they exist. One should further consider how the site might be affected during a flood.

Finally, looking beyond site boundaries is also important in determining how ecological systems are linked to other local or regional wetlands, open spaces and water systems. Significant and incremental insight into these questions can be gained through observation and data collection in both wet and dry seasons.

An assessment on existing conditions should include information on the following areas:

- Climate: Temperature, humidity, precipitation, wind speed, wind direction, wind duration, first and last yearly frostings, snow, frost, fog, temperature inversion, and the frequency of hurricanes, tornados, tsunamis, typhoons, etc.

- Geology: The age, structure and formation of rock patterns, geological profile and character, seismic activities, rock slide and mud slide activity, and bedrock

- Hydrology: Freshwater sources, tidal bores, waves, water volume, water quality and groundwater water level

- Topography: Regional contouring, topographical character, contour lines, cross-sections, sloping, slope aspect and digital topographical model

- Vegetation: Plant composition, communities, distribution and quantity of different species, and endangered species, history of fire outbursts, and evolution

▼ 图3-7：洪泛平原湿地恢复需与河道形态及其水文状况相对应。内蒙古呼和浩特市这片滨水区是可以运用易道为美国科罗拉多州斯坦普莱顿等地设计的恢复方案的理想场地

Figure 3-7: Restoration of floodplains needs to respond directly to channel topography and hydrology. This degraded riparian zone in Hohhot, Inner Mongolia (top) is an ideal location for adapting restoration approaches such as those that were used in EDAW's design at Stapleton, Colorado (bottom) in the United States

(摄影：Rowan Roderick-Jones)
(Photograph by Rowan Roderick-Jones)

(摄影：Dixi Carrillo)
(Photograph by Dixi Carrillo)

- 人类：聚落类型、土地利用现状、基础设施现状、经济活动、人口特征。

在上述内容中，并不是所有信息都必须由规划人员亲自去现场获取，这些资料的获取通常是一个长期而昂贵的过程，对于一个规划项目来说，通过若干次现场考察是无法一窥全景的，必须借助大量的文献查阅和专家咨询。这里所说的文献包括：政府发布的环境白皮书、环境质量公报以及类似公告；正规出版物提供的信息，尤其是学术期刊和各类地方志；由林业部门、环保部门、国土资源部门、海洋局、气象局等政府部门提供的各类技术档案。气候、地质、地形和人类特征通常可以通过这种方法来获取。此外，各学术团体的刊物、网站和论坛也通常能够提供非常及时的信息，尤其是野生动物和鱼类方面的资料。专家咨询不仅能够弥补资料上的空白，并且他们的经验可以用于分析湿地恢复的机会和限制因素。

但是，在湿地恢复之前，水文和生物方面的调查工作仍然是非常必要的，因为通常文献提供的信息都是基于大尺度调查的，对于小尺度的基地，许多有用的信息可能会被遗漏，许多待规划的区域也可能缺乏历史资料，或者资料陈旧，所以，在这种情况下，在湿地恢复之前所做的大量工作便能验证和补充所需信息。

应用激光雷达地形测量技术

传统测量是获取地形资料的传统方法，但是遇到大面积地区时就会出现效率低下的问题。有时，可用航空照片来推断高度信息，而在此基础上绘制的地形图，其清晰度要由地表植被来决定。在无法获取适当地形资料的地方，可运用机载激光雷达技术来实现。这种技术是用于获取地形资料最经济有效、精准可靠的方法之一，尤其是区域面积较大时。激光雷达技术是当前在北美洲和亚洲应用广泛的新技术（见图3-8）。作为"激光"雷达，这种技术的配置包括安装在小型飞机或直升机上的激光接收器及机载电脑和全球定位装置。接收器获取有关地球表面及其特征的三维资料，在较开放的区域，其精度在垂直方向为15~30cm，水平方向50~60cm。精度取决于地表植被类型，在植被稀疏的

- Wildlife: Habitat, types and quantities of different wildlife species, survey archives, endangered species, scientific and educational values

- Human activities: Type of human habitation, land use, infrastructure, economic activities and demographic attributes

Not all data from the above necessarily requires on-site retrieval by a design team. The collection of this is usually an expensive, long-term process. Even several on-site inspections are insufficient to provide a thorough overview for a design project. Extensive consultation using relevant documentation and expert opinions are essential for gathering and analyzing data. Appropriate documents include: environmental white books, environmental quality bulletins and similar official announcements issued by the government; information in publications, especially academic and local trade journals; technical documents obtained from government agencies in forestry, environmental protection, land resources, sea management and meteorology.

Data collection for climate, geology, topography and human activities is often done through the above means. Publications, websites and forums of different academic associations offer additional help in providing the latest information in relevant disciplines, such as wildlife biology and fisheries management. Expert opinions are valuable complements to collected data, as experts can offer specific experience in the analysis of the potential gains and limitations of a wetland restoration project.

It is similarly standard practice to undertake specific surveys in hydrology and biology before wetland restoration occurs. Data covered in available documents are usually based on large-scale surveys, and useful information regarding small-scale restoration sites may be unavailable. Many restoration sites have insufficient or outmoded historical documentation. Therefore, a thorough analysis on both larger and smaller scales should be undertaken to verify and supplement existing data before a restoration project proceeds.

Utilizing LiDAR for Topographic Surveying

Conventional surveying is the traditional method in obtaining topographic data, but can be inefficient for large areas. In some cases, aerial photos can be used to infer elevation information, from which topographic maps can be made with the resolution governed by the ground cover.

Where suitable topographical data are not available, the airborne Light Detection and Ranging (LiDAR) technology can be used to acquire data. This technology has proved to be

开放地区清晰度可达到最高。在树木密集的地区，激光光束穿透植被、探测实际地表高度的能力依据具体情况而各有不同，但总的来说要比航空照片的效果好。这组数码资料的资料点密度约为每平方公里450,000个数据点。凭此精度和数据点密度，准确地绘制出复杂地貌，以低于传统技术的成本再现自然及人工地面特征应不成问题。这种技术的成本与航空照片的成本差不多。资料获取速度约每天250~500km²。

one of the most cost-effective and reliable methods to acquire topographic data, particularly over a large areas. LiDAR is a new technology being used extensively in North America and Asia. The technology, the laser equivalent of radar, consists of a laser-based sensor mounted on a small airplane or helicopter, coupled with on-board computers and global positioning equipment. The sensor acquires three-dimensional data of the earth's surface and its features, with an accuracy of 15-30cm vertical, and 50-60cm horizontal in relatively open areas. The accuracy is dependent on the type of earth cover, and the best resolution can be achieved in open grounds with sparse vegetation cover.

In areas with a dense tree canopy, the ability of the laser beam to penetrate through the cover and detect the true ground elevation varies on a case-by-case basis, but is still generally superior to aerial photography. Data are digital and have a density of about 450,000 data points per km². With this accuracy and density, it is possible to produce precise renderings of complex topographies and natural and man-made features of land surfaces at reduced costs relative to more conventional means (the cost of the technology is similar to that of an aerial photo). The acquisition rate is about 250-500 km²/day. An example of a LiDAR image taken at the port of Houston in the United States is shown in Figure 3-8.

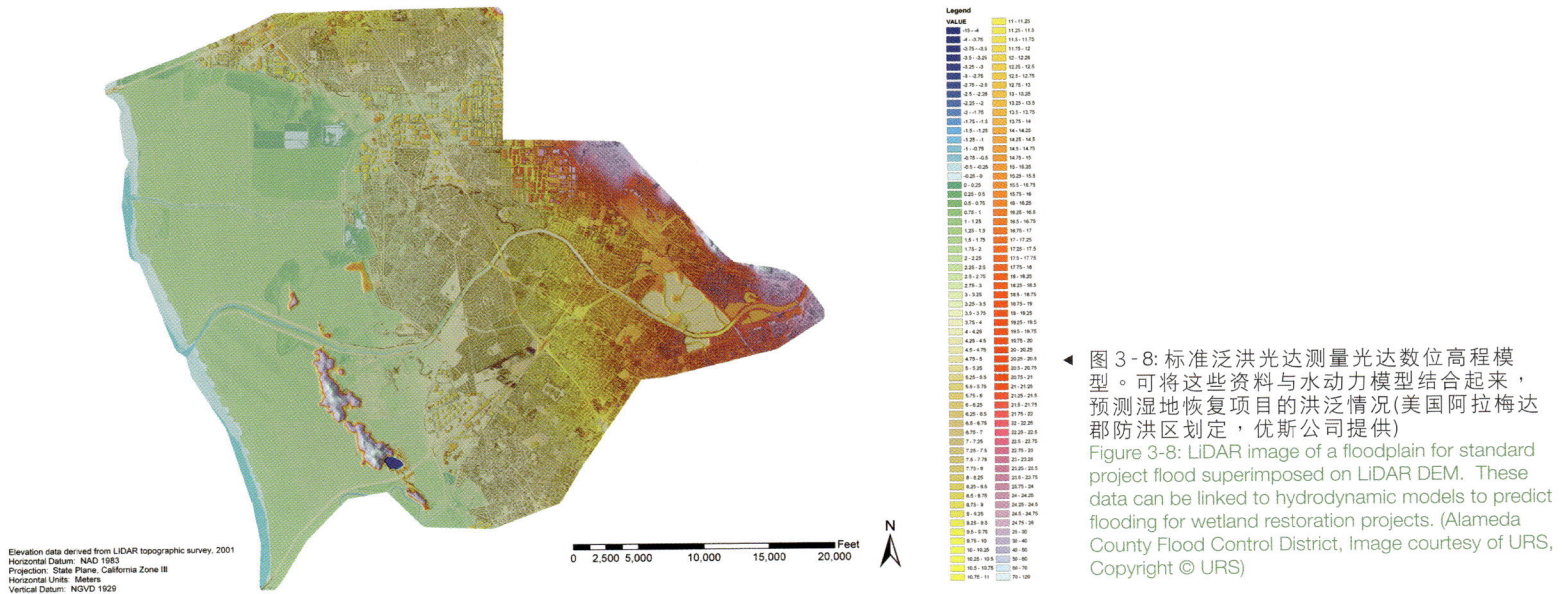

Elevation data derived from LiDAR topographic survey, 2001
Horizontal Datum: NAD 1983
Projection: State Plane, California Zone III
Horizontal Units: Meters
Vertical Datum: NGVD 1929

◄ 图 3-8: 标准泛洪光达测量光达数位高程模型。可将这些资料与水动力模型结合起来，预测湿地恢复项目的洪泛情况(美国阿拉梅达郡防洪区划定，优斯公司提供)
Figure 3-8: LiDAR image of a floodplain for standard project flood superimposed on LiDAR DEM. These data can be linked to hydrodynamic models to predict flooding for wetland restoration projects. (Alameda County Flood Control District, Image courtesy of URS, Copyright © URS)

水文学

水文条件通常是决定湿地形态和功能的主要控制机制（Mitsch and Gosselink, 2000）。保持生物作用和物理作用之间平衡的水文系统，包括了地表水和地下水的输入和输出、水位、土壤饱和以及洪泛的持续时间和频度。

最有用的水文学资料可能就是水文图。通常水文图描述了过去至少一年以上的水面高度。一些湿地类型，如苔藓泥炭地、淡水和咸水沼泽、有林沼泽和临时性池塘，都有各自不同的水位特征，可以据此确定该湿地类型的物理形态和生物物种。图3-9显示了不同湿地类型及大型、小型河流的水文图实例。

获取或制作项目范围内的精确水文图能明显提高规划效率。利用水文图这一工具，可以很容易确定不同植物群落的合适高度。举例来说，浮水植物可以承受一整年的水淹，而滨岸的湿生植物则根据物种不同只能承受一年中2周到6个月的水淹。

确定湿地恢复的主要水源至关重要。获取当地的地形图以及遥感影像（航片和卫星影像）可以让设计者了解到湿地是如何同水源（河流、小溪、湖泊、海湾、海洋等）相连的。如果湿地没有与地表水的联系，那么其水源补充则可能来自较隐蔽之处，如地下水（泉水、渗出水、高水位地下水）、雨水或非沟渠型地表径流。一些季节性湿地如春池或临时性池塘则完全靠降雨补充，仅在雨季时形成，到了旱季就干涸了（见图3-10）。

根据地图或航空照片所提供的信息，再进行实地验证是非常重要的，因为对航空照片中空间位置和光谱特征的误读是常有的事情，而且一旦照片过期，其反映的土地使用情况会发生很大的变化，尤其在中国东部地区，城市化过程非常快，土地使用情况需要及时更新。对航空照片的多年跟踪调查说明，有些因素如地形不会发生很大改变，而另一些因素，如土地使用、排水沟渠、道路及其它设施、植被群落，则有明显的变化。地形、排水量、径流量、降水量和水位资料可从当地林业和水务部门取得。有关地表径流量和土壤的资料可能不容

Hydrologic

Hydrologic conditions are very often the primary control mechanisms determining a wetland's form and function (Mitsch and Gosselink, 2000). Components of a hydrologic system that sustains the balance between biological and physical processes include surface and ground-water inflows and outflows, water levels, duration and frequency of soil saturation and flooding.

Perhaps the single most useful piece of hydrologic data is the hydrograph. A hydrograph depicts water surface elevations over time, usually over a one-year period. Wetland types such as bogs, freshwater and saltwater marshes, swamps and ephemeral pools each have distinct hydrographs which define the physical form and biological species of the wetland. Examples of hydrographs for various wetland types and large and small rivers are provided in Figure 3-9.

Obtaining or developing an accurate hydrograph for a restoration site can streamline the design process. With this tool, appropriate elevations for different vegetative communities can be easily established. For example, emergent vegetation can withstand submersion for the entire year, while depending on the species, riparian plants can withstand submersion for anywhere between two weeks to six months of the year.

Defining the primary source of water for a wetland restoration site is also essential. Obtaining maps of local topography and remotely sensed images (aerial and satellite images) will allow planners to observe how wetlands are associated with water sources such as rivers, streams, lakes, bays, and the ocean. Where wetlands exist without surface water connections, they may be sustained by less obvious sources of water, such as groundwater (springs, seeps, or a high water table) or rainfall and non-channelized surface runoff. Certain seasonal wetlands such as vernal or ephemeral ponds are supplied entirely by rainfall and exist only temporarily as wetlands, drying out during the dry season. Developing an accurate water budget and balance for the site will enable the designer to determine the appropriate restoration approach.

Ground-truthing of any information obtained by maps and aerial images is particularly important. The misinterpretation of spatial patterns and coloring on aerial images is common, and land uses may have changed significantly if images are outdated. Reviewing aerial photos taken over several years may show that some features, such as topography, have not changed much while others, such

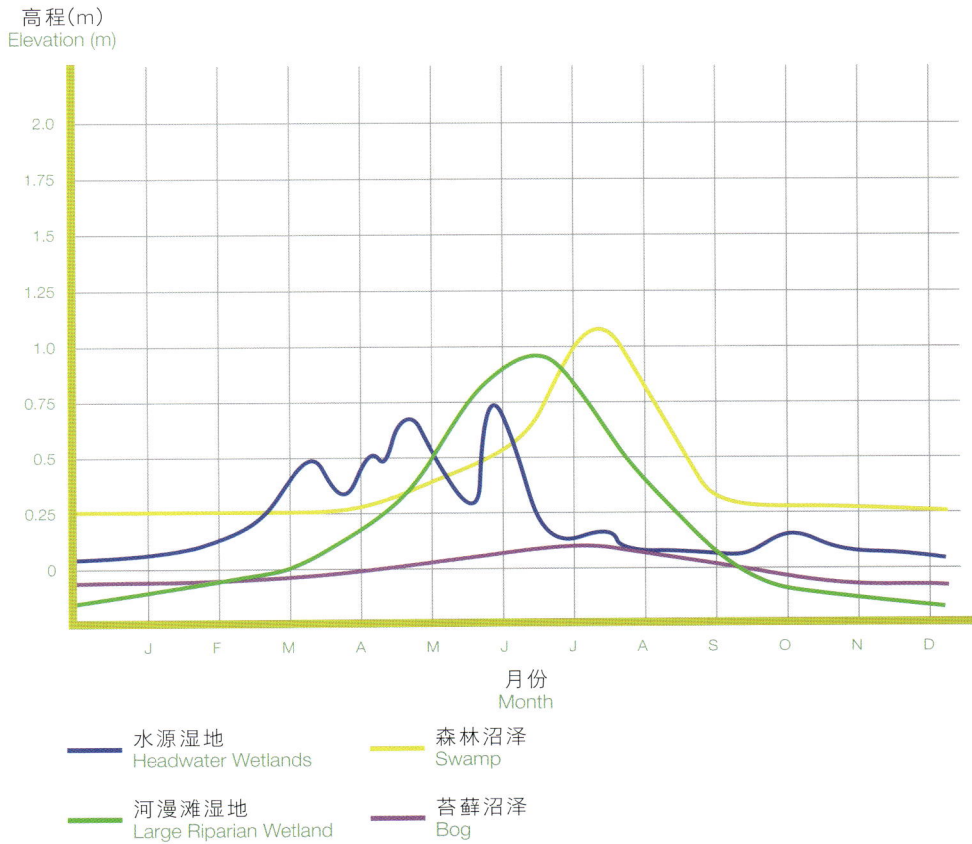

高程(m)
Elevation (m)

月份
Month

水源湿地
Headwater Wetlands

森林沼泽
Swamp

河漫滩湿地
Large Riparian Wetland

苔藓沼泽
Bog

◀ 图 3-9: 不同类型湿地的水位变化(绘制:易道)
Figure 3-9: Hydrographs for Various Wetland Types (Graphic: EDAW)

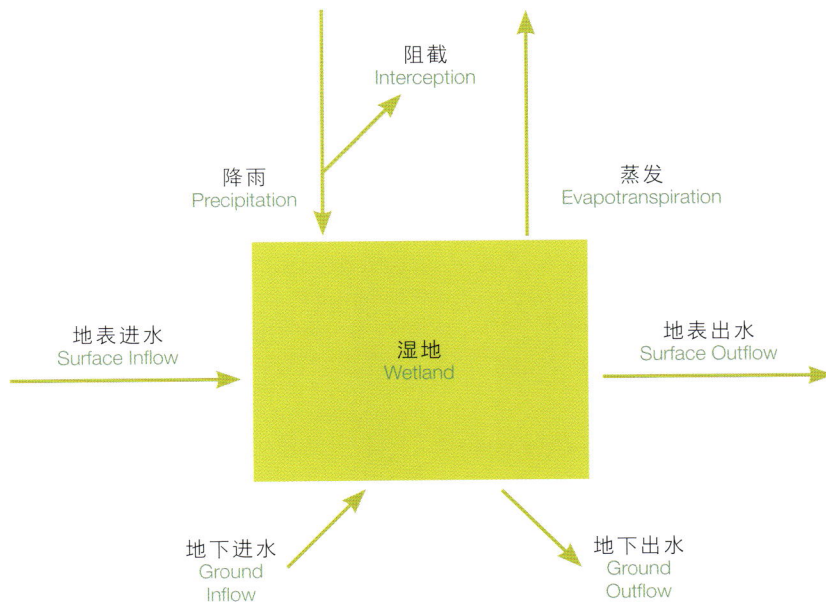

阻截
Interception

降雨
Precipitation

蒸发
Evapotranspiration

地表进水
Surface Inflow

湿地
Wetland

地表出水
Surface Outflow

地下进水
Ground Inflow

地下出水
Ground Outflow

◀ 图 3-10: 湿地进出水的常见组成情况(绘制:易道)
Figure 3-10: Components of a typical water budget (Graphic: EDAW)

易 获 取 , 但 为 了 有 效 规 划 湿 地 恢 复 工 程 , 这 些 资 料 是 必 不 可 少 的 。 在 大 多 数 情 况 下 , 购 买 上 述 资 料 是 完 全 值 得 的 。 流 域 或 地 区 的 土 壤 类 型 分 布 地 图 (可 从 当 地 林 业 部 门 或 国 土 资 源 部 门 取 得) 对 于 划 定 湿 地 土 壤 的 历 史 或 现 有 范 围 提 供 了 极 为 重 要 的 参 考 , 此 外 有 时 也 会 包 含 其 他 有 用 的 信 息 , 如 泉 水 、 池 塘 、 溪 流 和 排 水 沟 渠 的 位 置 。 一 旦 了 解 地 表 水 位 后 , 就 可 以 确 定 池 塘 的 深 度 以 及 是 否 需 要 衬 垫 材 料 。 如 果 需 要 全 年 有 自 由 表 面 流 , 则 可 将 地 形 构 建 为 盆 地 , 并 让 盆 地 最 深 处 低 于 预 期 最 低 水 位 。

对 于 退 化 严 重 或 特 征 不 明 的 湿 地 , 可 采 用 的 有 效 方 法 是 , 在 邻 近 地 区 寻 找 一 个 未 受 干 扰 的 、 各 方 面 情 况 相 似 的 现 存 湿 地 作 为 参 考 系 统 , 并 从 中 取 得 水 文 体 系 、 植 被 、 野 生 动 物 等 信 息 , 为 湿 地 重 建 做 好 资 料 和 信 息 准 备 。 如 果 在 当 地 或 本 区 域 无 法 找 出 这 样 的 参 考 湿 地 , 不 妨 参 照 世 界 其 他 地 区 相 似 气 候 和 水 文 条 件 下 的 案 例 , 不 过 需 要 注 意 的 是 , 有 些 因 素 如 植 物 种 类 和 土 壤 会 与 本 地 有 很 大 不 同 。

除 了 自 然 条 件 以 外 , 在 湿 地 恢 复 工 程 中 还 要 考 虑 人 类 活 动 和 影 响 。 现 有 的 和 规 划 的 道 路 、 停 车 场 、 排 水 沟 渠 、 水 库 及 其 它 设 施 , 可 能 会 对 湿 地 的 水 文 环 境 产 生 影 响 。 在 湿 地 恢 复 过 程 中 , 规 划 者 必 须 决 定 是 否 应 将 湿 地 纳 入 城 市 和 农 业 排 水 系 统 中 , 还 是 将 其 完 全 排 除 在 外 。 如 果 改 善 水 质 是 首 要 目 标 的 话 , 就 应 让 湿 地 接 纳 暴 雨 洪 峰 、 农 田 径 流 或 经 过 初 步 处 理 的 城 市 污 水 ; 为 此 , 必 须 向 有 湿 地 治 理 经 验 的 工 程 师 咨 询 , 以 规 划 湿 地 的 水 处 理 设 施 。

as land use, drainage ditches, roads and other structures, and vegetation communities, have changed significantly. Topography, drainage, runoff, rainfall and water level data should be available from local Water Authority or Forestry Bureaus.

Data on soil and groundwater levels are not commonly available but can be crucial when planning functional natural wetland restoration projects. In most cases, obtaining this data for a project is well worth the small cost involved. Soil maps for watersheds or districts may be available from the local Forestry Bureau and are invaluable references in locating existing and former wetland soils. Other useful information, such as the location of springs, ponds, streams, and drainage ditches is sometimes indicated on soil maps. Knowing the groundwater elevation will help in determining the depths required for the bottom of ponds, and whether liner materials may be required. If standing water is desired year round, restored wetlands can be excavated so that the deepest part of the basin is below the lowest anticipated water level.

It is also useful to obtain information from a nearby, undisturbed wetland with similar characteristics to the one envisioned for restoration. This is particularly true if the planned site area is significantly degraded, or if the historical characteristics of the wetland type under restoration is uncertain. The hydrologic regimes, biological complexity, and physical structure of the reference site can be mimicked at the restoration site. In the absence of a suitable, proximate or regional reference site, case studies from regions around the world with a similar climate and hydrologic characteristics can be utilized with the understanding that some characteristics such as plant species and soil may vary considerably.

Aside from natural conditions, it is important to consider human influences and impacts. Planned and existing roads, parking lots, drainage ditches, reservoirs, and structures may exert an influence on hydrology within the wetland. The planner should determine whether it is appropriate to either integrate or isolate the wetland from urban and agricultural drainage systems. If water quality improvement is a primary goal, it may be desirable to allow storm water or pre-treated municipal wastewater to enter the site. If this is the case, an engineer with wetland experience should be consulted to design the water treatment elements of the wetland.

有关水文学的问题

1. 怎样获取基本的水文资料？

2. 待恢复湿地的现有水文学特征是怎样的？

3. 待恢复湿地受到干扰前的水文学特征是怎样的？

4. 在待恢复湿地和参考湿地应测量何种参数？

5. 怎样的干扰活动造成了对湿地水文特征的影响（湿地是怎样并以何种原因失去或减少了水资源）？

6. 土地使用、道路、沟渠及其它人造设施所造成的改变是如何影响地表径流量和排水或成潭作用的？

7. 以上土地使用所造成的改变，以及现有的道路、建筑和其他人造设施，会以怎样的方式对所恢复、重建或改建湿地的能力发生影响？

8. 在本流域或本区域的哪些地方能够找到与待恢复湿地同样类型的参考湿地？

9. 湿地恢复工程会怎样影响下游的水文特征？

10. 地表和主要水源水位(地表、地面水)怎样影响湿地水文特征？

11. 在地形改变和规划因素上，可以采取什么措施来恢复地表和地下的水文特征？

12. 如何规划湿地，以使其能够应对洪峰？

13. 采取哪些软工程或生物工程方法，可以解决湿地的水文问题（包括水土流失）？

14. 是否存在一些限制因素（如建有堤坝或缺乏水源）将导致湿地的水文学功能无法完全发挥？

15. 重建工程的目标是否合理可行，同时使湿地完全发挥其水文学功能？

Questions to ask about Hydrology

1. Where are baseline hydrologic data found?

2. What is the existing hydrologic character of the site?

3. What is the historical pre-disturbance hydrologic character of the site?

4. What parameters should be measured at the restoration and reference sites?

5. What disturbances have affected the site's hydrology (how and why was water removed or diverted from the site)?

6. How might changes in land uses, roads, ditches, and other human-constructed features affect surface water runoff, and drainage or ponding patterns?

7. How might these changes in land use, and the presence of roads, buildings and other human constructed features affect your ability to restore, create, or enhance a wetland?

8. Where can reference sites for this wetland type be found in the watershed, or regionally?

9. How will restoration of a site's hydrology affect downstream areas?

10. How do the land surface and primary water source elevations (surface and ground water) affect a wetland's hydrology?

11. What physical landforms, changes and design elements can be made to restore hydrology both above and below ground?

12. How can the site be designed to deal effectively with extreme flood events?

13. What soft engineering or bioengineering methods are available to rectify the hydrologic problems, including erosion?

14. Is there any factor which alone (such as the existence of levees or lack of adequate water sources) might limit the ability to restore full hydrologic functions?

15. Are the project goals reasonable, feasible, and likely to result in the achievement of maximum ecological functioning possible for the site?

水 质

尽管水文状况是影响湿地恢复工程成败的首要因素，但是水质问题也同样不容忽视。具体来说，人们常误以为湿地能够吸收并处理所有类型的污染，并因此向湿地直接排放污水，这种做法会对野生动物和植被造成灾难性的破坏。来自周围环境的化学污染物质会毁掉湿地的净化水质能力，并改变湿地的特征，比如寒冷地区的高速公路旁常用盐保温，可是如果盐分渗入湿地的话，会对植物群落的生产力和结构造成不良影响，有可能引起有害物种的快速蔓延，而来自印染厂、造纸厂或化工厂的废水如果不经初步处理而直接排入湿地的话，其中的有毒有害物质会对湿地生物产生毒性，最终导致湿地植被的大量死亡（见图3-11）。又如，在美国加州Kestereson一个用于水质净化的湿地，由于来水中的硒含量很高，并且一度被忽略，因而导致很大范围内鸟类胚胎的突变。由此可见，详细分析水质对任何湿地恢复项目的成败都具有重要意义。

有关水质的问题

1. 是否存在水污染现象？是否有水华、赤潮、水生动物突然集体死亡的历史事件？

2. 可能的面源污染源在何处？

3. 盐度和酸碱度的变化情况如何？

4. 现有污染程度会怎样影响重建湿地的生物结构？

5. 湿地周围现在或以前有没有工厂所引起地下水或土壤污染？

6. 要解答上述问题1至4，应做何种测定？

7. 是请专业监测部门来做水质测定，还是通过简单的野外水质监测设备获得资料？

8. 污染问题是否可以解决，用何种方法来解决污染问题？

除了自然条件以外，在湿地恢复工程中还要考虑人类活动及其影响。现有的与规划的道路、停车场、排水沟渠、水库及其它设施，也可能会对湿地的水质、

Water Quality

While hydrology is most often the primary agent affecting the success or failure of a wetland restoration project, the influence of water quality can be equally important and thus should not be ignored. A common misconception is that wetlands can assimilate and treat all manner of pollution. This can lead to disastrous consequences for both wildlife and vegetation. Inputs of chemicals from the surrounding landscape can overwhelm a wetland's ability to improve water quality and can change the characteristics of the site. For example, salts are often used in cold climates along highways, and if they enter a wetland, can alter the productivity and composition of plant communities, possibly favoring pest species (Niering, 1989). In another case in Kesterson, California, a high concentration of Selenium (Se) in the source water for a treatment wetland was initially overlooked but later found to have caused widespread mutations in embryonic birds. Therefore, the rigor and level of detail shown in water quality data can determine the success or failure of any restoration project.

Questions to ask about Water Quality

1. Is water pollution evident?

2. What are the likely sources of pollution?

3. How will pollution levels impact a restored wetland's biology?

4. Are there or have there been industries at or near the site that could have caused groundwater and soil pollution?

5. What type of testing is necessary to obtain answers to question 1-4?

6. Do outside consultants need to be hired for testing or can data be easily obtained using simple field kits?

7. Will pollution problems require remediation, and what methods are available for fixing the problems?

8. What parameters should be monitored? How often should they be monitored and for how long?

Aside from natural conditions, it is important to consider human influences and impacts. Planned and existing roads, parking lots, drainage ditches, reservoirs, and structures may exert an influence on water quality, hydrology and the biology within a wetland. Adjacent land

水文环境和生物结构产生影响。附近土地的使用方式，如住宅、工业、商业、农业、畜牧、采矿、林业、开放水域、草原、为停车或娱乐目的的开放空间，会以多种形式对湿地产生影响。城市土地会向湿地排放油污、重金属、沉淀物和表面活性剂等污染物，农用土地则会因为大量排放杀虫剂、除莠剂和化肥而对湿地的生态系统产生不良影响，同时，要对工业设施进行全面调查，确定其是否达标排放，防止其向周围自然环境排放废水、废物、废气并危害当地的野生动植物。

uses such as residential, industrial, commercial, agriculture, grazing, mining, forestry, open water, grasslands, or park and recreational open space systems all affect a wetland in different ways. Urban landforms may discharge pollutants such as oils, heavy metals, sediment, and surfactants into the wetland, while agricultural lands can be a significant source of pesticides and fertilizers that could adversely affect a wetland's ecosystem. Industrial facilities should be investigated thoroughly to ensure that toxic materials and waters are not discharged into a natural area, where they could threaten wildlife.

◀ 图 3-11: 此处列举了众多湿地污染源中的几个，包括养殖池塘的排放物、工业废水和施工造成的侵蚀 (摄影 : Rowan Roderick-Jones)
Figure 3-11: A few of the many pollution sources to wetlands are represented here including (clockwise from left) discharge from industrial wastewaters, aquaculture ponds, and erosion caused by construction activities (Photographs by Rowan Roderick-Jones)

▼ 图3-12:扬子鳄和黑脸琵鹭是可以通过适当
恢复生态环境而得到保护的两种濒危动物
（图片来源：www.photos.com）
Figure 3-12: The Chinese alligator and black faced
spoonbill are two endangered species which can be
protected through restoration of appropriate habitats
in China (Source: www.photos.com)

生 物

湿地最初是因为水禽、涉禽栖息地而受到重视的，除了鸟类和鱼类之外，一些昆虫（如蜻蜓）、爬行动物（如鳄鱼、大鲵等）、水生哺乳动物（如水獭等）也应当根据历史记录和栖息地的适合度而成为湿地恢复中的目标保护物种。因此，在湿地恢复之前应该对生物资源有比较全面的了解，除了能够有针对性地开展栖息地恢复之外，也可避免因为湿地恢复而导致特殊生境的丧失（见图3-12）。

此外，许多大型哺乳动物如虎、鹿等都可能会将其领地及周边地区的湿地作为饮水地，因此，信息收集范围不仅包括湿地本身，还应尽量包括周边的栖息地。对于迁徙/洄游物种，还必须了解迁徙路线的位置和物种利用湿地的季节特征。林业部门通常备有大量相关信息，此外，还必须向经常在本地区进行动物学、植物学、生态学研究的科研人员咨询相关内容，对于缺乏研究资料的地区，向当地的渔民、猎户咨询通常能够得到大量有用的信息，尤其是水华或赤潮、兽类出没等与生产活动密切相关的事件，此外，还应明确他们对现有湿地的利用程度和依赖程度，比如湿地产品在他们总收入中的比例，这样的调查除了能够推测湿地资源现存量之外，还有助于在规划中将人为活动的控制和当地居民的福利考虑在内（见图3-13）。

有关生物的信息：

1. 当地的地带性植被类型和现有的植被类型？

2. 湿生植物和中生植物的分布格局及其优势度？

3. 濒危珍稀物种的种类、数量、种群动态及其利用湿地的季节特征？

4. 濒危珍稀物种的特殊生境需求？

5. 外来物种（尤其是已证实为入侵物种）、速生杂草的分布和群落特征？

6. 耐污物种、伴人物种的种类和数量？

Biology

The global awareness of wetlands is largely supported by their function as important water fowl and wading bird habitats. Apart from different species of birds and fish, some insects such as Libellulidae and Coenagrionidae dragonflies, reptiles and amphibians such as crocodiles and giant salamanders, or aquatic mammals such as otters can be identified as targeted protection species in wetland restoration projects, based on historical documentation and habitat compatibility. A comprehensive understanding of local biological resources can help restore a site in accordance to the specific habitat requirements of individual species and can prevent habitat loss during the restoration process.

Moreover, since large mammals such as tigers and deer tend to wander out of their territorial domains to drink water in nearby wetlands, it is advisable to collect biological data in areas beyond the restoration site. For migratory species, an understanding must be gained on migration routes and seasonal wetland use. National or local forestry agencies may be a good source of relevant information, and scientists conducting biological or botanical research in the region may also be consulted for their specific knowledge. In areas where biological data are scarce, useful information may be obtained from local fishermen and hunters, especially on agriculturally associated events such as algae blooms or red tides and wildlife populations. Through these sources, the degree to which a local population depends on existing wetlands can also be assessed, determining things like the proportion of wetland yields to overall income. These kinds of surveys allow a project designer to identify existing resources of a wetland site and, more importantly, make due consideration for controlling human activity and creating local community benefits.

Questions to ask about Biology

1. What are the regional vegetation types and existing vegetation types of the site?

2. What are the distribution patterns and domination levels of phreatophytes and mesophytes?

3. What are the types, quantities, community development and seasonal characteristics of wetland use by endangered species at the site?

4. What are the special habitat requirements of endangered species?

5. What are the distribution and community characteristics of foreign species (especially those

7. 野生动物，尤其是大型兽类的种类、数量、活动习性（是否对当地居民的生活生产带来危害）以及受人为干扰的影响程度（是否被商业性的捕猎或豢养？）

8. 资源生物的种类、生产方式、收获方式和资源现存量（包括有关的管理计划）？

9. 规划范围内的公益林、防护林、国家基本保护农田、商品林、防火通道、已承包的水产养殖池塘的范围、面积和群落结构（资源物种组成）？

10. 规划范围内是否已有全部或部分面积被纳入自然保护区或湿地公园的规划？规划要求如何？

11. 周边地区（100km范围内）是否具有自然保护区或国际重要湿地？自然条件是否相似？恢复湿地是否可能成为其补充栖息地？

在上述问题的调查中，可以明确湿地在生物栖息地方面的重要性。在针对物种保护的湿地恢复中，必须首先确定所要保护的目标物种，以便有针对性地采用生态恢复措施。确立依据如下：

1. 备选目标物种的濒危程度和种群规模；

2. 广泛查阅文献资料，以确定潜在的可恢复的物种种群；

3. 走访当地渔民、猎户、林业部门，调查野生动物现状，尤其需要关注是否具有保护物种的活动记录以及与人类活动的关系；

4. 分析湿地的水文和植被现状，推测可利用其作为栖息地的动物类型；

5. 了解基地周围的保护区、国家公园、山系、水系、迁徙或洄游路线等情况，以确立基地是否处于动物的活动廊道上，或者是否可能成为连接两个栖息地的补充廊道。

在确定目标物种之后，应该详尽了解其生境需求，可参阅已发表的学术论文和

identified as invasive species) and fast-spreading weeds?

6. What are the types and quantities of pollutant-resistant and human-associated species?

7. What are the types, quantities, habits (harmful or not to the life and production of the local community) and exposure to human disturbance (whether hunted or artificially bred for profit) of wildlife at the site?

8. What are the types, production modes, harvesting modes and resource pools (including relevant management schemes) of economic crops and animals?

9. What are the ranges, areas and biotope structure (composition of resource species) of commercial and non-commercial forests, protected forests, farmlands under national protection, fire buffer zones and commercial fishery ponds in the restoration site?

10. Is the restoration site totally or partially included in a planned natural reserve or national wetland park? If so, what are the requirements of the plan?

11. Are there natural reserves or Ramsar wetlands of international importance in the vicinity (within 100 km of the restoration site)? Are there similar natural conditions? Is there a possibility that the wetland will become a mitigation habitat for a nearby natural reserve or Ramsar wetland?

The answers to the above questions may determine the regional significance of the site as a biological habitat. If it is necessary to target specific endangered or protected species for the restoration project, the following steps can be taken to identify appropriate target species and their habitat requirements:

1. Determine the protected status and community size of potential target species;

2. Consult existing documentation to determine potential target species;

3. Consult local fishermen, hunters and forestry agencies to obtain the latest information on wildlife, especially records on the activities of potential target species and their interaction with humans;

▼ 图3-13：当地渔民，如天津附近团泊湖一带的居民，可提供与湿地生物多样性和具有商业价值的资源相关的宝贵生物信息（摄影：Rowan Roderick-Jones）
Figure 3-13: Local fishermen, such as these men from Tuanbo Lake near Tianjin, can often provide valuable biological information about a wetland's biodiversity and its commercially valuable resources (Photograph by Rowan Roderick-Jones)

专著、进行专家咨询以及前往相关的自然保护区。

在非针对物种保护的湿地恢复中，应该尽可能使恢复的湿地吸引更多的生物进入，提高当地的生物多样性（见图3-14，3-15），所依据的原则是：

1. 能够在当地提供一个人为干扰少的环境；

2. 能够提供水源、饵料（如果实、花蜜等）；

3. 植被本身具有较高的多样性，能提供不同层次的林冠或具有较好隐蔽性的草丛。

4. Analyze the site's hydrologic and vegetation status to determine the kinds of wildlife that may use it as a habitat; and

5. Establish an understanding of natural reserves, national parks, mountains, water systems and migration routes within the site vicinity, in order to determine whether it is located in the corridor of wildlife activities or can be used as a mitigation corridor to connect two habitats.

After identifying the target species, the project designer must obtain a detailed understanding of the species' habitat requirements by referring to published academic theses and monographs, by consulting with experts, and by visiting relevant natural areas.

If the restoration project is not for the protection of particular species, an alternative approach is to design the site to attract as many species as possible in order to increase biodiversity. The principles for this approach to site design are as follows:

1. The site should provide habitats with minimal human disturbance;

2. The site should provide a sufficient water source and forage crops (such as fruits and seeds); and

3. The vegetation should have high biodiversity and provide a multi-layered canopy or thick undergrowth as wildlife cover.

▶ 图3-14: 低潮时，慈溪的湿地恢复场地现出天然泥质浅滩(摄影：Rowan Roderick-Jones)
Figure 3-14: Low tide at Cixi exposes the natural mudflats near the restoration site (Photography by Rowan Roderick-Jones)

▶ 图3-15: 在全球环境基金宁波-慈溪湿地项目中，设计师们规划出大面积的泥滩湿地，将目标锁定为岸鸟，这些泥滩在此概念性场地规划图中显示为深棕色。其他生态环境类型有沼泽、季节性湿地和旱草甸等（图片来源：易道）
Figure 3-15: At the Ningbo–Cixi Global Environment Facility, designers targeted shorebird species by creating expansive areas of mudflat wetlands, shown as dark brown areas on this conceptual site plan. Other habitat types included marshes, seasonal wetlands, and dry meadows. (Source: EDAW)

研究中心(面积400平方米)
Research Center
(Indicative Footprint 400sqm)

机房
Pump House

入口池
Forebay Pond

教育中心和巡逻站
(面积100平方米)
Docent Center and Ranger Station
(Indicative Footprint 100sqm)

木径和观景台
Boardwalk & Viewer Stage

教育中心
(面积3,000平方米)
Education Center
(Indicative Footprint 3,000sqm)

湿地恢复手册 原则·技术与案例分析

社会与经济特征

长期以来，人们对湿地的社会经济特征的了解是非常模糊的，甚至在湿地恢复工作中缺乏专门的调查和研究。但是，城市和区域规划中的社会与经济特征评估以及上个世纪90年代以后兴起的湿地生态价值货币化研究能够为湿地恢复提供相关的调查依据和方法。

本书的第一章已经详细描述了湿地的生态价值与社会经济价值。但是，目前，除了可直接消费的物质较容易进行货币化之外，湿地的其他价值如气候调节、水文调节、科研与美学等价值如何进行货币化，以便能够使之更为有效地与市场、政策、管理、规划等各个环节相结合仍然是一个普遍的难题。

即使各方面都较相似的两块湿地在不同的社会、经济背景下也会导致生态价值估算的不同。比如，在较为贫困的地方，人们对湿地的生态旅游支付意愿肯定会低于经济较为发达的地方。问题在于，当同一块湿地具有变化较快的社会经济背景下时，就会产生令人困惑的问题。某些生态价值（如物质生产和水体净化功能）在经济发展初期得到了充分的挖掘，但是同时也产生了副作用或不良后果（如生物多样性下降、水质恶化等），当经济条件上升到一定条件时，人们对该湿地的其他生态价值（如美学价值等）产生了兴趣，却发现那些副作用已经导致人们的愿望无法实现，或者必须支付高额的代价来减缓和去除那些副作用。

社会经济状况是项目范围、当地及更大区域范围的内在社会经济因素，通常包括了人口模式、经济动力和文化政治背景的重要方面，因而每一块湿地的社会经济状况评估结果都会各不相同。调查当地社会经济状况对于明确规划用地是否适于开发以及湿地对当地未来的影响至关重要，使所做的规划既不能只着眼于当前最受重视的那些价值，也不能只在预测的未来经济发展趋势下评估和开发的湿地价值。

详细的经济调研可确保恢复工程的财务可行性以及给当地社会带来的效益，同时将项目成本及潜在的负面影响减至最小。社会文化因素还可以丰富湿地恢复的政治和社会层面。

Socioeconomic Conditions

For many decades, the socio-economic character of wetlands restoration lacked a systematic investigation and research process. Fortunately, beginning in the 1990s, new research methodologies for analyzing wetland values in monetary terms, as well as the socio-economic assessments of wetlands as they impact city and regional planning, have been extensively applied to wetland restoration projects.

Two wetland sites with similar environmental characteristics may be subject to very different valuations by communities with differing socio-economic conditions. Less developed communities may not place a high value on the environmental value of nearby wetlands. As such, wetlands can be subject to unmitigated exploitation through intense agricultural production, pollution or over-harvesting. As time goes on, and the community becomes more socio-economically developed, people begin to take a greater interest in other values of a wetland (such as aesthetic enjoyment). This greater interest in the environmental values of wetlands often serves to drive the restoration of wetlands that have been degraded by less developed communities in the past.

In this light, a rigorous socio-economic investigation is of critical importance in determining whether the intended site is suitable for development, and how this site may affect the future of the local region. Such investigations ensure that the planning focus rests not only on the most desired present values, but also on broad variables that consider the practical and healthy use of wetland values in a way that aids future economic development in the region, without destroying the wetland that will advance this development.

Analyzing socioeconomic conditions are critical to understanding the development suitability and impact of a wetland. Detailed economic research will ensure financial feasibility and benefit local communities, while minimizing costs and potential negative impacts of a project.

Site assessments should include a multitude of socioeconomic variables that are contingent upon the availability of quantitative and qualitative information. Significant data can often be obtained through local government bureaus, such as planning or statistical departments. Information must also be gathered through stakeholder interviews and on-site observations.

项目的社会经济现状评估包含多种社会经济指标。评估的效果取决于可获得的信息数量和品质。通常，大部分信息资料是通过当地政府部门获得的，如规划局或统计局（国家统计局有定期发布的统计年鉴）。另外，还必须通过与利益相关者的访谈和实地考察来收集资料。目前在规划中的社会经济评估包括以下几个重要范畴。

人口，移民和人口变化

- 当地人口数量是了解目前共有多少住户将受到开发影响的基础。它不仅直接作用于湿地的经济效益和复原计划，而且是管理部门制定住户动迁战略计划的重要依据。

- 移民与人口增长趋势被用于评估将来在开发过程中有多少住户会受到影响。在评估湿地的生态价值以及为解决环境保护问题需要修建什么设施时，这一点非常重要。

- 区域人口越多，将来对住房、民用设施和开发密度的需求也就越大。

经济

- 地区生产总值（GRDP）是指一个地区生产的商品和服务的总价值，可以体现该场地与其他地区相比的相对经济实力。它还从就业和增值两方面反映出该地区的主要就业情况。这些对于理解当地经济如何随湿地开发而改变以及对当地就业和产品输出的潜在影响很重要。

- 收入水准：城市居民可支配收入和农民纯收入是生活水准的两个指标，并且对评估以下主要因素必不可少：开发的经济效应；对某些自然资源的需求和消费意愿；人口搬迁和再安置战略。

  在中国，对收入和GRDP的统计资料从本地乡镇级别开始考虑。收集有关项目地区的具体经济资料时，必须与当地居民谈话、进行实地考察和社会调查。

社会文化

- 对环境资源的社会取向：不同区域的人们对湿地有着截然不同的社会

A socioeconomic analysis includes the following key categories.

Population, migration and population change

- The on-site population is the basis to understanding how many households will be influenced in the immediate term by the restoration project. This directly contributes to the household resettlement strategy (if needed), and to a wetland economic impact and revitalization strategy.

- Migration and population growth trends are used to estimate the number of households that will be affected by the course of development. This is important in determining the overall development impact of a wetland park, as well as the necessary facilities required to address environmental preservation issues.

- The greater regional population serves as a broad indicator of future demand for housing, civic facilities and development density.

Economic

- The Gross Regional Domestic Product (GDP) is the total value of goods and services produced by a region. It indicates the strength of the site's economy relative to other localities. It also shows the key employment activities on the site in terms of both employment and value-add. These are critical measures in understanding how the local economy will change upon development of the wetland, as well as its potential impact on local jobs and output.

- Income levels: Urban disposable income and rural net income are indicators of the standard of living. Both are necessary to assess key factors like the economic impact of the development, the demand and willingness to pay for certain natural assets, and in forming a relocation and resettlement strategy.

- In China, statistics for income and GDP are measured on a local township scale. For detailed economic information of the project site area, data collection must include interviews with local residents, site observations and community surveys.

Socio-cultural

- Social preferences for environmental assets also provide useful information. Different countries have contrasting societal standards for green open spaces for wetland parks. The majority of Chinese urban

▲ 图3-16: 农村湿地的应用：湿地住宅、农作物及一些农产品，如本地产的鱼类及野生和家养的禽类 (摄影：Rowan Roderick-Jones)
Figure 3-16: Rural utilization of wetlands: wetland dwellings, harvesting, and some of the numerous products including native fish species, and wild and domesticated birds (Photographs by Rowan Roderick-Jones)

▼ 图3-17:这些瓦缸里装的是有名的浙江绍兴产的黄酒。易道在为绍兴设计一座大型的湿地公园时，就考虑到生产这种黄酒所必需的纯净水 (摄影：Rowan Roderick-Jones)
Figure 3-17: These casks hold the famous yellow wine produced in Shaoxing, Zhejiang province. The necessity of clean water for the production of yellow wine was considered during EDAW's design of a large wetland park in Shaoxing (Photograph by Rowan Roderick-Jones)

▲ 图3-18: 易道在设计绍兴湿地公园时，将浙江绍兴的这个古码头作为文化古迹保留下来 (摄影：Rowan Roderick-Jones)
Figure 3-18: This ancient wharf in Shaoxing, Zhejiang province, was preserved as a cultural relic in EDAW's design plan for the Shaoxing wetland park (Photographs by Rowan Roderick)

标准。大城市里不少人都比较看重有湿地的地方，河畔、湖滨和海滨已经成为人们休闲旅游、观鸟摄影的最佳去处，湿地在城市绿地中的比重也日益增加。相反，有些农村地区却把河道、低洼地等湿地与害虫和贫困生活联系在一起（见图3-16）。因此，了解某地区的社会取向对于进行湿地恢复设计、可行性分析和让整个社会认可湿地很重要。

- 历史上，湿地是人与动物繁衍生息的地方，因此蕴藏着多种历史文化遗产。决定保护或恢复湿地时，必须既要考虑其机会成本，又要考虑到历史价值（见图3-17,3-18）。

在调查了社会经济状况后，可对被恢复的湿地进行生态价值的评估，以便分析哪些价值在体现过程中可能相互冲突，而哪些则是具有相互促进作用的。生态恢复可以具有多个目标，体现多个价值，但是往往只有一二个主要目标。作为主要恢复目标的生态价值应该不仅能够为当地居民提供持续性或替代性的生计，并且有利于当地经济发展后的湿地保育与开发计划。中国的许多学者都在这方面进行了探索。目前的评估方法包括：费用支出法（即支付意愿调查）、市场价值法、减轻损害费用支出法、影子工程法等。由于这方面的工作在国际上也属起步阶段，许多方法仍然存在争议，需要在实践中不断完善（欧阳志云等，1999；崔丽娟，2001；陆健健，2006）。

但是，不可否认的是，通过对湿地生态价值的研究，可以唤起人们对湿地的重视，避免设立过于脱离实际的目标或对湿地产生不良的影响。

areas place higher premiums on areas with wetland parks. In contrast, certain rural communities may associate a wetland park with pests and the creation of uncomfortable conditions. Hence, understanding community social preferences are important for the design, feasibility analysis and social acceptance of the wetland.

- As home to generations of animals and farmers, wetland areas often contain heritage assets of the local population. The decision to preserve or transform the wetland must consider the historical value and opportunity costs of development.

After a socio-economic profile is completed, an assessment of a wetland's ecological values must be conducted to determine which values are most desirable given the economic, social, natural and other site conditions. An ecological restoration project may have multiple goals and values, yet only one or two of these would be principle goals. Having an ecological value as the principle goal of a restoration project stems not only from a wetland's ability to support alternative livelihoods for local residents, but in its capacity to bolster development schemes around the wetland that would support and help sustain the economy over the longer term. A number of Chinese scholars have explored this area in recent years, including Ouyang Zhiyun et al, 1999, Cui Lijuan 2001, and Lu Jianjian 2006.

The following section explores the various assessment approaches currently practiced.

上多奇河湿地恢复
美国加州大浩湖
业主：加州公共服务部房地产服务部门加州大浩湖保护局
Upper Truckee River Wetland Restoration
Lake Tahoe, California
Client: California Department of General Services, Real Estate Service Division and California Tahoe
Conservancy

EDAW | AECOM

3.4 湿地恢复的若干选择

3.4.1 生态功能的预测

湿地恢复的多种选择

A) 一种是狭义的恢复，即调整并恢复湿地的水文过程和生物过程，可能需要采取原有群落恢复的措施，包括物种引入等，最终恢复生态系统的活力和自我维持能力。

B) 第二种是复原，通常针对若干目标种或生态系统的某一个服务功能，在较短的时期内通过调整生态系统管理策略来完成。

C) 第三种是重建，即构建一个新的湿地类型，这种情况通常发生在水文条件已经无法恢复的情况之下，比如，已经围垦的高潮带湿地通常很难恢复到自然的潮滩湿地，但是可以构建为淡水沼泽或林泽。

在着手进行湿地恢复之前，必须在这三种恢复选择中选择一种。选择的最主要依据就是湿地退化程度（见图3-19），即湿地退化的程度越轻，就越容易恢复为与原先一致的湿地，随着湿地退化程度的加重，通常就只能恢复部分结构和功能，但是需要付出更多的投入，包括人为引入目标物种，改造地形以恢复类似湿地的水文特征等。重建湿地是投入最多的恢复选择，所针对的湿地往往是已经极度退化或者无法满足原有湿地类型的外部条件，水文状况的根本性改变是促使管理者采用重建湿地的重要原因。重建湿地需要的时间是最长的，除了实施重建工程的投入之外，还需要长期的监测和管理。

此外，还要考虑的是对所恢复的湿地类型的选择。是自由表面流湿地还是潜流湿地？是季节性湿地还有永久性湿地？是受潮汐影响的湿地还是无潮汐影响的湿地？通常，退化越严重的湿地，可选择的类型就越多。

3.4 Alternative Analysis and Ranking

3.4.1 Ecological Function Forecast

Multiple options for wetland restoration

There are multiple categories of wetland restoration,including:

A) Restoration in the narrow sense, that is, adjusting and restoring the hydrologic and biological processes of the wetland site, attempting to achieve its original biological community (including introducing new species into the site), and eventually rebuilding the productivity and self-regulating aspects of the ecosystem.

B) Rehabilitation, or regulating the ecosystem in a limited time period to attain discernable improvements in certain target species or certain ecological services.

C) Re-creation, or re-creating a new wetland type for a restoration site, can be undertaken when the hydrological conditions have precluded the possibility of normal restoration. For example, high-tide zone wetlands already encircled by dykes and reclaimed as farmlands are difficult to restore as natural mudflats, but can be turned into freshwater swamps or marshes.

One of the above options must be chosen before the actual restoration, a decision based principally on the level of degradation at the site (see Figure 3-19). The less the site is degraded, the easier it can be restored into its original state. As degradation intensifies, the structures and functions of the site usually can be restored only partially, and this may generate high costs from target species introduction, land re-contouring and the re-establishment of wetland hydrology. Reallocation can be an expensive option and is usually used for extremely degraded sites or sites without any satisfactory external wetland conditions. A substantial change in hydrology may be enough reason for a project manager to adopt this option. Reallocation also takes the longest period of time to accomplish and requires long-term monitoring and management in addition to investment required for its construction.

In addition to the options mentioned above, the selection of wetland types to be restored should also be considered. Options include whether the wetland will be free surface flow or sub-surface flow, seasonally or permanently flooded, tidal or nontidal, etc. Generally, more severely degraded sites allow for more restoration options.

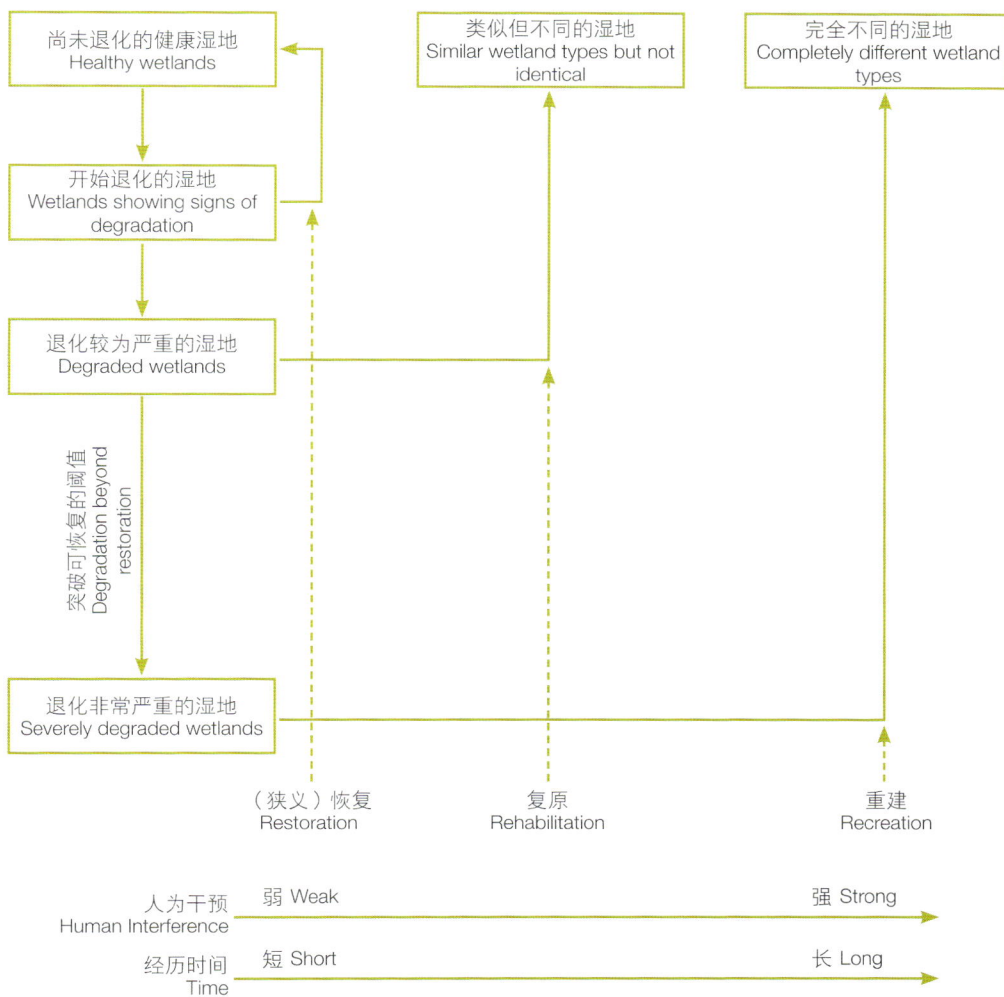

尚未退化的健康湿地
Healthy wetlands

开始退化的湿地
Wetlands showing signs of degradation

退化较为严重的湿地
Degraded wetlands

突破可恢复的阈值
Degradation beyond restoration

退化非常严重的湿地
Severely degraded wetlands

类似但不同的湿地
Similar wetland types but not identical

完全不同的湿地
Completely different wetland types

（狭义）恢复
Restoration

复原
Rehabilitation

重建
Recreation

人为干预
Human Interference

弱 Weak
强 Strong

经历时间
Time

短 Short
长 Long

◄ 图 3 - 19: 湿地恢复的三种选择 (改自：Simenstad 等，2006)
Figure 3-19: Three choices in wetland restoration (Adopted from Simenstad et al, 2006)

对最佳方案的判断

在决定湿地类型的时候，如果已经确定了需保护的目标物种，则通过该物种的栖息地要求来确定最佳方案，或者直接根据该物种的栖息地要求进行湿地恢复。除此之外，在恢复湿地的主要功能上也可能存在很多选择（见图3-20）。因为湿地恢复通常是多个专业交叉的工作，不同专业背景的人会有不同的想法，因此，在确立最终的湿地恢复方案前，会产生若干可供选择的恢复方案。任何湿地恢复项目都可能有几种途径，也许并没有完美的方案，但是，通过对比，可以获得最佳方案。以下是一般决定最佳方案的依据：

- 最可能达到预设的目标；

- 可能产生的生态风险最低；

- 如果必须引入非本地物种的话，防止其恶性扩散的机制最完备；

- 能为当地人提供可持续的生产或生活方式；

- 最低的投入产出比；

- 尊重并保护当地的文化遗产和民间习俗。

在比选最佳方案时，可采取的方法包括：案例参考、社会调查、基于GIS的情景分析等。通过案例参考可以确定湿地恢复的可操作性以及前景预测，社会调查则可确定湿地恢复的公众接受程度以及公众对未来湿地生态服务功能的期望程度，引导湿地恢复更加具有针对性，并做及时调整。

基于GIS的情景分析方法是目前越来越多采用的评价工具。GIS是可叠加各类信息的空间分析工具，它可以有效地管理具有空间属性的各种资源环境信息，对资源环境管理和实践模式进行快速和重复的分析测试，便于制定决策、进行科学和政策的标准评价。

此外，GIS还可以有效地对各个时期的资源环境状况及生产活动变化进行动态监

Selecting a preferred option

There are multiple options for restoring the chief functions of wetland site. Wetland restoration is an interdisciplinary task that often invites different opinions from stakeholders with different specializations, and several valid choices may be considered for a final restoration plan. If a target species to be protected has been selected, the habitat requirements of the target species may determine the best option and even the whole restoration process of a site. Multiple approaches to a given restoration project are inevitable. Although a perfect plan is conceivable, the best option is one that is achievable through the careful assessment of the available choices. A plan is considered a "best option" if:

- It is most likely to achieve preset goals (whether for habitat improvement, water quality, flood control, etc.);

- If the plan generates the least possible ecological risks;

- A complete mechanism is in place to thwart the unwanted proliferation of foreign species;

- Sustainable living and/or production conditions are provided to the local community;

- The plan warrants the best input/output ratio; and

- The cultural heritage and folk customs of the site locale are accorded sufficient respect and protection.

Available methods in the selection of a best option include case studies, social surveys and using a Geographical Information System (GIS), or GIS-based, scenario analysis. Case studies can help determine the feasibility and future of a wetland restoration, while social surveys reveal both public acceptance levels towards a restoration project and public expectations towards future ecological services of the restoration site. Both methods may lead to timely adjustments for a restoration plan, allowising the project to focus on appropriate goals.

GIS-based scenario analyses have recently become more widely available and popular as an evaluation tool for alternative analysis. GIS is a space analysis tool with overlays of different datasets which may include among other information, vegetation cover, slope, soils, groundwater and surface water systems, and monitoring locations. It is used to efficiently manage resources and environmental information with spatial attributes, conduct fast, repeated analytical tests on the administration and the practice of resource and environmental management, and to provide the basis for decision-making

方案一: 兴建跨河道堤坝断面 B:B
Option 1: In-Line Check Dam. Section B:B

海拔(m)
Meters Above Sea Level

淡水
Fresh Water

方案二: 疏浚现有河道断面 B:B
Option 2: Dredge Existing Channel. Section B:B

海拔(m)
Meters Above Sea Level

淡水
Fresh Water

方案三: 控制潮汐波动断面 B:B
Option 3: Controlled Tidal Fluctuation. Section B:B

海拔(m)
Meters Above Sea Level

咸水
Brackish Water

方案四: 控制中央湖泊水位断面 B:B
Option 4: Controlled Central Lake. Section B:B

海拔(m)
Meters Above Sea Level

水泵，提水到中央湖泊
Pump to Central Lake

淡水
Fresh Water

淡水
Fresh Water

优点/限制
Benefits/Constraints

泵水的长期运行成本较低
Lower long term operation costs for pumping

跨河道堤坝的建设成本较高
Higher capital cost for construction of check dam

中央湖泊及内河道的地下水位较高
Higher water table at central lake and internal canals

兴建跨河道堤坝需要有关部门批准
Permitting maybe required to construct check dam

需要即时洪水监看
May require real-time flood monitoring

提供多个灌溉方案
Multiple irrigation options

方案一: 兴建跨河道堤坝
Option 1: In-Line Check Dam

水泵与水渠
Pump and Pipe

地面流
Surface Flow

净化湿地
Treatment Wetlands

市政供水用于灌溉
Municipal Water Supply for Irrigation

可调节的跨河道堤坝
Inflatable Check Dam

高尔夫球场的灌溉
Golf Course Irrigation

新堤坝
New Levee Wall

现有水闸
Existing Water Gate

新水闸
New Water Gate

Fresh Water Lake

Fresh Water Inflow

断面 A
Section A

海拔(m)
Meters Above Sea Level

淡水
Fresh Water

~1.5 m

虹吸管/水泵或没有闸的暗渠(引力流)
Culvert with gate - gravity flow

淡水 Fresh Water.

流回水道(引力流)
Gravity flow back to canal

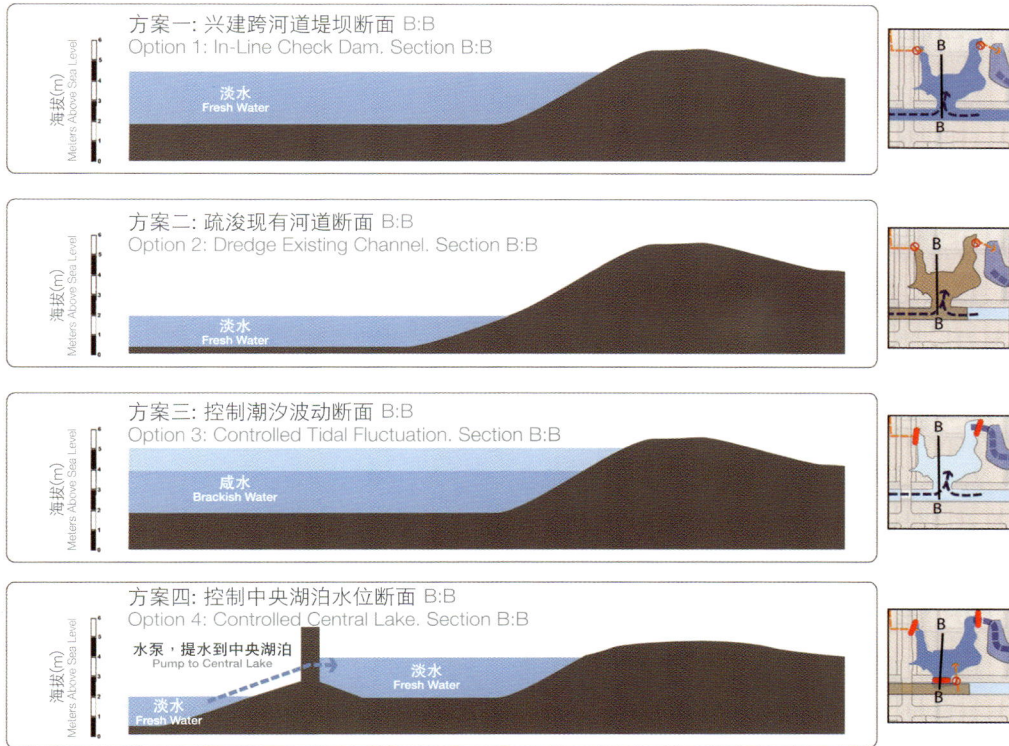

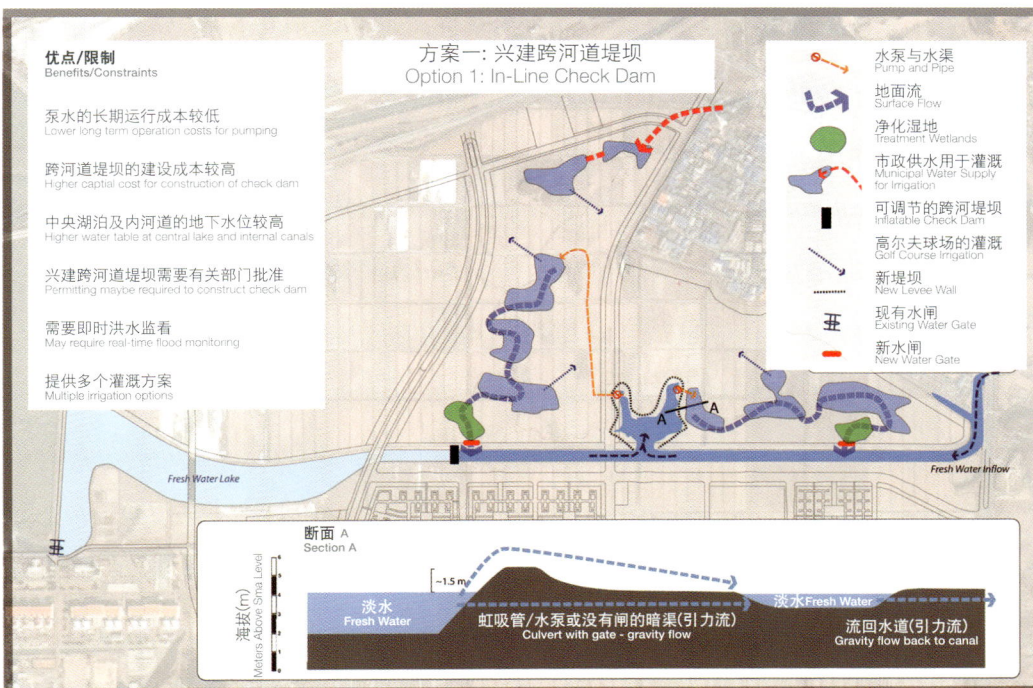

图 3 - 20: 易道在韩国做的 Cheongna湿地项目可供选择的多个的水文环境恢复方案包括潮汐和淡水方案。此处所示的方案1运用外环水道水库，提供合适的水文环境（图片来源：易道）
Figure 3-20: Options for restoring hydrologic conditions suitable for wetlands at EDAW's Cheongna project in Korea included tidal and freshwater schemes. Shown here is Option 1, which utilized an in-line check dam to provide appropriate hydrology. (Source:EDAW)

测和分析比较，即除了评价以往人类活动对自然资源的影响之外，还能在虚拟状态下评价以后不同资源开发模式可能产生的生态与环境影响，以便遴选出最合适的开发模式。目前，景观生态学的发展也多采用GIS技术，在栖息地形状、面积、数量和完整性判断上，重要自然资源的空间分布及其变化等方面具有极其显着的视觉化和可量化的优势（见图3-21）。通过GIS，可综合资料收集、空间分析、决策过程和布局规划等各个过程，为不同专业的工作人员提供可共用的平台，是一个发展迅速且强大的技术支撑。

情景分析

比如，在下面的简化案例中，假设可选的情景有三种，分别是栖息地价值恢复、防洪与休闲价值，水质改善中有较佳的表现。

根据当地的规划需求，在栖息地价值、防洪与侵蚀控制、水质、休闲价值和经济性等五个评价因数分别给予评价，并且每一个评价因数都有相应的权重（表3-1为例），各情景在每个评价因数上体现的价值越高，则分数越高（通常最高可设为5分），累计得到综合评价，分数最高的情景三可以被认为是最佳方案。

当然，各因数的权重可在现场评估的基础上采用专家评估的方法进行修正，对于靠近自然保护区或国际重要湿地的区域，栖息地价值应该得到更高的权重，对于人口密集且工业发展迅速的地区，水质和休闲价值可设较高的权重，对于洪泛平原和有过洪水记录的地区，则赋予防洪与侵蚀控制功能较高的权重。所有因数的权重相加为1。

and a standardized evaluation thereof. Furthermore, GIS analysis can provide a dynamic monitoring and analytical comparison for resource and environmental conditions and economic production over certain periods of time.

Apart from an actual evaluation of the historical impacts of human activities towards natural resources, GIS also offers a virtual evaluation on how different resource development modes may impact ecosystems and environments in the future. Currently, GIS is widely used in the development of ecosystem planning, because it has the advantage of both visualization and quantification in determining the shape, size, number and integrity of habitats and the spatial distribution and evolution of key natural resources. Data collection, spatial analysis, decision-making and layout planning are all integrated in the uniform platform of GIS, which can easily be shared by project members of different specialization. GIS has and will continue to be a fast-developing and powerful technical tool for wetland restoration.

Typical Scenario

The following hypothetical cases illustrate a mechanism for selecting preferred restoration alternatives:

Table 3-1 provides three scenarios whose outcomes indicate a good overall performance for each one. Evaluation standards fall into five categories: habitat value, flood and erosion control, water quality, recreational value and economic value, all according to restoration requirements. A weight coefficient is attached to each evaluation category. The scores each scenario receives in five categories are multiplied by the weight coefficients and summed up to reach the overall value. As scenario III comes up with a highest final score, it may be considered the "best option".

The weight coefficients in all categories are subject to modifications based on site evaluation and expert judgment. In areas adjacent to natural reserves or Ramsar wetlands, the habitat value should be weighted higher. In densely populated, industrially developed areas, the water quality and recreational value should be weighted higher, and in floodplains and areas with a frequent flood history, the flood and erosion value should be weighted higher. All weight coefficients must come to a sum of 1.

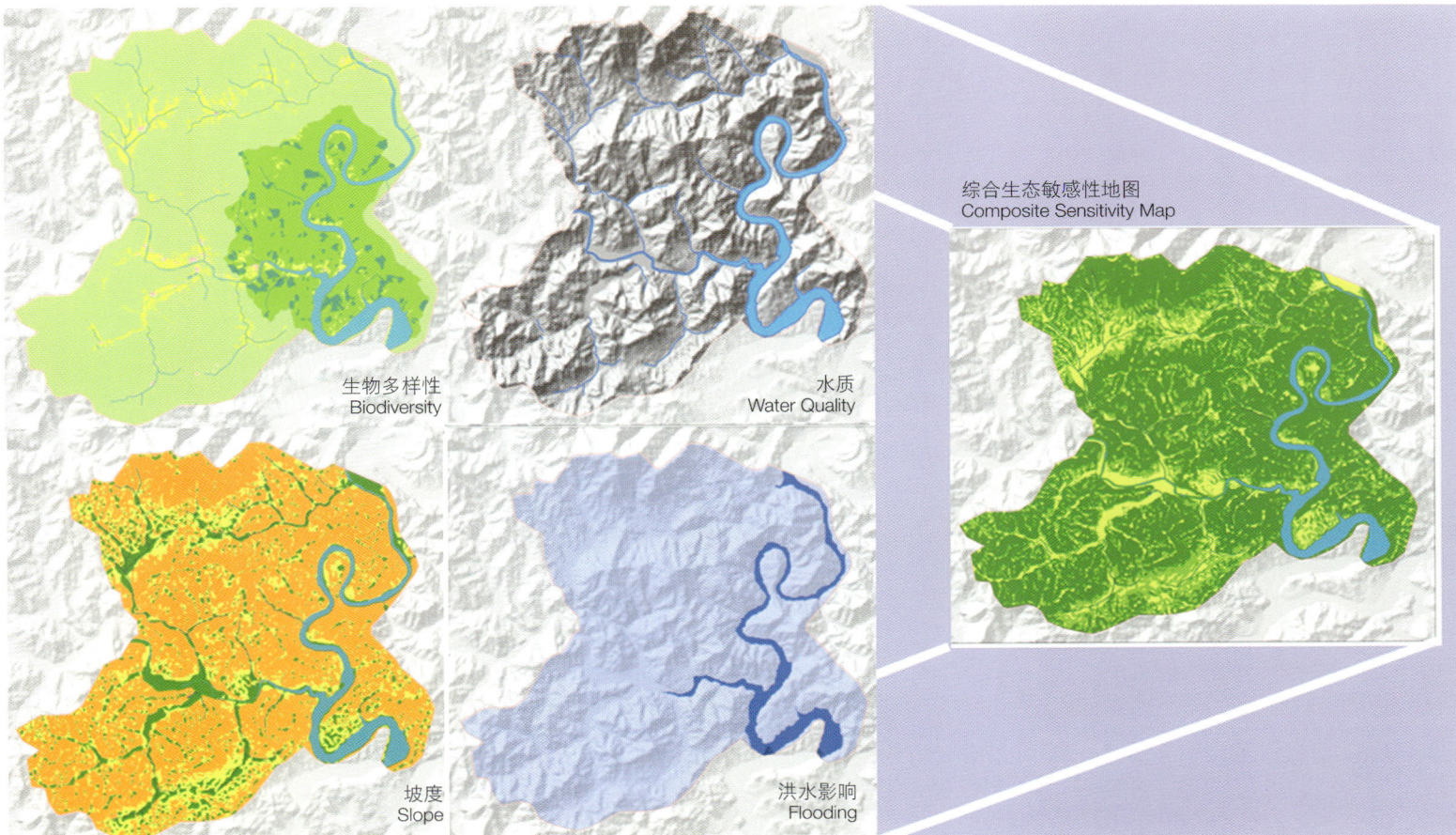

生物多样性
Biodiversity

水质
Water Quality

综合生态敏感性地图
Composite Sensitivity Map

坡度
Slope

洪水影响
Flooding

▲ 图 3-21:易道在辽宁青山沟的总体规划研究中,用GIS综合生态敏感性分析图确定了合适的
恢复点而不是开发点(图片来源:易道)
Figure 3-21: GIS composite sensitivity maps were used to determine appropriate restoration versus development
locations in EDAW's masterplan studies at Qingshangou, Liaoning Province (Source: EDAW)

▼ 表 3-1:情境分析表在根据当地的规划需求上的五个评价因数分别给予评价,综合评价因数上体现的价值最高,可
以被认为是最佳方案
Table 3-1: The scores each scenario receives in five categories are multiplied by the weight coefficients and summed up to reach the overall value

	栖息地价值 Habitat value	防洪与侵蚀控制 Flood & erosion control	水质 Water quality	休闲价值 Recreational value	经济性 Economic value	综合评价 Overall value
权重 Weight	0.3	0.2	0.2	0.1	0.2	
情景一 Scenario I	5	3	3	2	2	3.3
情景二 Scenario II	3	5	4	5	3	3.8
情景三 Scenario III	4	4	5	3	4	4.1

3.4.2 社会经济预测

场地初征

在选择最佳方案的过程中，衡量各个方案的社会经济状况是一个重要的步骤，对恢复湿地的管理和可持续发展具有重要意义。

社会经济预测是对未来经济乃至社会状况的分析。其依据是历史资料、已知市场走向以及对未来发展的分析判断。社会经济预测是一项重要的设计工作，它提供有关社会变革以及这些变革如何与开发形成互动方面的重要信息。

根据当地的特色和所得资料，对湿地项目的预测有很多种。预测是在统计资料和市场调研基础上所做的定量评估（见图 3-22）。然而，许多对发展中国家的社会预测事实上是通过分析访谈及考察结果对社会趋势所做的定性评估。

通常，湿地项目预测要考虑的最重要的问题是以下几个社会和经济因素：

人口

- 项目范围内、当地乃至更大区域范围的人口在过去 10 年中是如何变化的？促使这些变化发生的内在动力（例如道路设施、经济增长等）是什么？随着这些驱动力继续发展，未来人口增长将会发生怎样的变化？

- 预计 20 年后的人口情况会如何？未来人口将刺激当地的住宅、民用或休闲以及商业或零售用地的需求，因此在做湿地恢复的规划时必须予以考虑，尤其是当该湿地恢复项目是以增强其美学和休闲功能为目的的时候。

- 城市人口的增长情况很重要，因为它可以反映出崇尚高水准生活的中产阶级数量及其发展状况，而农村人口的增长则对当地自然资源的利用规模具有重要意义。

- 固定人口与流动人口之间的比例。流动人口不仅包括游客，更包括一些追随自然资源的渔民和猎户，他们虽然对居住条件的要求很低，甚

3.4.2 Socioeconomic Forecast

Initial Characterization

Socioeconomic forecasts are analyses of future conditions in the economy and society at large. These estimates are based on historical data, known market trends and the analysis and forecasting of future development. As such, socioeconomic forecasts are a critical design aspect that provides vital information regarding changes in society and how these changes will interact with development.

A wide range of forecasts are possible for wetland projects, depending on the site's unique characteristics and the availability of data. Forecasts are quantitative assessments determined from statistical data and market research. However, many social forecasts in developing countries are in fact qualitative assessments of societal trends using analysis of interviews and observations.

The following social and economic factors are considered to be the most important for forecasting wetland projects.

Population

- How have the project site, regional and greater regional populations changed in the past ten years and what are the underlying forces (such as transportation networks, economic growth, etc.) driving this change? How will population growth change in the future as these forces continue to develop?

- Ultimately, what is the estimated population in 20 years time? The long-term population will drive demand for residential, civic, recreational and commercial or retail land in the local community, and must be considered in planning a restoration project.

- Additionally, growth in urban populations is significant because it indicates the size and growth of a middle class that demands higher living standards.

- Migrant populations are not limited to tourists alone. Included in this category are fishers and hunters who regularly use natural resources. Most do not have requirements beyond basic living conditions, with some even relying on their fishing boats as a means of shelter. Their production and living circumstances in relation to a wetland, however, directly influence a wetland's bio-ecological environment. Therefore, helping these groups find an alternative livelihood and managing their production, should be a long-term goal of wetland restoration

至以船为生，但是由于他们的生产和生活都直接影响湿地的生物和环境，对他们的生计引导和生产管理应该是湿地恢复后长期管理的重要内容。

消费者及社会取向

- 消费者对公共开放空间的取向和态度将出现怎样的变化？随着收入增加，越来越多的家庭需要高品质的生活环境，因而愿意出钱在社区里建设精美的公共绿地和水景。

- 广大民众希望在他们生活的社区里有何种住宅和旅游开发项目？城市内的湿地恢复存在哪些重要的社会机会与限制？

- 通过采访主要的利益相关者和政府官员并对当地居民进行民意调查，确定消费取向和社会取向。

游客到访量

- 目前这个地区接待多少游客？增加或限制游客数量的因素是什么？随着地区经济和政府各项事业的发展，游客访问量会起什么变化？湿地恢复项目是否能吸引国际、国内和当地的游客？预计游客量是多少（见图3-23）？

- 是否需要根据野生动物的栖息特性确定游客的季节性控制和总量控制？

- 游客访问量预测使规划者能够设计出与此相应的旅游设施和环境保护工程，在保证旅游业效益的同时，保持湿地的环境品质。

经济结构变化

- 在过去这10年中，是否出现了哪些重大的经济结构变化？国家统计局把各镇第一、二、三产业的经济统计信息制成表格，可用于评估当地就业状况和经济发展方向，得出在今后的10到20年间，经济发展的走向如何？

- 经济结构变化可以体现出当地人口的生计和适当的再安置政策以及未来当地的就业情况和湿地恢复的经济机会等方面的重要情况。

Consumer and societal preferences

- How are consumer preferences and attitudes changing with regard to public open space? As incomes grow, an increasing number of households demand higher quality living environments and are willing to pay a premium for a well-designed public green space in their community.

- What kind of residential and tourism development does the community prefer? What are the significant societal opportunities and constraints that a wetland park provides?

- Consumer and societal preferences are determined through interviews with key stakeholders and government officials, and through surveys of the local population.

Tourism visitation

- How many tourists does the area currently receive and what is currently driving or constraining these figures? How will tourism visitation patterns change as the regional economy and government initiatives develop? Will the wetland park be attracting international, national, or local visitors?

- Is it necessary to implement seasonal control and total volume on the number of visitors allowed, based on how the wild animals make use of the habitat?

- Tourism visitation forecasts allow planners to design the appropriate tourism facilities and environmental protection structures that will maintain the quality of the wetland park while still capturing the benefits of tourism.

Structural changes in the economy

- What are the key structural changes in the economy over the last 10 years? And how will the economy continue to evolve in 10-20 years? The China National Statistics Bureau tabulates economic statistics of the primary, secondary and tertiary industries for each township. These can be used to assess the employment and growth direction of the local economy.

- Structural changes in the economy provide important information regarding the livelihoods of local populations and appropriate resettlement policies, as well as future local employment and economic opportunities provided by a wetland park.

▼ 图 3-23: 在易道的浙江绍兴镜湖湿地公园设计方案中，旅游业起着重要作用（摄影：徐志剑）
Figure 3-23: Tourism played a major role in EDAW's design of the Shaoxing Wetland Park, Zhejiang Province (Photograph by Eddie Tsui)

成本估算

成本估算对于经济影响评估至关重要，在进行湿地设计方案比较时，是颇见成效的一个评估因素。地点不同，成本常常也相差很大。这是因为成本是由许多变数决定的，如当前的环境品质、劳动力的可获得性和基础设施资源等。因此，做成本估算必须深入调查当地的劳动力和资源市场，并且通常要与本地的估算师协作完成。

成本分为四大类：场地准备费用、施工费用、再安置费用以及持续性年度维护费用。

场地准备

场地准备是指为提供适合湿地改善的条件而必须进行的土地处理。通常，场地准备费用占项目总成本的10%。湿地恢复的场地准备工作一般包括以下几个方面：

- 清除或削减多余植被，包括入侵物种、杂草等；

- 土壤处理；

- 林相改造；

- 摧毁或移走污水管及其他人造工程；

- 地面建筑物搬迁；

- 动员项目团队：设计和可行性研究。

施工费用

施工费用是将一块土地转变为达到项目目标的功能性湿地所需的实际工作费用。通常，此费用在总成本中的份额最大，约占整个资金成本的30%到50%。施工费用一般包括：

土地施工
- 挖掘和运输；

- 平整、压实和运输废渣；

- 筑堤；

- 流量控制设施；

Cost Estimates

Cost estimates are critical to any economic impact assessment and provide an effective means for comparing alternative wetland design schemes. Costs often vary significantly per locality because they are based on a wide range of variables, such as existing environmental quality and the availability of labor and infrastructure resources. For these reasons, cost estimates require research into the local labor and resource market, and are typically performed in cooperation with local quantity surveyors and design institutes.

Costs are categorized into four main types: (1) Site preparation (2) Construction costs (3) Resettlement costs; and (4) Recurring annual maintenance costs.

Site preparation

Site preparation involves the ground treatment necessary to provide conditions suitable for wetland restoration. Site preparation typically constitutes approximately 10% of the total cost of the project. The following aspects are common in wetland site preparation.

- Clearing/slashing of unwanted vegetation

- Soil treatment

- Removal/relocation of heritage trees

- Demolition/relocation of sewage pipeline and other built structures

- Relocation of above-ground structures

- Mobilization of project team: design and feasibility studies

Construction costs

Construction costs are those related to the practical work for converting a piece of land into a functional wetland that achieves project goals. Construction costs are often the most significant proportion of the total costs, ranging from 30% to 50% of overall capital expenditures. These costs typically include:

Land construction
- Excavation and hauling

- Grading, compacting and transporting land fill

- Embankment construction

- 进水工程（挖方和填方）；

- 排水池；

- 筑堤坝和防洪墙

交通基础设施
- 路障；

- 公路或其他访问点和入口。

环境准备
- 湿地植被种植：播种、扦插、移植等；

- 必要的疏林工作；

- 以景观美化为目标的栽培。

动迁费用

动迁是指将该区域的当地居民搬迁到新居点并提供财政补偿的过程。由于土地征用破坏了住房、社区结构、社会网络和社会服务，因而常会产生动迁成本。做财务补偿时，必须考虑到生产资产，包括土地和收入来源。一般来说，以湿地公园为主要目的的湿地恢复项目动迁费用要占项目总成本的30~50%。但是，对于原本就处于荒野状态的地区，如自然保护区的湿地，动迁费用可能是非常低廉的。

中国的动迁政策是依据亚洲发展银行的标准制定的。该政策很全面，包含许多方面：法律顾问、经济评估、社区参与、投诉程式、组织机构责任和未来监督。但是，该政策主要在大型基础设施项目中得以落实，遇到小型项目时，实施起来常常就没有那么严格。

动迁费用通常因项目目标不同（特别是当地居民是否继续在项目地区从事生产活动）而有很大差异。许多湿地在设计时是为了让村民继续在湿地上耕作，或通过新的商业或旅游机会继续在湿地上稳定谋生。与动迁相关的大项费用见表3-2。

- Flow control facilities

- Inlet structures (cut and fill)

- Outlet basin

- Laying dams and floodwalls

Transport infrastructure
- Traffic barriers

- Roadways or other access and arrival points

Environmental preparation
- Wetlands planting: propagation and harvesting

- Replanting

- Landscaping

- Relocating heritage plant species

Resettlement costs

Resettlement is the process of transporting and financially compensating local populations on the site to new settlements. Resettlement losses most often arise because of land acquisitions that disrupt housing, community structures, social networks and social services. Productive assets must be considered in the financial compensation package, including both land and income sources. In general, resettlement costs for wetland parks in China range anywhere from 30% to 50% of the total cost of the project. However, relocation costs can be considerably lower at sites that were previously wilderness areas, such as wetlands located in natural reserves.

China's national resettlement policy is based on standards set forth by the Asian Development Bank. It encompasses a comprehensive plan with numerous aspects: legal counsel, economic valuations, community participation, grievance procedures, organizational responsibilities and future monitoring. However, these policies are primarily implemented for large-scale infrastructure projects, while smaller projects often have less stringent regulations.

Relocation costs vary according to project goals and are especially dependent on whether or not local residents are engaged in agriculture, aquaculture or industrial production at the site. Many wetlands can be designed in a way that allows rural residents to maintain stable livelihoods on the wetland through continued farming activities and new commercial or tourism opportunities. Table 3- 2 provides a comprehensive list of the major costs associated with resettlement.

▼ 表3-2: 动迁费用因项目目标不同，以下为相关的大项费用及细目(来源：易道经济组；《亚洲发展银行移民手册》)
Table 3-2: Comprehensive List of Major Costs Associated with Resettlement (Source: EDAW Economics; ADB Handbook on Resettlement [1998])

类别 Category		费用项 Cost Items
移民准备与补偿 Resettlement preparation and compensation		· 对有关人员和编目资产的普查与调查成本 · 信息咨询费用 · 对损失资产的赔偿：土地、建筑物等 · 重置土地的费用 · Cost of census and survey of affected persons and inventory assets · Cost of information and consultation · Compensation for assets lost: land, building structures, etc. · Costs of replacement land
搬迁与过渡 Relocation and transfer		· 搬运可搬运物件的费用 · 重置房产的费用 · 场地和基础设施开发和各项服务的费用 · 过渡期间的最低生活保障金 · 重置营生和停工的费用 · Cost of moving and transporting movable items · Cost of replacement housing · Cost of site and infrastructure development and services · Subsistence allowances during transition · Cost of replacement businesses and downtime
收入恢复计划 Income restoration plans		· 对收入恢复计划的成本估算 · 增益服务(卫生、教育等)成本 · 环境改善一揽子计划(森林、土壤保持、牧业用地等) · Cost estimates for income restoration plans · Cost of incremental services (health, education, etc.) · Environmental enhancement packages (forestry, soil conservation, grazing land, etc.)

▼ 表3-3: 湿地的经济效益 (来源：易道经济组)
Table 3-3: The Economic Benefits of Wetlands (Source: EDAW Economics)

使用价值 Use Values	直接使用价值 Direct	· 农业(稻米、蔬菜、树和其他作物) Agriculture (rice, vegetables, trees, other crops) · 渔业 Fishing · 食品 Food · 休闲 Recreation · 建筑材料 Building material · 薪柴 Fuel · 交通 Transport · 其他原材料 Other raw materials
	间接使用价值 Indirect	· 防洪减灾 Flood/storm prevention & protection · 有机物和营养物的储存与循环利用 Storage and recycling of organic matter and nutrients · 防腐与生物防治 Erosion and biological control · 废水处理 Wastewater treatment · 为当地居民和庄稼灌溉提供水源 Water supplies for local population and/or crop irrigation
	可供选择的价值 Option Values	将来可以用湿地造福社会的方案：药物研究、工业、休闲、纯净水等 Future potential applications of wetlands that will benefit society: Pharmaceutical research, industrial, leisure, clean water, etc.
非使用价值 Non-Use Values		· 文化价值 Cultural value · 美学价值 Aesthetic value · 文物价值 Heritage value · 遗产价值 Bequest value
正外部效应 Positive Externalities		· 房地产价值的提升 Increased real estate values

持续性年度维护费用

湿地恢复是一项长期的工作，不仅仅是初期的设计和施工过程。因此，在进行湿地恢复规划时，必须将长期的管理和维护成本考虑在内。持续投资湿地维护对于恢复造福后代的具有可持续发展性的高品质湿地极其重要。湿地维护费用占总资本支出的10%~30%，具体取决于湿地恢复的类型。

湿地维护包括以下方面：

1) 生物监测和对自然资源的保护：植物群落、野生动物群落、鱼类、害虫、枯枝落叶、沉积物和水质。
   - 植物效用、健康状况和密度；
   - 虫害，水中和陆地上的杂草；
   - 肥料和除草剂的应用；
   - 对濒危和稀有物种的栽培；
   - 沉淀物和污染物含量；
   - 水的径流量和质量；

2) 对海岸线、河堰或堤坝、进水口或排水口及水流系统的结构保养。

3) 对当地人造工程的保养：苗圃、环境实验室、教育设施、行政大楼、公路及人行道。

4) 执法和安全向导。

5) 应急反应。

经济效益分析

湿地恢复项目通常能为当地居民乃至广大社会产生重大经济效益。这些效益以各种价值和正外部效应表现出来，参见表3-3。然而，湿地恢复设计的不同会大大提高对经济的正面影响，就旅游与商业领域出现的新商机而言尤其如此。本节着重介绍对于分析和选择其他湿地设计方案最重要的货币和非货币经济效益。

非货币效益：条件评估与支付意愿

湿地创造的使用价值和非使用价值与利润或支出无关，是通过使消费者满意的产品的非货币效益体现出来的。虽说是非货币效益，但也有许多方法将这些价值量化。一般来说，评估的依据是一旦湿地无法再造福于人，人们愿意付出多少来换回湿地的好处与服务。

Recurring annual maintenance costs

Wetland restoration requires a long-term commitment. The initial design and construction process is the beginning of a responsibility to maintain the ecological and functional value of the wetland over time. As a result, recurring management and maintenance costs must be taken into consideration during the planning stages of the wetland restoration. Maintenance costs range from 10% to 30% of total capital expenditures, depending on the type of wetland development.

Wetland maintenance includes the following aspects:

1) Biological monitoring and maintenance of natural assets: flora, fauna, fish, pests, litter, sediments, and water quality and includes:
   - Plant performance, health, and density
   - Pest infestation, aquatic and land weeds
   - Fertilizer and herbicide applications
   - Stocking of endangered and rare species
   - Sediment and pollutant content
   - Water runoff quantity and quality

2) Structural maintenance of shorelines, weirs/dams, inlet/outlets and water flow systems

3) Maintenance of built structures on-site: nurseries, environmental laboratories, educational facilities, administration buildings, paths and walkways

4) Enforcement and security navigation

5) Emergency response

Analysis of economic benefits

Wetland parks generate significant economic benefits for local residents and the greater community. These benefits are reflected in different types of valuations and positive externalities, as summarized in Table 3-3. Different park designs, however, can create dramatic positive economic impacts, particularly in terms of the potential for new tourism and commercial opportunities. This section introduces the monetary and non-monetary economic benefits that are most useful for analysis and selection of alternative wetland designs.

Non-monetary Benefits: Contingent Valuation and Willingness-to-Pay

Use and non-use values created by the wetland are not related to profits or expenditures, but are instead reflected in the non-monetary benefits of the product that contributes to a consumer's satisfaction. Although these

▼ 图 3 - 24: 您愿意出钱在您家附近建设这样的湿地吗？(摄影：Dixi Carrillo)
Figure 3-24: Would you be willing to pay to have this wetland near your home? (Yolo Bypass, California. Photograph by Dixi Carrillo)

条件评估法是评估支付意愿最常用的方法。这种方法运用调查的形式确定当地民众对湿地的好处与服务功能的明确偏好。条件评估法调查在美国和欧洲已有很长的历史，包括澳大利亚的卡卡杜国家公园、美国阿拉斯加州 Exxon Valdex 石油泄漏事件等，因此这套方法经过时间的千锤百炼，已能够将调查与反应偏差控制在最小限度。条件评估问卷通常包括：

- 您对这20hm²湿地的破坏索要多少赔偿？

- 为保护湿地中的某某遗迹，您愿意支付多少钱？

- 您愿意额外支付x元让您的社区拥有x吗（见图3-24）？

条件评估法可供经济学家对湿地恢复项目的"增量产品"进行评估，包括表3-3中列出的非使用价值、间接价值、可供利用的价值和正外部效应。此外，该方法还是得到广泛认可的评估直接使用价值的手段，如住户在得知附近有个湿地公园时的满意程度。虽然如此，条件评估法由于存在各种调查偏差，并且一次周密的调查需要花很多时间和金钱，因此也受到很多批评。

具体应用此法的广度和深度应视场地特征及项目目标而定。设计和方法对于条件评估结果的统计准确至关重要。

货币效益：商业效应分析
湿地公园对当地居民和商业也有着可量化的货币效益影响。分析这些影响时的预测过程涉及两大领域：(1)估计农业生产力的变化；(2)找出并预测新商机。

农业生产力
湿地公园从种植农作物的种类和数量方面影响着该场地的农业总产量。当腾出耕地作其他开发用途时，这种影响就是负面的；而当环境品质提高生产力时，即会出现正面影响（见图3-25）。评估农业生产力对于了解当地农户受到的经济影响以及该区食品安全受到的全面影响很重要。

鉴于因土壤质量、害虫、水及其他农业

are non-monetary values, there are numerous methods to quantify them. In general, valuations are based on what people would be willing to pay for the wetland goods or services, if the wetland were no longer capable of doing so.

The contingent valuation method is the most commonly used means to estimate willingness-to-pay. This method uses surveys to determine a local population's stated preferences for the wetland's good or service. Since contingent valuation surveys have a long history of usage in the United States and Europe, including Kakadu National Park in Australia and the Exxon Valdez oil spill in Alaska, US, the methodologies have evolved sharply over time to minimize different kinds of survey and response bias. Contingent valuation questions typically read as such:

- How much compensation would you demand for the destruction of 20 ha of wetland park area?

- How much would you pay to preserve the "X" heritage buildings in the wetland park?

- Would you be willing to pay "X" extra dollars in order to have "X" nearby your community?

The contingent valuation method allows economists to estimate the economic value of incremental wetland park outputs, including the non-use, indirect, option use values and positive externalities in the table above. The method is also widely accepted as a means to estimate direct use values as well, such as the satisfaction a family enjoys knowing that a recreational park exists in their neighborhood. Nevertheless, criticisms of the contingent valuation exist regarding different types of survey biases as well as the cost and time needed to perform a rigorous survey.

The scale and depth of its specific usage should be based on site characteristics and project objectives. The design and methodology are critical for statistically accurate contingent valuation results.

Monetary Benefits: Business Impact Analysis
A wetland park will also have a quantifiable monetary impact on local residents and businesses. The forecasting process of analyzing these impacts involves two main areas: (1) Estimating changes in agricultural productivity, and (2) Identifying and forecasting new commercial opportunities.

▼ 图 3 - 25: 在美国加州 Yolo Bypass这个多功能洪水控制、野生动物保护和农业区上，农业与湿地并行不悖 (摄影 : Dixi Carrillo)
Figure 3-25: Agriculture and wetlands go hand in hand in California's Yolo Bypass, a multipurpose flood control, wildlife and agricultural zone (Photograph by Dixi Carrillo)

因素差异而产生的不可预知的影响力，量化未来的农业产量非常之难，因此，对农业进行全面评估时，有关植物种植、土壤和病理学方面的问题应向农作物专家咨询。评估使用以下办法：

- 确定现有农产品：土地面积、农作物种类和产值；

- 完成湿地开发后估算农业用地面积；

- 根据新环境状况估算新的农产量，作物种类多样和增值；

- 找出可种植的新农作物。

新商机

好的湿地恢复项目能够为该社区创造新商机。例如，湿地常会成为附近居民户外运动的场所，如徒步游行和赏鸟等。这些旅游活动会从各各方面增加收入，例如门票和停车费、食品饮料销售以及教育慈善捐赠。此外，当环境得到改善时，湿地产品（如渔产品、农产品和衍生的旅游纪念品）可被赋予更高的价值，增加当地的收入。如果周边的农村因此而能够得到绿色食品基地、无公害产品基地的认证，带来的经济收益将是非常可观的（见图3-4~3-6）。

估算这些收入必须采取三大措施：

- 找出可能存在的新商机；

- 就每种商机（如酒店、野营等）对该区域当前的竞争情况展开评估：搜索不同的定价和选址；

- 根据当前本地市场及湿地的环境状况，预测湿地的游客到访量和收入情况。

Agricultural Productivity

Wetland parks impact the overall agricultural output of a site in terms of the type and amount of crops grown. This impact can be both negative, whereby farmland is lost to other development, or positive, whereby environmental quality increases productivity. Assessing agricultural productivity is thus important in gaining an understanding on the economic impact of local rural households, as well as the overall effect on regional food security.

Quantifying future agricultural outputs is a significant challenge because unpredictable variables and impacts arise from differences in soil quality, pests, water and other agricultural factors. A comprehensive agricultural assessment will include crop specialists in plant growth, soil and pathology. These assessments include:

- Identifying existing agricultural outputs: land area, crop type and output values

- Estimating agricultural land area upon completion of the wetland development

- Estimating new agricultural output based on new environmental conditions, disaggregated by crop type and value-added

- Identifying potential new crop outputs

New Commercial Opportunities

Successful wetland restoration project have the potential to create new commercial opportunities for the community because of their popularity as destinations for outdoor sporting and recreational activities. These tourism activities generate significant expenditures from a variety of revenue sources, such as entrance and parking fees, sales on food and beverages, and educational charity donations. Scenarios that need to be considered include:

- Identifying new potential commercial opportunities

- Assessing existing competitive regional activities for each type of commercial opportunity (such as hotels, camping, etc.), and scanning for different price points and locations

- Forecasting wetland visitation and revenues based on the park's current local market and environmental conditions

▼ 表 3-4: 所示农业调查范例取自加拿大多佛尔圣克雷尔湖湿地恢复项目农作物预测（来源：圣克雷尔湖湿地恢复项目议案，2001）
Table 3-4: Below is a sample format for an agricultural survey, as taken from the Lake St. Clair Wetland Restoration project in Dover, Canada.
a. 1996年英国多佛 (Dover)小区农作调查资料
a. 1996 Agricultural Census Data for the Dover Community
b. 全部的蔬菜量设定为4年内轮耕的农作物产量，包括2年内耕种大豆和玉米
b. All vegetables are assumed to be grown in a four-year rotation that includes two years in soy beans and grain corn.
c. 谷类和豆类作物设定耕种在黏土沃土壤中
c. Grain and bean crops are assumed to be grown on clay loam soils.
d. 玉米地的英亩数从玉米种子成长市场理事会取得（全部的玉米假设为耕种于黏土沃土壤中）
d. Seed corn acreage from the Seed Corn Growers Marketing Board (All seed corn is assumed to be grown on clay loam soils).
e. 加工农产品数量为主
e. For processing rather than fresh produce.
[来源：圣克雷尔湖湿地恢复项目议案(2001)]
[(Source: A Proposed Wetland Restoration Project, Lake Saint Clair (2001)]

农作物 Type of Crop	农作物收获量成长值 Total Crops Grown[a]		干土农作物和肥料(沙地) Crops on Sand and Muck[b]		黏壤土农作物(黏土) Crops on Clay Loam Soil[c]	
	公顷 hm²	百分比 %	公顷 hm²	百分比 %	公顷 hm²	百分比 %
玉米 Grain Corn	8384	32%	1227	25%	7157	32%
玉米种子 Seed Corn[d]	2833	10%	-	0%	2833	13%
大豆 Soy Beans	11215	41%	1227	25%	9988	45%
小麦 W. Wheat	2318	9%		0%	2318	10%
马铃薯 Tomatoes[e]	3,475	5%	1406	29%	-	0%
豌豆 Peas-	1406	2%	542	11%	-	0%
红萝卜 Carrots[e]	259	1%	259	5%	-	0%
甘薯 Sugar Beets	53	0.2%	53	1%	-	0%
胡椒 Peppers[e]	39	0.1%	39	1%	-	0%
其他 Other	155	0.6%	155	3%	-	0%
总计 Total	30137	100%	4908	100%	22296	100%

▼ 表 3-5: 列出加拿大多佛尔圣克雷尔湖湿地恢复项目的预计休闲开支。估算是在对当地商业和旅游设施的 调查基础上做出的（来源：圣克雷尔湖湿地恢复项目议案，2001）
Table 3- 5: The below shows estimated recreational expenditures for the Lake St. Clair Wetland Restoration project in Dover, Canada. Estimates are based on local surveys of the commercial and tourism facilities. (Source: Calculated and abridged from Kreutzwiser [1981]. Source: A Proposed Wetland Restoration Project – Lake Saint Clair [2001])

	观光者 Nature Viewers	渔夫 Fishermen	鸟禽猎者 Waterfowl Hunters	次使用者 Secondary Users
旅行 Travel	$21	$26	$33	$43
住宿 Accommodation	$3	$11	$11	$44
食物和饮品 Food and drink	$11	$20	$51	$96
其他 Other	$3	$17	$43	$31
总计 Total	$38	$74	$137	$213
当地花费 Local spending	$18	$35	$65	$108
平均人数 Mean party size	2.8	2.8	2.0	3.3

第1步：社会经济预测 主要领域： ・人口预测 ・居民需求预测 ・旅游和酒店需求预测 ・商业、零售和市民需求预测 方法： ・对当地利益相关者进行的访谈或调查，如居民、业主和政府官员 ・对场地周围较大区域的旅游、住宅和商业用地做竞争性评估 ・文献查阅 ・统计资料查阅 ・实地考察	Step 1: Socioeconomic forecast Key areas: • Population forecast • Residential demand forecast • Tourism and hotel demand forecast • Commercial, retail and civic demand forecast Methodologies: • Interviews/surveys with local stakeholders such as residents, business owners and government officials • Competitive assessments of greater regional tourism, residential and commercial sites • Literature review • Review of statistical sources • Site observations
第2步：成本估算 主要领域： ・场地准备费用 ・施工费用 ・动迁费用 ・续生性年度维护费用 方法： ・分析项目设计的几大成本领域 ・调查当地项目的成本资料 ・与本地估算师协作进行	Step 2: Cost Estimate Key areas: • Site preparation costs • Construction costs • Resettlement costs • Recurring annual maintenance costs Methodologies: • Analyze project design for key cost areas • Survey local projects for costing data • Collaborate with local quantity surveyor
第3步：经济效益分析: 主要领域： ・非货币效益 ・货币效益：实务影响分析 ・农业生产力 ・新商机 方法： ・条件评估调查 ・与农业专家合作，估计农业生产力的变化 ・市场调查并采访利益相关者 ・对新商机进行竞争性评估和可行性研究	Step 3: Analysis of Economic Benefits Key areas: • Non-monetary benefits • Monetary benefits: Business impact analysis • Agricultural productivity • New commercial opportunities Methodologies: • Contingent valuation survey • Collaborate with agricultural specialists to estimate changes in agricultural productivity • Market research and interviews with stakeholders • Competitive assessments and feasibility study of new commercial opportunities

3.5 湿地生态恢复的工程设计

3.5.1 被动与主动的湿地恢复

如上所述，湿地生态恢复通常有多种选择，在栖息地价值、防洪与侵蚀控制、水质、休闲价值和经济性等各个方面都有所不同。从恢复途径看，也可分为两种：被动的湿地恢复与主动的湿地恢复。

被动的湿地恢复

一种偏重于借助自然系统本身的自我维持和自我恢复能力，比如，在湿地恢复初期，尽可能清除系统内的人为干扰（封山禁牧、消除并处理点源污染、禁止收割芦苇和渔猎湿地动物、拆除海岸围堰等）并恢复湿地的水文状况，然后通过植被的更替来逐步恢复湿地（见图3-26）。

这个恢复途径与目前大多数自然保护区采用的方式相同，在国外如美国也较多采用此种方法，优点是直接用于生态恢复的费用低廉，充分利用当地和邻近湿地的种子库，经过一段时间后，自然植被得以恢复，不足之处在于这种方法不适于中国大部分人类活动频繁的地区，置换土地和迁移人口都需要高昂的经济代价，并且恢复需要的时间很长，在恢复初期，其生物多样性功能和休闲价值是无法体现的。这种方法通常用于尚存有湿地特征或可用简单低科技手段重建的场地，其目的在于用最小代价重新建立湿地自然生态，所借助的是自然循环过程。

主动的湿地恢复

另一种生态恢复途径则较为积极主动，除了恢复湿地水文状况外，还要恢复湿地的基底和植被，甚至进行关键物种的再引入。严重退化或湿地类型难以识别的待重建湿地就需要这种方法；在有些情况下，湿地及其所有的结构和功能都必须在一个全新的场所创建出来。这种恢复途径包括改变地形轮廓，通过水流控制设施来管理水位和水量，种植各种水生、湿生和中生植物，清除侵入性和非本土物种，以及根据具体物种生存需求来添加合适的土壤。因为对规划、工程、园艺、建筑、人力、物力等方面的要求较高，所以主动方法通常成本较高。

3.5 Engineering Design and Specifications

3.5.1: Passive and Active Restoration

As stated above, there are multiple options available for a wetland restoration project, each with variable degrees of emphasis on habitat value, flood and erosion control, water quality, recreation, and economic value. However, wetland restoration approaches can be divided into two general categories: passive and active.

Passive

The passive approach aims to revive the self-preservation and self-restoration abilities of the wetland ecosystem, or the wetland's natural ability to achieve dynamic equilibrium. The critical first step in passive restoration is to remove the factors of human disturbance. In some cases, this may consist of simple actions such as fencing off land to exclude livestock (see Figure 3-26), removing and treating point source pollution, preventing the harvesting of reeds and wetland animals, breaching berms to the sea, or redirecting river water flows from rivers.

At the same time, wetland hydrology can be affected in such a way that vegetation succession may gradually rehabilitate the wetland. This approach is widely adopted in most Chinese nature reserves and in the US. Low restoration costs are a major advantage of the passive approach, since local and nearby seed reserves are the primary source of new natural vegetation which gradually accumulates over time. However, passive approaches have seldom been used in the more densely populated areas of China, because land replacement and resident relocation would require high economic investments, and adequate biological and recreational values are not present in the early stages of the restoration.

The passive approach aims at re-establishing natural wetland processes with minimal effort and is generally chosen for sites which retain some basic wetland characteristics, or which can be restored using simple low-tech solutions. In general, the passive approach requires the aid of natural processes.

Active

The active approach is predictably more intensive and usually consists of restoring wetland hydrology, re-establishing the base and vegetation of the site, and introducing key target species. Severely degraded wetlands, or sites which are not identifiable as wetland ecosystems, generally require this approach. In some instances, the wetland and all of its forms and functions must be created in full from a raw site. Some active interventions include re-contouring, using flow control structures to manage water levels and flow rates, planting

▲ 图 3 - 26: 水滨篱栅是防止河流受到家畜不良影响的常见被动方法 (摄影 : Dixi Carrillo)
Figure 3-26: Riparian fencing, such as the one shown above at EDAW's North District Flood Control Project in Sacramento, CA, is a common
passive method for protecting streams from the detrimental impacts of cattle (Photograph by Dixi Carrillo)

但是，人为活动对湿地的影响可以通过交通设计、污水收集和处理工程设计、旅游设施设计等来控制，并且可以通过移植植株、清除外来物种、动物种群恢复等方式来缩短湿地恢复到理想状态所需要的时间。此外，渔业等人类活动可以根据生态系统内各种群的动态进行适度的开展，比如，在退渔还湿的地方，除了湿地恢复外，可以开展垂钓等生态旅游，通过对资源生物的中度干扰来维持该种群持续增长。

根据中国的国情，我们认为，后一种恢复途径适合于大部分湿地，尤其是中国东部人口稠密、经济发展迅速的地区。该恢复途径需要多种技术。

hydrophytes, phreatophytes and mesophytes, the removal of invasive and non-native species, and the addition of appropriate soils depending on species' specific requirements. Active restoration approaches are typically more costly due to higher dependence upon design, engineering, horticulture, and construction labor and materials.

In practice, human disturbances to wetlands are controllable through the careful design of transportation, waste water collection and treatment, and tourist facilities. Restoration can be further accelerated to the desired functional state through plant relocation, removal of non-native species and the restoration of a site's biological community. These actions may be combined, depending on a species specific dynamics in the ecosystem, with human activities such as fishing and biomass harvesting. For example, in places where fishery ponds are being restored back into wetlands, ecological tourism such as angling may be conducted simultaneously, wherein a moderate disturbance to an economic species may stimulate its continual growth.

It is believed that the active approach to restoration is applicable in most Chinese wetlands, especially in densely populated, fast-developing eastern and coastal areas in China. The approach requires a series of replicable technologies, including those necessary for an active intervention, and is described herein.

(摄影:Rowan Roderick-Jones)
(Photographs by Rowan Roderick-Jones)

(摄影:Dixi Carrillo)
(Photograph by Dixi Carrillo)

(摄影:Rowan Roderick-Jones)
(Photograph by Rowan Roderick-Jones)

湿地恢复手册 原则·技术与案例分析

116

3.5.2 湿地生物恢复技术

植被

主要包括物种选育、培植技术、群落结构优化配置与组建技术、群落演替控制与恢复技术等。植被是湿地的特征之一,各湿地类型有不同的植被特征(见图 3 - 2 7),因此,不论在哪种湿地类型选择植物种类时,必须考虑所选物种的地带性、耐淹性、耐盐性、耐寒性等特性。本节将根据不同湿地类型进行分述。

对于沼泽湿地而言,通常需要开展的工作包括:

- 利用湿地的天然种子库进行湿地植被恢复,包括湿地内的种子库和孢子库、种子传播、植物繁殖体。这个方法可用于刚刚退化的湿地,或者周边具有发育良好,并可为其提供种子库的湿地(见图3-28)。

- 如果湿地退化厉害,并且距离邻近湿地较远,则可通过播种大量乡土植被的种子、营养体扦插、移植等方式进行湿地植被的恢复。移植的植株可从类似湿地中采集,但是采集的强度不能太高,以免对其他湿地产生不良影响。运用组织培养技术,可在短期内获得大量幼苗,并且不会对其他湿地产生不必要的干扰。

- 采用适度放牧方法控制或阻止木本植物的入侵,这是一种低成本且有效的植被管理方法,适合尤其以保护鸟类生境为主的湿地。食草动物能为湿地留下粪便等,补充湿地土壤的有机物含量,并且它们的活动能够增加湿地土壤的微生境,如蹄印等。

对于河流湿地而言,通常需要开展的工作包括:

- 根据河流水位变化情况设计不同植被的分带格局,包括沼生植被带(含湿生草本植物和耐湿的木本植物)、挺水植被带、漂浮植被带(或浮水植被带)和沉水植被带。为保证扦插和移植的存活率,木本植物应该设计在平均最高水位以上,并且应设计林下带。根据所采用的草本植物的

3.5.2 Ecological Restoration

Vegetation

Vegetation is an essential component of wetland habitats, with different wetland types supporting distinct vegetation communities. Decisions regarding vegetation planting and management must take into consideration the wetland type of the restoration site, and recognize the species' characteristics selected, such as tolerance to submersion, salinity, temperature and other environmental factors. This section summarizes appropriate methods for vegetation restoration and its management in different wetland types.

Marsh Wetlands

- Where the restoration site is only slightly degraded, or is adjacent to existing wetland habitats, marshland vegetation will often regenerate quickly with minimal intervention from wetland managers. Vegetation can be produced from the seed bank in the soil of the restoration site itself, or from seeds/spores dispersed from adjacent sites.

- If the site is heavily degraded or isolated from other wetland habitats, a more proactive approach to vegetation restoration is required. Methods that can be adopted in this instance include the seeding of native plants, taking cuttings or transplanting whole plants. Specimens for transplantation can be collected from similar wetland sites, but if this approach is used, care must be taken to avoid or minimize damage to the donor wetland site. A preferred approach is to culture the required number of plants in a nursery prior to transplantation.

- Marshland habitats are dynamic, early successive habitats, and regular control and management of their vegetation is required to maintain conditions and prevent the colonization and succession by woody plants. Various techniques can be adopted to maintain marshland vegetation, including cutting, controlled burning and application of herbicides. A novel and cost-effective approach to marshland vegetation management is to use grazing animals such as cows and Water Buffalo. In addition to grazing and controlling vegetation, dung produced by herbivorous animals replenishes the organic content of wetland soil and their activities (trampling, for example) helping to increase the diversity of a habitat within the marsh, creating small, open-water pools.

River Riparian Wetlands

- The riparian habitats of rivers are characterized by narrow zones of different types of vegetation with

▲ 图 3 - 28: 典型的淡水沼泽 (摄影 : Dixi Carrillo)
Figure 3- 28: A typical freshwater marsh in the Yolo Bypass (Photograph by Dixi Carrillo)

不同耐淹性来决定它们在最高水位和最低水位之间的位置。不具有浮叶的沉水植物应设计在最低水位以下的浅水区内，以利于吸收光能。当河流具有足够的宽度，并且在不妨碍航运等其他水上活动时，可设计漂浮植物群落。

- 滨岸植被必须具备一定的宽度。保持河岸岸边30m以上的植被带会起到有效的降温、过滤、控制水土流失等作用，提高生境多样性。60m以上宽度的河岸植被带可以满足动植物迁移和生存繁衍的需要，并起到生物多样性保护的需要（见图 3-29）。

- 常用的栽培技术包括：播种、移植、扦插等。在植被恢复之前，应采用机械或人工办法去除杂草，尤其是律草等蔓生杂草。对于土壤有机质含量低的滨岸，可通过覆土、施加有机肥的方式进行弥补，但是，应该避免在雨季和汛期进行。

对于湖泊湿地而言，通常需要开展的工作包括：

- 适当控制大型沉水植物的生长。采用的方法包括收割、引入特定的食

specific tolerances to inundation. These include a submerged plant zone, an emergent plant zone, a facultative plant zone (including submersion-resistant woody plants), and finally, a zone composed largely of terrestrial plant species. It is therefore essential that the species planted in each zone are carefully selected for their tolerance to inundation. In general, woody plants should be placed above the highest average water level, grasses and reeds should be placed between highest and lowest water levels according to their resistance to submersion, and submerged plants should be positioned below the lowest water level. Additionally, submerged plants without floating leaves should be placed in shallow water to ensure that they receive enough light.

- Riparian habitats of rivers perform a wide range of environmental services too: they help regulate river temperatures, filter run-off, control erosion, and act as an important wildlife habitat and migration route. How well the riparian habitat performs these functions is largely dependent on its size. It is recommended that the vegetation zone alongside each riverbank be at least 30 meters – and preferably over 60 meters – in width.

- As with all wetland habitat restoration projects, riparian habitats usually require improvement or modification prior to vegetation planting. For example, invasive climbers (e.g., Mikania micrantha) must be

▼ 图 3 - 29: 美国乔治亚州内一处沿河的河岸区 (摄影 : Dixi Carrillo)
Figure 3-29: Riparian zone along a river in Georgia, 9(USA (Photograph by Dixi Carrillo)

草昆虫或鱼类、下调水位、沉积物覆盖等。

滨岸植被的恢复方法与河流湿地相似，自陆地向开阔水域建造湿生植被带、挺水植被带、扎根浮叶植被带、漂浮植被带、浮游植被带、沉水植被带以及开阔水域的浮游植被带等7个湿地植被带（见图3-30）。

- 采用机械或人工收割方式控制生长过于密集的挺水植物，尤其是当植被已经阻碍水流，导致湖泊淤浅，并成为湖泊的内源性营养物负荷。

对于滨海地区的草本盐沼湿地而言：
由于同时具有盐度和水淹的胁迫，因此，可选的物种远远少于淡水湿地。建议种植乡土物种，在干扰严重的地方，应首先减缓水力冲刷的影响，并移植先锋物种。在环境条件较好的地方，可种植演替晚期的物种。为了增加种植的成功率，可采用单种群落恢复的方法，减少和控制种间竞争（见图3-31）。

对于红树林湿地而言
除了减少红树林（见图3-32）受到的人为干扰，让受干扰红树林进行自我恢复外，其他恢复技术包括：

- 胚轴造林技术：将胚轴直接插入土壤基质，插入深度约为胚轴深度的1/3~1/2，过深胚轴容易腐烂，过浅容易被海浪冲走。此法简单易行、适宜大型繁殖体造林，但受胚轴成熟季节性约束较大。研究表明，秋茄造林采用胚轴插植，方法效果较好，深度为种苗长度的1/3~2/3。秋茄根系海绵状容易折断，应该使用裸根小苗造林选择半年内生的幼苗，一年生或多年生的小苗和幼树不易成活。同时，截断胚轴造林时，胚轴越短，成苗率越低；越靠近顶端，成苗率越高（廖宝文，2003）。此种方法一般选择大型繁殖体种类，例如红海榄、秋茄、木榄等，费用低、技术简单，可进行大规模造林和次生林改造。

- 容器苗造林技术：用聚乙烯薄膜袋（育苗袋，装满营养土石，直径7cm，高20cm）。在海上苗圃进行人工育

removed through mechanical or manual means before planting, and habitat restoration areas with low organic content may be improved through soil replenishment and the application of organic fertilizers.

Lake Wetlands

- The restoration of riparian vegetation at the periphery of lakes is similar to that of riverine wetlands: species planted in each zone must demonstrate a tolerance to the level of water inundation predicted in a particular zone.

- Submerged and floating plants in lakes can spread rapidly and adversely impact the ecology of a lake, preventing light from penetrating to its lower levels, and potentially leading to anaerobic conditions. Submerged or floating plants require control by means of harvesting, introducing specific herbivorous insects or fish, or by lowering water levels and sediment cover.

- Emergent plants may also require control by means of mechanical or manual harvesting, especially when plants are dense enough to impede water flow and cause stagnation.

Brackish Water Wetlands

Salt-Marsh Wetlands
Salt-marsh wetlands are a severe habitat for plant species due to rapid changes in their hydro-period and salinity. As such, there is a rather narrower selection of plants available for salt-marsh restoration than for freshwater wetlands. As with all wetland habitats, native plant species are the primary choice for salt-marsh restoration projects. In heavily degraded sites, the impact of water scouring must be mitigated before the introduction of advance plants is possible. In less degraded sites, plants in later stages of ecological succession may be introduced.

Mangrove Wetlands
When selecting sites for mangrove wetland restoration, it is particularly important that infringement on nearby mudflat habitats is avoided. Although they can appear suitable for mangrove planting, mudflats are themselves ecologically valuable habitats that should be preserved. Key restoration approaches specific to mangrove wetlands include the following:

- Propagule planting: Many species of mangrove are

图 3‑30: 湿地恢复区展示了湖滨区的植被分布状况，最前面是湿地植被，中间地带是滨水灌木，后面是高地松树 (摄影：Dixi Carrillo)

Figure 3-30: The Upper Truckee wetland restoration site demonstrates vegetation zonation adjacent to a lake with wetland vegetation in the foreground, a band of riparian shrubs in the middle ground, and upland pine trees in the background (Photograph by Dixi Carrillo)

图 3‑31: 图中的盐沼恢复是盐池恢复工程的一部分。请注意图的右侧未恢复的盐池以及左侧连接沼泽地和潮汐水道 (摄影：Dixi Carrillo)

Figure 3-31: Restored salt marshes as part of the San Francisco Salt Pond Restoration project. Note the contrast between the un-restored salt ponds on the right and the tidal channel on the left that connects the restored marshes to the San Francisco Bay. (Photograph by Dixi Carrillo)

▲ 图 3‑32: 香港米浦湿地保护区被潮汐淹没的红树林 (摄影：Rowan Roderick-Jones)
Figure 3-32: A flooded mangrove forest at the Mai Po Wetland Reserve, Hong Kong (Photograph by Rowan Roderick-Jones)

苗，培育出相应规格的容器苗进行定植造林。此法技术含量高，造林品质好。适合小型胚轴种类及特殊生境的造林。此种方法一般选择隐胎生种类如白骨壤和桐花树等，常用于逆境造林、补植和科学实验。

- 天然苗造林技术：直接从红树林群落中挖取天然苗来造林，但容易伤根，并对原红树林群落发展造成负面影响。此种方法一般为繁殖体未成熟季节时的应急造林或种苗短缺时的补植，与上两种方法比较而言造林效果差，不宜作为主流造林方法。

- 林相改造技术：对于次生红树林，如果是大面积纯林，可采用先间伐后引进其他红树林种的方法来进行恢复（李玫等，2003）。

大型底栖动物

大型底栖动物是湿地食物链上的重要环节（见图3-33），也是吸引鸟类的重要因素，尤其是涉禽。此外，双壳类动物通过过滤水的方式获取水里的食物，能提高水体透明度。

目前在大型底栖动物群落恢复方面没有特别的技术，所采取的方法为：人工投放和控制捕捞强度。进行人工投放时，需考虑所投放的底栖动物是否是当地的乡土物种、是否适应湿地的盐度和沉积物或基底条件以及是否会侵占其他物种的生态位。

鱼类

在湿地恢复中，鱼类群落的动态可以用来监测湿地水质。水质越差，鱼类的多样性就越低，反之亦然。对于鱼类群落恢复，所采用的方式有：投放鱼苗、建造人工鱼礁以及控制捕捞强度。

控制捕捞强度并非就是完全禁渔，而是在鱼类的繁殖、育幼期以及索饵洄游期间，降低捕捞强度或者禁渔，但在其他时间，可以通过捕捞成体来维持种群处于增长阶段。

投放鱼苗是一种快速增加鱼类数量的方法，除了人工投放之外，最好的办法就是与附近水体（如海洋）恢复水系通畅，

▼ 图3-33: 深水大型底栖动物有图中所示保卫着米浦保护区上宝贵的泥质浅滩的招潮蟹等螃蟹和小蜗牛，但不包括像弹涂鱼这样的底栖鱼类（摄影：Rowan Roderick-Jones）
Figure 3-33: Benthic macroinvertebrates include species such as this fiddler crab and small snails, but exclude bottom dwelling fish like these mud skippers, shown here defending their valuable mud flat territory at the Mai Po reserve (Photograph by Rowan Roderick-Jones)

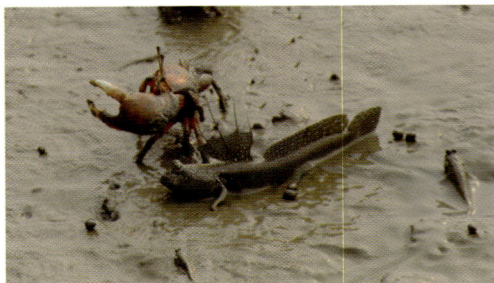

viviparous to some extent, with seeds germinating and developing into seedlings (commonly called propagules) prior to their release from the parent plant. Planting mangrove propagules is a simple, effective method of establishing mangrove habitats. The propagules can be planted directly into a soil base to a depth of about 1/3-1/2 of a propagule's length. Due to its low costs and simple nature, this method is used widely in the creation of large-scale mangrove restoration and enhancements.

- Planting cultured saplings: Culturing viable saplings is possible for many mangrove species. Saplings can be grown in small polythene bags filled with nutrient soil before being planted at the restoration site. This method requires certain technical information regarding the growth requirements of different mangrove species, and is generally easier when plants with small diaspores (such as Aricennia marina and Aegiceras corniculata).

- Planting saplings taken from existing mangroves: Transplanting saplings directly from existing mangrove communities to the restoration site is another approach to mangrove restoration. Transplanting saplings has a lower success rate than other planting methods, as sapling roots are often damaged during transplantation. Likewise, sourcing saplings from existing mangrove habitats can adversely impact the donor site. As this method often creates comparatively unsatisfactory effects, it must not be considered as a primary means of foresting.

- Secondary mangrove enhancement: Previous attempts at mangrove restoration have often involved creating large stands of a single species of mangrove. The species and structural diversity of such habitats can be improved through targeted clearance of mangroves and the subsequent introduction of other mangrove species (Li Mei, 2003).

Benthic Macroinvertebrates

Benthic macroinvertebrates refer to larger faunal species (>0.5mm in size) that occur in the bottom sediments of wetlands. Macroinvertebrate communities are a primary link in the wetland food chain as an important food source for larger predators including birds, fish and amphibians.

Benthic macroinvertebrates will usually colonize suitable habitats quickly, as many species have one or more highly mobile stages in their life cycles. For example, a number of freshwater macroinvertebrates (such as stoneflies or

湿地恢复手册 原则·技术与案例分析

在涨潮或调水的同时，自然纳入鱼苗，这种方法有助于湿地内恢复自然的鱼类群落。如果采用人工投放的话，就需要考虑所投放鱼类的食性，根据它们在食物链上的位置来决定投放的必要性。在大型沉水植物或藻类过度生长的水体内，可以投放草食性鱼类。底栖杂食性鱼类则可以减少沉积物再悬浮和内部营养负荷。

人工鱼礁多用于近海生态系统，可能包括一些潮下带湿地。其原理是投放对水质无害的大型固体（通常是废弃船体、混凝土块等），增加水底的生境复杂性，为鱼类的繁殖和栖息提供场所。这个方法已经成为大型固体废弃物的一个处理方法。但是，采用这个方法之前，必须确定：投放地点的沉积物表面没有珊瑚礁等生物礁体，投放材料长远不会向水体释放有毒有害物质（见图3-34）。

两栖动物
两栖动物是一类典型的湿地动物，很多种类在其生活史中，有些阶段是无法离开水体的。中国的许多两栖种类都因为栖息地丧失或破碎化、过度捕捞等原因而成为珍稀濒危物种。对这类动物的保育和恢复需要仔细研究它们对栖息地的要求。有的物种如蛙类能适应稻田、滨岸等不同湿地环境，但有的物种如大鲵则只能适应非常洁净的水体（见图3-35）。目前，这类动物的保育措施有：控制捕捞强度或禁捕、栖息地恢复和重新引入。在重新引入前，如果引入的个体为人工繁殖个体，则必须进行谨慎的野外放养试验。

鸟类
营建适宜水鸟生活的栖息地通常是湿地生态恢复的主要目标之一。由于鸟类的活动性很强，并且多为迁徙物种（见图3-36），因此，必须首先确定需要吸引和恢复的目标物种，方法见本章"湿地的现状评估"的相关内容，然后再仔细研究目标物种对生境的需求，并进行针对性的栖息地恢复。这些工作通常包括：

提供充足的鸟类饵料
对于植食性的水鸟，应恢复那些能够提供其食用的种子、块根、块茎的植物群

dragonflies) have flying adult stages, and some mollusk species have planktonic larval stages. The key to restoring macroinvertebrates to a site is creating a suitable habitat for the targeted species or communities. Various factors affect the suitability of wetland habitats for macroinvertebrates, including water quality, substrate type, and the presence or absence of vegetation and flow rates.

Fish
The dynamics of fish communities are good indicators of a wetland's water quality and the integrity of its habitat, with poor quality habitats generally indicating a low biodiversity of fish, while the converse is also true. Restoration practices for fish communities include fry supplementation, the construction of artificial reefs, and controls on recreational or commercial fishing.

Fry supplementation is used to quickly establish or enlarge fish communities. Although artificial supplementation is usually successful, a more desirable approach is the natural recruitment of fry during high tides or water exchanges, provided that there is a reliable water flow between the restoration site and nearby water bodies. With artificial supplementation, it is important that the habitats to which the fry are introduced are capable of supporting them. For example, herbivorous fish may be supplied into wetlands with an abundant supply of submerged plants or algae, and benthic omnivorous fish may be supplied as a means to reduce sediments deposition and internal source eutrophic burden.

The construction of artificial reefs is usually applicable in near-shore, sub-tidal ecosystems. This approach consists of placing large, inert structures onto the bottom sediments in order to increase habitat complexity and to provide a suitable area for fish reproduction and breeding. It is also a useful method of displacing large marine structures, such as redundant ships. Care must be taken in placing artificial reefs to ensure they do not damage any natural reefs (such as rock or coral reefs) in the deployment area, and that there is little or no risk of toxic discharge from the artificial reef into the surrounding waters.

Fishing controls do not necessarily mean an indiscriminate ban on all fishing, but may refer to a reduction or temporary cessation of fishing during key periods of the year when target fish are reproducing, breeding or migrating. At other times, adult fish can still be harvested in order to stimulate the growth of the fish communities.

▼ 图3-34: 健康的淡水湿地通常都盛产鱼类，为渔民和当地住户提供了娱乐和生计
Figure 3-34: Fish are usually abundant in healthy freshwater wetlands and provide recreation and sustenance for fishermen and local communities

（图片来源：www.photos.com）
(Source: www.photos.com)

（摄影：Rowan Roderick-Jones）
(Photograph by Rowan Roderick-Jones)

图 3-35:两栖动物，如鲵和蛙，都是湿地群落和食物链上的重要成员(图片来源：www.photos.com)
Figure 3-35: Amphibians such as salamanders and frogs are important members of the wetland community and food chain (Source: www.photos.com)

落；对于肉食性的水鸟，则应恢复上一个营养级的种群或群落，一旦得以确定，可采用上述的相应技术。由于许多水鸟都可能以具有较高经济价值的大型底栖动物为饵料（见图3-37），因此，需要谨慎评估那些资源生物的生物量，以及适宜的人为捕捞强度。对于迁徙物种，当它们停留在湿地觅食时，应该严格控制湿地的人为活动，除了防止捕捞导致饵料缺乏外，也应避免直接干扰鸟类。此外，有些鸟类会在水体盐度方面有特殊的要求。鸟类的生境需求可查阅相关的专业书籍，或者寻找参考系统（如被同种水鸟利用的自然保护区等）来获取必要的信息。

提供水鸟觅食所需的水动力条件
有些水鸟会喜欢沿着潮汐的前缘觅食，其中的一个推测是因为适当的水动力能够冲刷滩面，使底栖动物暴露出来，降低捕食难度。对于这类水鸟，恢复湿地的潮汐影响是最好的方法。

提供隐蔽的繁殖或栖息场所
天然的洞穴、植被都可能是水鸟的繁殖或栖息场所。但是，在植被恢复时，必须注意生境的复杂性，具有斑块状明水面的芦苇群落比人工湿地的芦苇床具有更高的栖息地价值。

为了恢复一个具有自我维持能力的湿地生态系统，人为的辅助设施不应该成为吸引并留住野生动物的主要手段。传统的动物园豢养方式不适用湿地恢复，因为被折翅或剪羽的方式对恢复鸟类的自然种群几乎没有任何意义，并且所营建的不是自然状态下的鸟类群落。在恢复初期，可采用投放底栖生物的方式来吸引涉禽和游禽，但是，投放的目的应该立足于恢复当地的底栖动物群落，以便减少日后的投饵数量，并能维持较为稳定的涉禽或游禽群落。又如，可以通过搭建人工鸟巢来吸引鸟类，但是，应该注意鸟巢的多样性，以便吸引不同种类的鸟，并且控制人工鸟巢的数量，保留自然状态下的树杈、草丛、倒木等吸引鸟类自行筑巢。

珊 瑚 礁
珊瑚礁是一类特殊的湿地类型（见图3-38）。除了减缓对珊瑚礁的影响，使珊瑚礁进行自然恢复外，目前采取的恢

Amphibians

Most amphibians are dependent on wetlands during certain stages of their life-cycles. Many amphibians in China are considered endangered because of the loss and fragmentation of their wetland habitats, the impact of pollution and their over-collection for food. Restoration of amphibian populations to wetland habitats requires a detailed understanding of the habitat requirements of the species concerned. Some amphibian species, such as the Common Asiatic Toad (*Bufo melanostictus*) can tolerate a wide range of wetland ecosystems, while others like the Giant Salamander (*Andrias davidianus*) have very specific habitat requirements, being limited to cold, unpolluted, fast-running mountain streams and lakes. Preservation and restoration measures for amphibians include harvest controls or bans, habitat restoration and species re-introduction. For species re-introduction, a careful field survival test must be conducted if the intended species are artificially cultured.

Birds

The main ecological goal of many wetland restoration projects is providing suitable habitats for waterbirds. Since birds are highly mobile animals (especially waterbirds, many of which are migratory), habitat restoration usually consists of identifying one or more target species (see the "Wetland site assessment" section of this chapter), studying the habitat requirements of these target species, and designing a wetland restoration program based upon that knowledge. Actions for bird habitat restoration usually include the following practices.

An adequate food supply
A restored habitat should provide sufficient amounts of food to attract target species. Plant communities provide edible seeds, root and stem tubers provide food for herbivorous birds, and re-established faunal communities attract insectivorous, piscivorous and other carnivorous birds. Since many waterbirds eat economically valuable species such as crustaceans and fish, the biological resources and appropriate degree of harvesting these economic species must be carefully assessed. This is especially critical when a large influx of migratory birds arrive at the site: human activities must be strictly controlled to prevent food shortages due to harvesting as well as unnecessary disturbance to the birds. Professional publications and reference sites – such as natural reserves that support similar species – should be consulted in order to determine the specific habitat requirements for different birds.

▲ 图 3 - 36: 一大群鸟来到淡水沼泽地（图片来源 : www.photos.com）
Figure 3-36: A dense flock of birds take flight at a freshwater marsh (Source: www.photos.com)

复措施有：

珊瑚移植技术：珊瑚移植包括收集整个或部分健康的珊瑚礁生物群体，然后将其移植到一个与其环境条件相似的退化的珊瑚礁。移植成熟的生物群体可以加快自然恢复的速度，因为它回避了群体生命周期中具有较高死亡率的幼年期，避免了引入不适合的幼体。珊瑚移植的研究主要有两类，包括整个珊瑚礁生物群体的移植和生物群体片断的移植。

人造珊瑚礁技术：为珊瑚虫等造礁生物提供合适的栖息地，原理类似于人工鱼礁。人造珊瑚礁可以采用塑胶、竹子、砾石以及混凝土等材料，要具有一定高度，并且建造在垂至于大海盆的方向上。人造珊瑚礁内部要提供各种尺寸生物栖息的空间（Rilov and Benayahu, 1998）。注意此方法不适用于大面积的珊瑚礁恢复，虽然能够有效增加海洋生物量，但其不能成为海洋生态系统的有机部分，不能生长，而且建筑地点要考虑不能影响正常的航线。此种方法花费时间和经费较高，同时可能造成污染，因此在选择恢复地点以及材料选择上要慎重。

珊瑚养殖技术：结合珊瑚虫产卵时间，利用珊瑚接合体进行珊瑚培育，少数种

Improving hydrodynamics for food-hunting
Different species of water birds tend to hunt for food at specific water depths, with many preferring to hunt along tidal-lines (where hydrodynamic conditions and clear surface mud expose benthic creatures). Restoring the tidal influence of the wetland site is a good way to attract such birds.

Providing suitable reproduction/breeding habitats
Natural caves and patches of dense vegetation are favored reproduction and breeding habitats for many species of wetland birds. However, a habitat's complexity must be considered during vegetation restoration. Reed communities interspersed with patches of open water provide a much more valuable habitat than reed beds with no open water system.

Coral Reefs
The following restoration approaches have been adopted to reduce negative impacts towards coral reefs and to facilitate their natural rehabilitation.

Coral transplantation: Transplantation refers to collecting a healthy coral community completely or partially, and moving that community to a degraded reef with similar environmental conditions. The successful transplantation of coral communities may significantly speed up natural restoration processes, because planktonic stages of corals, which provide natural recruitment, have high mortality rates.

▼ 图 3 - 37: 在加州的 Yolo Bypass，一只红腿鹬受这里的浅水吸引，想在这里寻找无脊椎动物为食（摄影 : Dixi Carrillo）
Figure 3-37: A red-legged stilt at the Yolo Bypass in California is attracted by shallow water where it can forage for invertebrates (Photograph by Dixi Carrillo)

类的珊瑚幼虫是可以成功饲养的，并且费用较低。

除了针对上述各生物类群的恢复途径外，与生物有关的恢复策略还包括：湿地附近的土地使用方式中，如城市公园、高尔夫球场或农田可以成为自然或经恢复的湿地与高密度的城市或工业用地之间的缓冲带；湿地附近的农场和农业林带，如果和湿地之间尚有未开垦的缓冲区的话，可以提供有价值的山地栖息地。

如果待重建的湿地附近有主要道路、铁路干线或工业设施，就必须采取措施减轻噪音污染和视觉污染。种植浓密植物的围堰和隔离墙通常足以阻挡上述干扰。

为了应对附近土地的使用方式，一定要取得当地的分区规划档，同时对私人开发也要有所了解。待恢复湿地及其附近地区的现有情况不能与未来5至10年的规划方案冲突。规划和建设部门是上述信息的最好来源。

Construction of artificial coral reefs: Artificial reefs similar to those used to attract fish can also be deployed to encourage coral formation. Such reefs can be made from plastics, bamboo, gravel or concrete. Reef blocks must be sufficiently large and placed perpendicular to the seabed (Rilov and Benayahu, 1998). While artificial reefs can provide a number of ecological benefits, this approach can be costly if applied on a large scale, as it can affect navigation routes, while still other types of artificial reefs (such as old ships) can be a source of pollution. Sufficient caution must therefore be attached to the decision of construction sites and materials.

Coral culturing
Juvenile corals can be collected during the coral spawning period and cultured in controlled conditions, before being released back into the restoration site. Some types of coral polyp larvae can be successfully cultured at low costs.

Aside from the restoration approaches mentioned above for individual biological communities, there are other restoration strategies related to biology.

Land uses adjacent to wetlands include urban parks, golf courses or farmland, all of which can provide a buffer zone between a natural or restored wetland and higher density urban and industrial uses. Farms and agro-forestry zones are capable of providing valuable adjacent upland habitats if there are uncultivated buffer areas between the wetland and the fields.

Noise pollution and visual disturbances should also be mitigated if the proposed wetland is adjacent to large roads, rail corridors or industries. Berms and walls planted with dense vegetation usually provide the best protection from such disturbances.

When considering adjacent land uses it is important to obtain local zoning and planning documents as well as information on proposed private developments. Often what exists at and around the site is not what is in the plans for the site five to ten years in the future. Planning and Construction Bureaus are the best sources for this kind of information.

▼ 图 3-38: 珊瑚礁作为湿地尚未完全得到认可，但湿地公约将其纳入湿地的范畴（图片来源：www.photos.com）
Figure 3-38: Coral reefs are not always considered wetlands, but are defined as such by the Ramsar Convention (Source: www.photos.com)

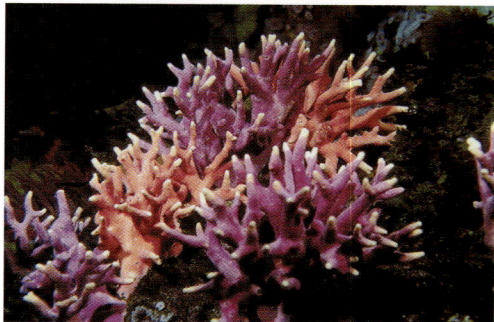

3.5.3 水文学设计

大多数待恢复的湿地都因为人类活动对水文系统造成干扰而有或多或少的退化现象。人类活动干扰的后果有多种表现，如洪水泛滥、干旱和湿地生产力降低，这些又进一步造成了地形变化。在美国加州萨克拉门多河三角洲，历史上为了进行农业生产，人们用堤坝把河中湿地岛屿与河隔开，结果在接下来的50年中，湿地上称为"泥炭"的沉积有机物质不断氧化、退化，导致湿地沉降达3m。到了今天，由于重建需要填埋上亿立方米的泥土，该处已几乎不可能恢复自然水文系统。

以上是个极端的例子，说明在土地对湿地的支持作用中，地形和水文状况发挥了非常大的作用。湿地原有功能和生产力的恢复，通常需要建立水平衡和水文模型，还要获知水资源分配和排放的具体资料，以类比自然水文系统。直接干扰应由有经验的水文学家来建立，并在模型中包括降雨量、地表水渗透、水气蒸发量及其它所有流入、流出量。

正如上文第3.3.2节（现场特征）中提到的，光达技术可用于大面积区域的地形测量，从而结合水力模型预测湿地恢复场地的洪泛周期（见图3-39）。

恢复工程中对湿地进行水文控制的部分可包含堤坝和土地工事、沟渠和水道、水流和水位控制设施（如堰和截水管）、防洪旁路和溢水设施等。这些设施的建设有利于创建良好的土壤和水环境，为持续发展湿地植物和吸引野生物种创造条件。

堤坝和土木工程

在由地势差异而形成的湿地中，构建围堰是很有效的一种方式，可以保持比湿地原始状态高的水位，但高水位在沟壑水中的营养物质高时，容易造成淤塞。围堰可以改变自然水的供给机制，可以缓解由于供水减少产生的干旱缺水，对于其他原因引起的缺水，则需要工作人员在这种次有的补水方式和不采取补水方式上进行选择。

堤坝位置

为了在恢复湿地内达到最大程度的土地

3.5.3 Hydrology

Most sites proposed for wetland restoration have been degraded to some degree due to the alteration of the hydrologic regime by human activities. These alterations can have an adverse effect on wetlands, causing over-flooding, drying out, and reducing a wetland's productivity. These changes in turn can affect landforms. In the Sacramento River Delta of California, historic wetland islands were isolated from the river by the construction of levees for agricultural production. Over the next 50 years, the land subsided by as much as three meters as accumulated organic materials like peat, oxidized and degraded in the wetlands. Today, restoring natural hydrology is near impossible, or a Herculean task, as it would require billions of cubic meters of fill soil.

While extreme, the above case illustrates the importance of landforms and hydrology in determining the land's ability to support wetlands. Restoration of original wetland functions and productivity often requires the development of water balance and hydrology/hydraulic models, as well as details such as water distribution and discharge systems to emulate natural hydrologic regimes. Hydrologic models should be prepared by experienced hydrologists and should account for rainfall, groundwater infiltration, evapotranspiration and all other inflows and outflows. In addition, the hydrologic model should be coupled with proper hydraulic models to develop restoration scenarios that take into account wetting and drying cycles of the wetland and allow for adjustments to the proposed internal earthwork design. Hydraulic models can vary in their sophistication depending on the level of details required. Typical models in use today include the hydrodynamic, one-dimensional and two-dimensional models.

As previously mentioned in section 3.3.2 (Initial Site Characterization), Light Detection and Ranging technology (LiDAR) can be used to perform topographic surveys on large geographic areas and can be coupled with hydrodynamic models to predict flooding periodicity at wetland restoration sites.

Project elements which exert hydrologic control over a site might include levees and earthwork, channels, waterways, and outfalls for water supply, flow and water level control structures, such as weirs and stand pipes, and bypass and overflow structures for flood control. Construction and spatial placement of these structures are important in creating soil and water conditions for the establishment and sustenance of wetland plants and the attraction of wildlife.

▼ 图3-39：易道在中国海南做的一个项目中，表明水质和潮汐起伏的二维水力模型输出图（图片来源：易道）
Figure 3-39: Output of a two-dimensional hydrodynamic model showing water quality and tidal fluctuation at an EDAW project in Hainan Island, China (Source: EDAW)

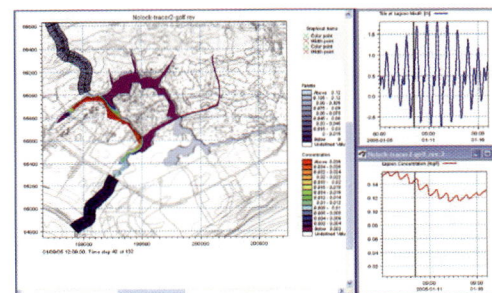

图 3-40: 该堤坝防止长江水泛滥时灌进历史上的湿地区，这里如今是村庄和农田(摄影: Rowan Roderick-Jones)
Figure 3-40: This levee prevents the Yangtze River from flooding into a historical wetland area which is now occupied by villages and agricultural fields (Photograph by Rowan Roderick-Jones)

图 3-41: 创建该湿地的目的是通过在自然排水区上修建小型堤坝，截住并处理暴雨径流
(图片来源: www. photos.com)
Figure 3-41: This wetland was created to capture and treat stormwater runoff by the placement of small levees across a natural drainage (Source: www. photos.com)

漫灌，人们常修建堤坝以建立大块浸水区域，但却不考虑其内部的水位梯度；这样形成的大片开放水域，常因深度问题使湿地植物无法繁殖，使鸟类无法觅食、栖息。堤坝的位置应纳入土地的自然轮廓，其形成的浸水区域应能够使开放水域和湿地的效用达到最大化（见图3-40,3-41）。

轮廓线堤坝系沿地形等高线设立，可以让管理者精确地控制湿地内部水位和植被分布。堤坝所依据等高线的间距因地形而有所不同，在较平整的土地上，因严格需要草本植物繁殖，等高线间距可设为0.2m，而在地形较复杂的土地上，因有开放水域、浮水植物、河滨区域和山地区域等多种栖息地都需纳入湿地场所，登高线间距可设为2m。

总之，越平整的土地上，等高线间距越小。湿地恢复规划者应在建设成本和功能可行性两者间作出平衡，以决定堤坝最终的数量。应充分考虑所有待建的堤坝都应受到充分考虑，使其对现有湿地栖息地的影响减至最小，对创建新湿地的效用最大。

堤坝设施
堤坝不管大小，其结构都必须符合关于稳定性、承重量、水土流失控制和永久性的工程规格，规划中必须考虑建设原料、滑坡、植被、宽度和高度。

为了保持堤坝的长期稳定性，选用合适的土壤、斜率和压缩度非常重要。有些土壤类型是不适合修筑堤坝的，如粗糙的沙土容易流失和被水侵蚀，有机土会慢慢分解致使堤坝缩小甚至溃塌；相反，低缩水膨胀率、高压缩度的土壤，如黏土、淤泥土和壤土，则很适合修筑堤坝。由于湿地土壤大多含有有机成分，所以不能现场取用。总之，黏土、淤泥土和壤土有高压缩度、低缩水膨胀率，因而最适合作为堤坝材质（Kelley et. al. 1993）。当地有工程能力的规划部门都应该能够提供意见，包括有关堤坝规划的工程规格。

为了减少地表径流、风浪等给泥土堤坝带来的水土流失，植被覆盖是非常必要的。非木本植物如多年生、一年生草本植物可以很容易地播撒或栽植于堤岸

Levees and Earthwork

Constructing levees is an effective way of maintaining water levels higher than their original state for wetlands on graded terrain, and as a means of mitigating water shortages that result from supply problems. The potential drawbacks of levees include possible sedimentation through reduction of flow velocities and the alteration of natural water supplies and discharge mechanisms that could affect areas outside of the project site. If water shortages in a degraded wetland are not anthropogenic, but rather caused by natural processes, alternative water supply methods, or no water supply at all may be the preferred restoration option.

Levee Placement

Often, in order to maximize inundation of land within a restoration site, levees are built to create large water impoundments with little regard for interior elevation gradients. This can result in the creation of large open water areas at depths that would prevent the establishment of wetland plants, and subsequently reduce foraging and habitat values for birds. Levee placement should therefore be integrated with the area's natural land contours and create impoundments which maximize the benefits of open water and wetland vegetation.

Contour levees run along topographic contour lines, and allow managers precise control over internal water levels and thus over vegetation structures throughout the wetland. Contour intervals on which to construct levees can range from 0.2 meters for very flat land where herbaceous growth management is strictly desired, up to two meters on topographically variable lands where a wide range of habitats including open water, emergent vegetation, riparian and upland areas will be integrated into the landscape.

In general, the flatter the land is, the fewer levees required. The restoration planner will need to balance construction costs and functional feasibility with the number of levees on-site. All potential levees should be considered in ways that minimize their detrimental effects on an existing wetland habitat, and for maximizing the ability to create or enhance new wetlands.

Levee Structure

Structurally, whether large or small, levees should be designed to comply with engineering specifications relating to stability, weight bearing capacity, erosion control, and permanence. Design considerations include construction materials, side slope, vegetation, width and height.

湿地恢复手册 原则·技术与案例分析

砂砾层
Gravel Trail

参考种植设计
Planting as Specifed

2:1斜坡，防侵蚀保护层，
耐湿植被
2:1 Slope, Erosion Blanket,
Seed & Plant

置石加固
Riprap Stone Toe

夯实土壤
Compacted Native Soil

池底防漏层
Compacted Native Soil

10-25mm 碎石
10-25mm Grovel

夯实土壤
Compacted Native Soil

碎石加固模块(成品)
Gabion Structures Retaining Works

管路
Culvert

池底防漏层
Pond Liner

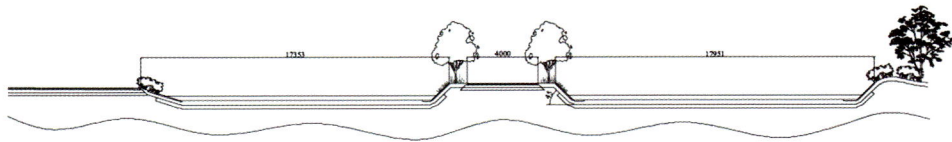

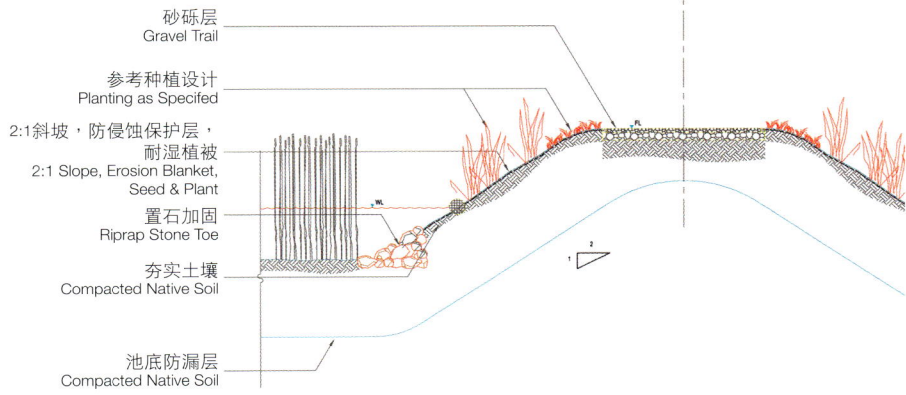

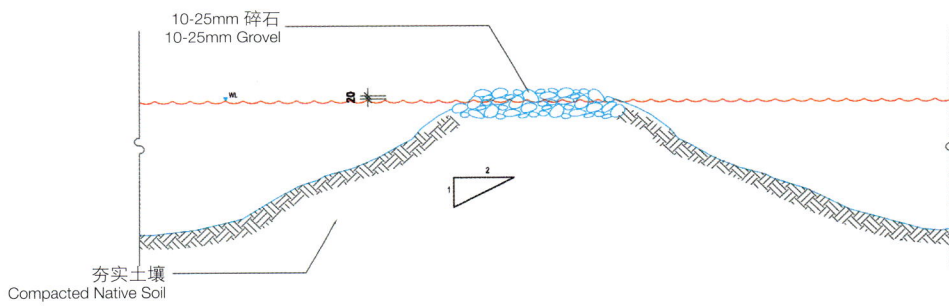

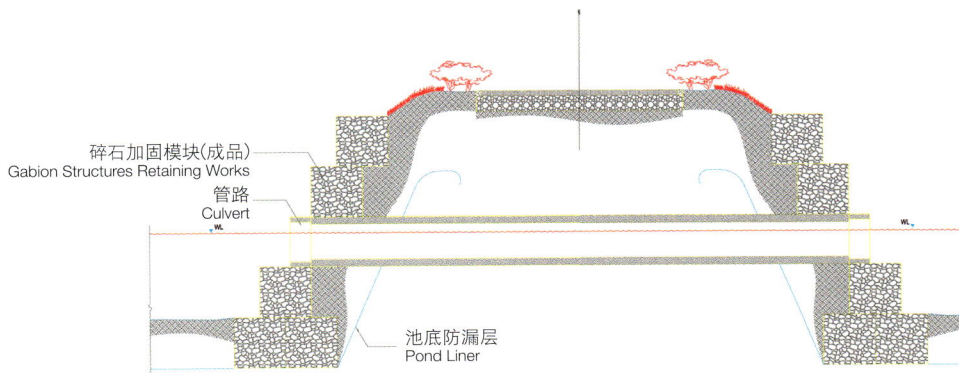

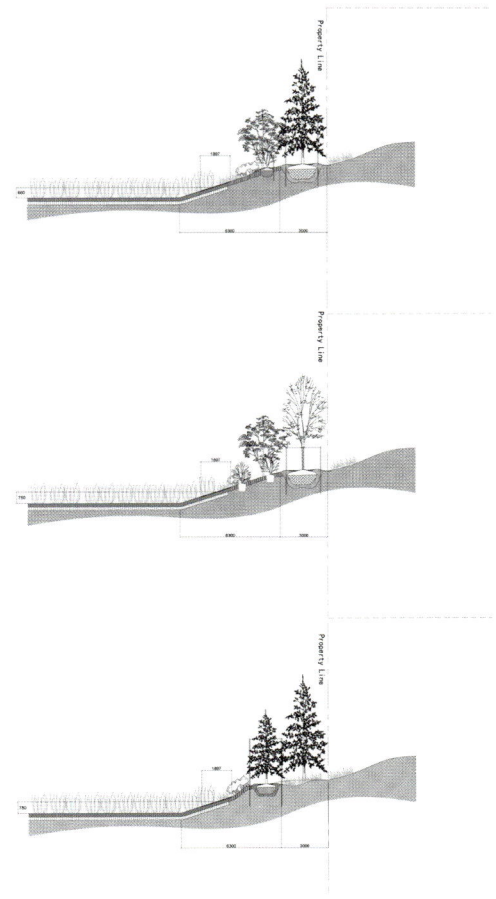

上，这些草本植物的根很浅，不会像乔木或灌木造成堤坝稳定性问题，而是在地表覆上一层浓密植被，可以固定土壤，稳定堤坝（见图3-42,3-43）。

如果吸引野生动物是首要目标的话，那么不妨种植有可食用种子的草本植物。堤岸斜度和土壤决定了地下水位深度，由此也决定了堤岸上可以成活的植物类型：如果斜坡较陡峭、土壤沙砾较多，植被会很快转化成河滨和山地类型；如果斜坡较平缓、土壤毛细管丰富，地下水位会离地表很近，湿地植被就会继续向斜坡上部繁殖。最适合的植被类型会根据地区不同而有差异，因此不妨咨询当地的生态学家或园艺家，征求其对物种选择的意见。

为了使堤坝能够支援用于植被维护和制造坡度的设备，必须对其进行工程改造。在小型恢复场地，小型拖拉机和割草机就足以应付；在较大的恢复场地，就要用到重型机械如推土机和反铲挖土机。这些机械设备部分决定了堤坝的三维尺寸，如宽度和堤岸斜度。堤岸应平缓升高，以供方便维护并防止某些哺乳动物挖洞破坏堤坝；通常3:1到5:1的边斜率就足够了。不过平缓的堤岸占用更多空间，需要更多的填埋土和建筑材料，因此成本较高。在较小的恢复场地或线性走廊（如河边），堤坝只能向空中发展，修建较陡峭的堤岸将有助于恢复工程。如果恢复场地确实需要较陡峭的堤岸，如1:1到3:1的边斜率，那么可以用石笼修筑整段堤墙，也可用散石增修稳定性坡脚。

堤坝高度可用下列经验公式算出：

堤坝高度 = 最大洪水深度 + 0.5m + 48h内最大降雨量

一个典型的浮水沼泽湿地，水深为0.6～1m，堤坝高度就可根据降雨量不同而设为1.5～2m；季节性湿地水深通常只有0.2～0.5m，堤坝高度就可设为1～1.5m。如果河边常受到强而短的洪水袭击，很快被洪峰淹没的低堤会比在洪峰中仅部分受到冲击的高大护堤少受破坏，因而更为可取。如果湿地直接承受来自洪峰的径流或是作为洪峰的临时储水区，必须在堤坝规划中设计紧急泄洪道，以保护

It is critical to use appropriate soil, slope and compaction levels in order to ensure the long term stability of a levee. Not all soils can be used to build levees. Coarse, sandy soils are generally inadequate because of their high erosion potential and susceptibility to water infiltration. Organic soils may decompose over time, causing shrinkage or even the collapse of the levee. Soils with low shrink swell capacity and high compaction potential such as clays, silty clays, and loamy clays are generally most suitable for levee construction (Kelley, et al 1993). Because many wetland soils are high in organic material, they may be unsuitable for use in levees on-site. Local Design Institutes with engineering capabilities should be consulted to provide assistance in obtaining engineering specifications for levee design.

Vegetation is desirable on earthen levees to reduce erosion from surface runoff, wind and waves. Non-woody vegetation, such as perennial and annual grasses can be easily broadcast or hydro-seeded along levee banks. The roots of these plant types are not deep enough to destabilize levees – which might occur with trees and shrubs, as they instead tend to form dense mats close to the surface, binding the soil and protecting the levee.

If the attraction of wildlife is a primary goal, grasses with edible seeds can be planted. Bank slope and soil determine the depth to water table ratio and therefore the vegetation that is appropriate for growing on the banks. For steep slopes and sandier material, vegetation can rapidly transition to riparian and upland types, while on shallower slopes with high capillary potential, the water table will remain closer to the surface, and wetland vegetation might have the ability to exist further up the slope. Appropriate vegetation types will vary regionally, and consultation with a local ecologist or horticulturalist is recommended for species selection.

Levees should be engineered to provide support for the equipment used in the maintenance and grading of vegetation. At smaller restoration sites, small tractors and mowers might be used, while in larger wetlands, heavier equipment such as bulldozers and large backhoes should be anticipated. The desired maintenance equipment will partially determine levee dimensions such as width and bank slope. Bank slopes should be gradual to allow safe maintenance access and to deter potential damage by burrowing mammals which prefer steeper slopes. Side slopes with ratios between 3:1 and 5:1 are usually satisfactory. Levees with shallow slopes occupy more surface area, require more fill and construction material, and are thus more costly. On small sites, such as linear

堤坝结构并调整洪水水位。沿河堤坝的修筑和移动通常需要征得当地水务部门或有关机构的同意或批准。

堤坝内部的受淹围堰有助于改善水质，是湿地恢复工程的常见特征。这种设施通常较低矮，其设计就是为了让水浅浅地漫过。受淹围堰用预设的优先通道减少了流量，相当于使湿地"短路"，其作用在于，让每一单位水都能有机会在湿地中留驻相同的时间，该时间相当于一平均单位水的留驻时间或经计算得出的湿地留驻时间。这样，受淹围堰就起了一个预设挡板的作用，改善通过湿地的水流。

直接供水
对于由于缺少水供给而干涸的湿地，可以通过直接补水来进行初期的湿地恢复。可以利用现有的河渠作为输水管道，也可以铺设专门的给水管道进行湿地直接输水。除了从其他流域调集外，也可以利用雨水进行水源补给。雨水输水可以通过引力作用排水的方式实现（包括通过梯田式的阶梯补水、排水管网或泵）（见图3-44，3-45）。利用泵排水可能导致水溶物质增加。

水流控制设施
湿地需要相关设施和建筑以向湿地内提供水源、为点收集设施供水、控制水位和减小洪涝灾害。如果正确设置和使用水流控制设施，就能够最大限度地类比自然水文系统。让水依靠重力自然流入优于用水泵送水，因此如果可行的话，就应采用这一方法以减少长期的经济成本；可考虑用湿地上游的拦截坝和分流建筑来营造重力自然水流的良好环境。排水调节设施应设置在储水区的最低位，并应有足够大的容量保证水能够完全、迅速地排出（Kelley et. al, 1993）。

湿地恢复工程有多种水流控制方法，其中大多数都包括调节水位、排水，以及在湿地区域或区域内小区块间截断或导向水流。有些水流控制设施的结构很简单，如原木拦截堰、固定V槽堰等（见图3-46）。其他设施有自动水流监测、可调节的机动化水闸等。V槽堰的设计和建造十分简便，并能测量水流量，因此在湿地恢复中应用很广，可以从许多公司购买统一型号的水闸和堰，

corridors along rivers, this might mean compromising a high aerial proportion of the site. In these situations, steeper levee banks may be desirable in order to maximize restoration. If site conditions mandate steeper slopes, between a ratio of 1:1 and 3:1, gabion can be used for the entire wall or additional toe stabilization can be applied using loose rip-rap.

In general, a levee's height is calculated using the following rule of thumb:

Levee Height = maximum depth of flooding + 0.5 meters + maximum 48 hour rainfall.

For a typical emergent marshland, or wetlands with water depths of 0.6–1 meter, this would amount to levee heights between 1.5–2 meters depending on rainfall. With seasonal wetlands, which typically have shallower flooding depths between 0.2–0.5 meters, levees might only be 1–1.5 meters tall. Adjacent to rivers prone to severe but short term flooding, lower levees which are quickly overtopped and submerged are more desirable as they often sustain less damage than large, protective levees which are only partially breached during a flood. For in-line wetlands which receive direct runoff from storm water, or which are used as temporary impoundments for flood control, emergency spillways can be incorporated into the levee design to maintain the structural integrity and appropriate flooding heights. When adjacent to rivers, placing or moving levees generally requires permits and authorization from local water resource bureaus or similar management authorities.

Internal, submerged berms are a common feature for wetland projects which also aim to improve water quality. These structures typically have very low elevations and are designed to be submerged just enough to allow for a shallow overflow. Submerged berms decrease flow through preferential flow pathways, thus "short-circuiting the wetland". In essence, berms ensure that every unit of water has a similar chance of staying in the wetland for a time period equal to that of the average unit of water – or what is called the calculated wetland residence time. By doing so, submerged berms act as engineered baffles which improve plug flow through the wetland.

Direct Water Supply
All wetlands require a reliable, predictable water source. For some, groundwater and rainfall may be sufficient sources while others may require surface connections such as rivers and waterways. For degraded wetlands that may have dried out due to human or natural

▼ 图3-44: 美国DTC新月公园的这块洼地将开发区的水沿陡坡输送到暴雨湿地(摄影 : Dixi Carrillo)
Figure 3-44: This swale at the DTC Crescent Park in the US delivers water from a development zone down a steep slope towards a storm water wetland (Photograph by Dixi Carrillo)

▼ 图3-45: 陡坡上可用散石来防止侵蚀和坍塌(摄影 : Dixi Carrillo)
Figure 3-45: Rip-rap can be used on steeper slopes to prevent erosion and slope failure (Photograph by Dixi Carrillo)

这比根据场地定制设施要便宜得多；选择水流控制设施的合适型号时，要在工程目标和设施的操作难易度、成本与维护要求两者间作出平衡。

湿地常会因浮水植物和湿地植物而产生很多有机碎屑，有时在洪涝时，整株植物会被连根拔起随水冲往下游。因此，水流控制设施通常需要同时使用碎屑栅，以防止水管、水口堵塞。与为水处理而修建的湿地不同，自然湿地并不一定需要在湿地区域内平均分配水流；但是，如果水处理能力被包含在工程目标中的话，那么便应规划额外的水流分配管道，以实现在湿地内水流面积的最大化分配。

衬垫

自然湿地恢复场所和用于处理、储存城市洪水径流的湿地，通常并不需要铺设不透水的衬垫。相反，用于处理城市、工农业废水的湿地，以及用于保护地下水资源免受污染影响、确保水处理有效时间的湿地，则必须铺设衬垫。如果湿地恢复工程有创建野生动物栖息地和处理废水的双重目标，就必须进行对废水品质和保护地下水资源的必要性的批判性评估，以确定是否需要铺设不透水的衬垫。

一个湿地如果地下水位较低、土壤沙化、易渗水，其恢复工程常倾向于采取铺设不透水衬垫的办法来减小水分流失，却不去考虑水质要求。其实，湿地本身是一个自我封闭的系统，即便在沙土质场所也不会有很快的水分流失；就像一个经工程改建过的渗透盆地，经过一两年的细菌繁殖、微粒沉淀和有机物堆积后，该湿地底部的毛细管空间就被堵塞了，减小了水分在其底部的渗透率。

一个水力深度为0.6m的成熟湿地，水分渗透率可能只有3mm/d或0.5%/d左右，因此该湿地的水力留驻时间至少应有200d，然后该系统才会干涸；当然，在最初一两个成长季节内的水分流失速度会高得多。总之，在决定是否采取铺设不透水衬垫时，必须首先考虑水质要求和地下水保护，其次再考虑水量和渗透。

常用的衬垫材料有高密度聚乙烯、膨润土、聚氯乙烯和聚丙烯。选择衬垫材料时，应考虑到覆盖面积、成本和材料耐久性。

disturbances, one of the first steps in restoring hydrology is providing a water supply. Serviceable rivers and canals can be used as channels for water transportation, and special pipes for direct water delivery to a wetland are also an option where stormwater can be collected from watersheds. Rainwater supplies can be tapped through gravity drainage, including land grading, and drainage network and pumps.

Flow Control Structures

Wetlands require facilities and structures that can deliver water to them, distribute it to and from point collection facilities, control water depths, and mitigate flooding impacts. The correct placement and use of water control features will enable the natural hydrologic regime to be simulated to the extent possible. When feasible, gravity flow into the wetland is preferred over pumping, and should be utilized to avoid long term economic costs and feasibility. Check dams or diversion structures upstream of the wetlands should be considered to create favorable conditions for gravity flow to occur. Structures regulating water discharge should be at the lowest elevation in the impoundment and be large enough to permit complete, rapid dewatering (Kelley et al, 1993).

Numerous flow control solutions are available to the wetland restoration planner. Most include some ability to regulate water levels, to drain the wetland, and to arrest or redirect flows between various wetland areas and "cells". Some flow control structures are very simple, such as log-stop weirs, or non-adjustable V notch-weirs. Others offer capabilities such as automated flow monitoring, and adjustable and motorized gates. V-notch weirs are used for many wetland restoration projects and are preferred for their simple design and ability to measure flow rates. Pre-fabricated gates and weirs are available from a number of companies and are often cheaper to obtain than constructing custom designed structures on-site. When selecting the appropriate type of flow control structure, the ability to meet project goals should be balanced with ease of operation, cost, and maintenance requirements.

Wetlands typically create large amounts of organic debris from floating and wetland plants. Occasionally, entire plants may be uprooted and flow downstream during large storms. Debris screens are commonly required in conjunction with most flow control structures to prevent the clogging of pipes and openings.

In contrast to constructed wetlands for water treatment, natural wetlands do not necessarily require even flow distribution over a wetland area. If water treatment capabilities are desired, than additional distribution pipelines should be designed to maximize even sheet flow across the wetland.

Liners

Natural wetland restoration sites and wetlands used to treat and store urban storm water run-off do not commonly employ impermeable liners. These structures are, however, a general requirement for wetlands used to treat municipal, industrial and agricultural wastewaters, used both to protect groundwater resources from potential contamination and to ensure sufficient treatment contact times. If a wetland restoration project will serve the dual purpose of creating a wildlife habitat and treating wastewater, a critical assessment of the wastewater quality and the need to protect groundwater resources should be undertaken to determine whether an impermeable liner is required.

At sites with low groundwater tables and sandy, permeable soils, a restoration designer may be tempted to limit water loss by employing an impermeable liner, even when water quality considerations do not mandate using one. Wetlands are in fact, self sealing systems that, even on sandy soil sites, will not lose water rapidly. Similar to an engineered infiltration basin, after a year or two of bacterial growth, particle sedimentation, and organic material buildup, pores at the bottom of wetlands become clogged, reducing the ability of water to pass through the wetland bed.

For a mature wetland with a typical hydraulic depth of 0.6 meters, water infiltration rates may be as low as 3 mm/day, or 0.5% per day. Hydraulic residence times in such a system would have to approach 200 days of consistent infiltration to cause the system to dry out. Of course, water loss during the initial two growing seasons may be much greater. The use of liners should therefore be considered more for water quality considerations and the protection of groundwater, and secondarily with regards to water volumes and infiltration.

A number of liner materials are available from manufacturers such as Firestone and Reef Industries, among others. Common materials include High Density Polyethylene, Clay Bentonite, Polyvinyl chloride, and polypropylene. Selection of liner systems should take into account the area to be covered, cost, and liner durability.

3.5.4 水质

自由表面流湿地（FWS湿地）通常用于处理各种废水，包括已初步处理的工业和生活废水，以及雨水。湿地已经被用于处理工业废水，如石油化学物质，纸浆厂废水，纺织工厂初步处理的废水和填埋场的沥出液。FWS系统尤其适合于净化低污染物浓度的大量废水。

不同的人工湿地在污染物去除能力上是不同的，但是它们都能很好地去除各种污染物，包括N、P营养盐、BOD、COD、石油类、悬浮物、重金属（Fe, Ni, Cd, Mg, Mn, Pb）、挥发性有机氯化物、病原体、杀虫剂和油脂类（Bastian and Hammer, 1993; USEPA, 1993）。物理沉降过程、植物的还原过程和过滤作用是最重要的去除过程（见表6）。这些过程能有效去除90%~98%的悬浮物质，颗粒有机物（颗粒态BOD）和依附于沉积物的营养盐和重金属。通过围隔、光降解和微生物过程能有效去除95%~99%的油脂。同样，人工湿地对病原体的去除率也很高，达到了70%~95%，去除途径主要有沉降和过滤、自然死亡和UV降解。

溶解态有机物、氨氮和正磷酸盐等液体物质的去除率相对较低，为60%~70%。大部分溶解态有机物都是在有氧条件下被水体里的细菌、附着在植物上的菌藻共生体以及沉积物表面的微生物所降解的。氨氮主要通过需氧的微生物硝化过程、植物吸收和挥发等过程去除，硝态氮则主要通过反硝化过程和植物吸收来去除。厌氧过程相对较为缓慢，需要较长的时间。磷的去除率为45%~75%，主要通过土壤吸收过程、植物同化过程和碎屑物沉积等过程来实现。金属的去除率为60%~99%，主要通过吸收和有机物络合作用来实现。各种污染物质的去除率见表3-7。

微生物易化过程很大程度上取决于水温、氧化还原电位势和pH值。虽然冬季时大型植物基本停止了生长，但是去除率仍然可以保持在较高的水准上。工厂产生的热能可用于增加FWS湿地的水温，在最冷的季节里提高微生物的降解能力。

3.5.4 Water Quality

Free water surface (FWS) constructed wetlands are commonly used to treat a large variety of wastewaters including domestic and industrial secondary effluents, and stormwater. While subsurface wetlands – also called gravel bed wetlands (with underground flow) – are also frequently employed for water treatment, this text briefly discusses the capabilities of treating polluted effluents using FWS wetlands.

Wetlands have been applied for treatment of secondary municipal and industrial wastewaters including petrochemical, pulp mill, and textile industry secondary effluents, and landfill leachate. They have also been employed to treat numerous agricultural and recreational facility wastewaters, including those from wineries, animal husbandry operations (hog farms, cattle feed lots), golf courses and other facilities. FWS systems are particularly well suited for "polishing" high volumes of waters with low pollutant concentrations and often follow up-front treatment of highly concentrated wastewaters. While constructed wetlands vary in their pollutant removal capabilities (see Table 3-7), they are still capable of serving as excellent treating mechanisms for a variety of pollutants including nutrients (nitrogen and phosphorous), BOD, COD, petroleum residuals, sediment, heavy metals (Fe, Ni, Cd, Mg, Mn, Pb), chlorinated volatile organics, pathogens, explosives, pesticides, and oil and grease (Bastian and Hammer, 1993; USEPA, 1993). Among the most important removal processes are the physical processes of sedimentation via reduced flow velocities and filtration by vegetation. These processes account for the high removal rates for suspended solids (90-98%), the particulate fraction of organic matter (particulate BOD), and sediment-attached nutrients and metals. Oil and grease are effectively removed (95-99%) through impoundment, photodegradation, and microbial action. Similarly, pathogens show good removal rates (70-95%) in constructed wetlands via sedimentation and filtration, natural die-off, and UV degradation.

Dissolved constituents such as soluble organic matter, ammonia and ortho-phosphorus tend to have lower removal rates (60-70%). Soluble organic matter is largely degraded aerobically by bacteria in the water column, plant-attached algal, through bacterial associations, and microbes at the sediment surface. Ammonia is removed largely through microbial nitrification (aerobic), plant uptake, and volatilization, while nitrate is removed largely through denitrification (an anaerobic process) and plant uptake. Biologically mediated anaerobic treatment steps

表 3-7: 自然表面流湿地中各种污染物质的去除率 （来源：Kadlec and Knight, 1996, Horne, 2005)
Table 3-7: Removal Rates of Various Pollutants in FWS Constructed Wetlands (Sources: Kadlec and Knight [1996], Horne [2005])

项目 Constituent	去除率 Removal	去除限值 (mg/L) Removal limit (mg/L)
TSS	61-98%	1 mg/L
BOD	70-97%	1 mg/L
NH_3	45-85%	0.1 mg/L
TN	85-95%	0.01 mg/L
TP	40-85%	0.1 mg/L
自由表面流湿地中的重金属去除率 Removal Rates of Heavy Metals in FWS wetlands		
Cu	63-96%	
Cd	70-99%	
Al	33-63%	
Fe	58-80%	
Mn	43-98%	
Pb	65-83%	
自由表面流湿地中的易燃物质去除率 Removal Rates of Explosives in FWS wetlands		
TNT	79-99%	
RDX	50-99%	
TNB	99%	
HMX	50-99%	
24 DNT	58%	
26 DNT	61%	

一级去除方程式（BOD、COD、NH$_3$ 和 TN）

根据基于栓塞流假设的一级反应动力学方程式，建立 BOD、COD、NH$_3$ 和 TN 的去除模型。此报告利用 Kadlec 和 Knight(1996) 所述的设计方法。在水文计算中，流出物浓度根据流速的变化而调整。基本计算方式如下：

$$C_{out} = ((C_{in}-C^*)exp(-K_{Tb}/q)) + C^*$$

其中：

C_{out} = 流出物浓度 (mg/L)
C_{in} = 流入物浓度 (mg/L)
C^* = 背景浓度 (mg/L)
K_{Tb} = 根据温度而定的速度常数 (m/y)
T_b = 平衡点水文
q = 水压负荷率 (m/y)

$$K_{Tb} = K_{20} \theta^{(T-20)}$$

其中：

K_{20} = 20°C 时的恒定常数
θ = 无量纲温度系数

are relatively slow and require longer residence times. Phosphorus removal (45-75%) occurs through soil sorption processes and through plant assimilation and burial as litter. Metals removal (60-99%) is achieved largely through adsorption and complexation with organic matter (Table 3-7).

Microbial processes are highly dependent upon water temperature, reduction-oxidation (redox) potential, and pH. Despite negligible macrophyte growth during the winter, removal rates can remain relatively high. Utilization of waste thermal energy from industrial facilities, for example, should be investigated as a means of increasing water temperatures in constructed wetlands, thereby improving microbial degradation rates during the coldest months.

First order removal of BOD, COD, NH$_3$ and TN

The removal of Biological Oxygen Demand (BOD), Chemical Oxygen Demand (COD), Ammonia (NH$_3$), and Total Nitrogen (TN) in free surface wetlands can be modeled according to first order reaction kinetics based on plug flow assumptions. This design approach is presented in further detail in Kadlec and Knight (1996). The basic design equation is as follows:

$$C_{out} = ((C_{in}-C^*)exp(-K_{Tb}/q)) + C^*$$

Where:

C_{out} = the outflow concentration (mg/L)
C_{in} = the inflow concentration (mg/L)
C^* = the natural wetland background concentration (mg/L)
K_{Tb} = the rate constant adjusted for temperature (m/y)
T_b = the water temperature
q = the hydraulic loading rate (m/y)

$$K_{Tb} = K_{20} \theta^{(T-20)}$$

Where:

K_{20} = the rate constant at 20°C
θ = dimensionless temperature coefficient

磷

因为含有磷的湿地通常都有复杂的化学反应，能够准确计算磷的保留量的预测模型极其少。美国佛罗里达州大沼泽湿地的一个试验区表明，磷的同化能力约为1g/(m²·a)。这至少比公布的硝酸盐去除率慢了两个数量级。"磷的1克法则"对于估算磷的去除率是相对简单和有用的方法，且被用于工业标准。

由于磷不存在气态形式，所以其主要存在方式为吸附到黏土和土壤颗粒上，与铁或其他金属元素结合在一起，作为有机体存在于植物上或以有机物的方式底埋在沉积物里。如果持续保持缺氧状态，磷就将保留在沉积物中。但是，沉积物对磷的持留有一个容纳极限，几年之后，磷的去除率将下降。在含氧量较高的条件下，磷将从土壤中释放出来，增加其在湿地中的浓度。

化学需氧量(COD)

用人工湿地处理工业过程中的COD已经被许多学者描述过了。Knight等人(1999)调查了五个处理石化工业流出物的表面流湿地，发现COD去除率在38%~86%之间。Vhovsek等(1996)发现在一个处理食品工业废水的高负荷湿地，COD去除率为92%。这个COD包含许多溶解剂、清洁剂和残留的植物物质(包括蛋白质、缩氨酸、氨基酸、醛类、酒精、弱有机酸，醛和酮)。Billore等(2000)在处理酿酒厂高浓度(高至8000 mg COD/L)流出物的过程中，COD去除率达到了64%。

有机物

因为湿地里含有有机物的化学反应都很复杂，能够准确计算有机物转化途径的预测模型极其少。在处理有毒有机物的过程中，自由表面流湿地最根本的设计考虑是处理物件必须在系统中滞留足够时间(例如足够的半衰期)，以便目标物质被降低到预期的水准。这些物质必须转化成新的无毒的形式、永久地沉积在沉积物、挥发到空气里，或者被植物吸收并被收割。

Phosphorus

Due to the complexity of chemical pathways involving phosphorus in wetlands, few predictive models exist which can accurately assess Phosphorus (P) retention. The threshold P assimilative capacity based on a test site in the Everglades (Florida, US) is approximately 1 g/m² yr. This is at least two orders of magnitude slower than published nitrate removal rates. The "one gram rule for P" has proven to be a relatively simple and useful rule of thumb for estimating P removal and has been used as an industry standard.

Because P has no gaseous phase, the main pathway for retention is adsorption to clay and soil particles, binding with iron and other metals, uptake by plants and burial as organic matter. If anoxic conditions are maintained, P will remain in the sediments. However, sediments will have a finite capacity for P retention and after several years, P removal rates may decrease. P release from soils during periods of high oxygen presence can increase concentrations in the wetland.

Chemical Oxygen Demand (COD)

Treatment of COD in constructed wetlands for industrial and agricultural operations has been described by numerous authors. Knight et al, (1999) reported values of between 38% and 86% for COD removal from five free surface wetlands that treated petroleum industry effluents. Vhovsek et al, (1996) showed an average COD removal of 92% in a high load wetland designed to treat wastewater from the food industry. This COD consisted of numerous solvents, detergents, and residual plant substances (including proteins, peptides, amino acids, aldehydes, alcohol, weak organic acids, aldehydes and ketones). Billore, et al (2000) reported that they were able to achieve 64% removal of highly concentrated (~ 8000 mg COD/L) distillery effluent.

Organics

Because of the complexity of chemical pathways involving organic pollutants in wetlands, few predictive models exist which can accurately assess their fate. In treating toxic organic constituents, the fundamental design consideration for a constructed wetland is that the material must be retained within the system for a sufficient amount of time (an adequate number of half-lives) for the targeted constituent to be reduced to the desired level. Constituents can be transformed into new non-toxic forms, adsorbed permanently in the sediment, volatilized to the atmosphere, or incorporated and harvested from plant biomass.

Volatilization half-life times can be used for some volatile

重金属

水生系统中大多数重金属的行为和对环境的影响已有详细的研究，目前普遍认为，湿地的确具有持久的金属去除性能。在美国，湿地被用来处理含有高浓度重金属、对植物或野生动物几乎没有毒害的出水（如酸性矿井排水）。（Pascoe et. al., 1994）。

决定金属在湿地中行为的作用有很多：化合作用，含水金属氧化物、黏土或有机材料的吸附作用，阳离子交换，不溶性物质（如金属硫化物和生物积累）的形成，絮凝作用和沉积作用。湿地植物根系周边的空斑形成是许多金属元素转化的另一个重要途径。如果来水中出现高浓度金属元素，高浓度硫化物将有助于形成金属硫化物，从而去除金属元素在湿地中形成的许多重金属种类都是非生物可利用的，而且可能在土壤中矿化。其他金属元素，如汞（Hg）、硒（Se），都能引发湿地中的生物积累和毒害。

pH值

由于工业污水有极高或者极低的pH值，所以在进入自然湿地处理系统之前要先进入传统的酸度处理系统。处理后pH在6.5~8之间，适合生物反应（比如硝化反应）。起到酸性缓冲作用的物质包括$CaCl_2$、H_2SO_4和NaOH。

$CaCl_2$是一种价格便宜的钙离子源，在废水处理中起pH调节作用。主要用于处理废油和无机化合物，包括氟化物、磷酸盐以及重金属。在废水处理中采用$CaCl_2$的包括铝、钢、玻璃、肥料、石化产品的生产废水处理以及生活污水处理。

3.5.5 休闲设施

湿地恢复工程可在湿地内修筑休闲设施，供人类使用（见图3-47）。即便是严格用于栖息地保护的湿地场所，也能提供某种人类活动方式。人类活动设施多种多样，从简单的散步道、观鸟隐棚到大型、多功能的参观中心，如已成为国际著名旅游胜地的香港湿地中心和伦敦湿地中心。除了建立湿地中心外也可以有各类其它设施，如当地或区域性的观鸟社团会所、湿地科学家的研究设施、教师中心和公共信息亭。

▼ 图3-47：旧金山市国家野生动物保护区用小型木板桥来限制游客数量，而绍兴却建造大型木板桥以便接纳更多的游客(摄影：Dixi Carrillo)
Figure 3-47: While small boardwalks are utilized at the San Francisco National Wildlife refuge to limit visitor numbers (top), at the Shaoxing Wetland in China (bottom), large boardwalks were built in anticipation of greater tourist activity (Photograph by Dixi Carrillo)

Heavy Metals

The environmental fate and effects of most heavy metals in aquatic systems have been studied extensively and there is general agreement that wetlands are capable of significant permanent metals removal. In the US, wetlands are used to treat effluents (such as acid mine drainage) containing high concentrations of some heavy metals with little or no toxicity effects to plants or wildlife (Pascoe et al, 1994).

Many processes determine the fate of metals in wetlands: complex formation, sorption to hydrous metal oxides, clays or organic materials, cation exchange, formation of insoluble species (for example, metal sulfides, and bioaccumulation), and flocculation and sedimentation. Plaque formation around wetland plant roots is another significant transformation pathway for many metals. Many heavy metal species formed in wetlands are not bioavailable and will remain mineralized in the soil indefinitely. Other metals, such as Mercury (Hg) and Selenium (Se) are known to cause bioaccumulation and toxicity in wetlands.

pH

Because certain wastewaters, particularly industrial waters, and certain agricultural wastewaters can have extremely low or high pH values, conventional pH buffering treatment may be required prior to entering a wetland treatment system. Attaining a pH level between 6.5 and 8 is optimal for certain biological reactions such as nitrification. Methods of pH buffering include addition of Calcium Chloride ($CaCl_2$), sulfuric acid (H_2SO_4), or sodium hydroxide (NaOH).

Calcium chloride ($CaCl_2$) is an inexpensive source of calcium ions and pH adjustment in wastewater treatment. Specific uses include treatment of oily wastes and the removal of various inorganic compounds, including fluorides, phosphates and heavy metals. Industries using $CaCl_2$ in wastewater treatment include aluminum, steel, glass and ceramics, fertilizers, petrochemicals, and municipal wastewater.

3.5.5 Recreational Facilities

Goals set for human usage of wetland restoration projects will determine the appropriate recreational facilities for the site. Even sites strictly designated for habitat preservation are capable of accommodating some form of human interaction. The spectrum of facilities range from simple

规划休闲设施时，首先要平衡野生动物栖息地条件和人类活动的需要。人们参观湿地的目的主要是观鸟、休闲散步和徒步旅行、（主要面对中小学生的）科普教育和自然风景观光。一个湿地场所如果常年没有原始风光、鸟语花香，那么就没有人会有兴趣前来参观了；正因为此，规划休闲设施时必须保证其不对自然景观造成破坏，并能控制游客流量以防止对野生动物造成干扰。噪声和视觉干扰在很远距离外就会惊吓鸟类，因此应尽可能将参观者控制在离主要栖息地至少100m远的地方。另外，在规划小径和观景平台时，应将其隐藏在鸟类和其他野生动物视野以外，比如在淡水沼泽湿地可将小径专门修建在厚实的芦苇床中，这样可在少数关键点提供广阔的观景空间，而同时又用观鸟隐棚将小径隐藏起来（见图3-48）。

boardwalks and bird blinds to large, multipurpose visitor centers that are internationally recognized as major tourism destinations, such as the Hong Kong Wetland Center and the London Wetland Center. Between these extremes are other facility types such as clubhouses for local or regional bird watching societies, research facilities for wetland scientists, lecture centers and information kiosks.

When designing recreational facilities, balancing wildlife habitat requirements with human activities is a primary consideration. The main reasons visitors come to wetland restoration sites include bird watching, recreational walking and hiking, education (particularly for school children), and enjoying the natural scenery. Without the pristine feel of a site and an abundance of wildlife present for at least part of the year, much of the visitor experience will be lost. Because of this, facilities should be designed to be non-intrusive forms in the landscape, and to control human traffic to minimize disturbance to wildlife. Noise and visual disturbances will frighten off birds, even at a great distance. When possible, humans should be at least 100 meters removed from primary habitat areas. In addition, wherever trails and viewing platforms are planned, they should be hidden from view from birds and other wildlife. In freshwater marsh wetlands, one option is to isolate trails to thick reed beds, creating expansive views only at key points, and always hidden by bird blind structures (Figure 3-48).

▼ 图3-48: 此效果图显示出天津附近的一座湿地公园中观鸟台的作用（图片来源：易道）
Figure 3-48: A conceptual diagram illustrating the role of bird blinds at a wetland park near Tianjin (Source: EDAW)

3.6 建设后的监测和评估

3.6.1 成效评估

湿地恢复和创建工程最令人头痛的地方莫过于评估工程是否成功，因为目前对此尚无普遍标准。原因是多方面的，如缺乏成文的任务目标和长期监测（Kusler and Kentula, 1990b），评估者的主观观点（Roberts, 1993）等。经恢复后的湿地绝大多数还不到10年，在生态学意义上是非常"年轻"的，而对于生态成熟的湿地恢复工程，则普遍知之甚少，这导致科学家们难以对恢复湿地能否在功能上替代自然湿地作出预测。不过，专家判断和长期的研究成果已可以为选出较可能成功的恢复工程提供有价值的参考。

为了建立湿地恢复工程成功与否的标准，科学家们已作出了不少努力。最早的标准是，只要湿地条件适于湿地植被繁殖，那么其他生态功能就已存在或会慢慢培养出来；如今人们普遍认为，一个重建场所"一片绿荫"并不意味着工程成功，评判恢复工程的标准应以湿地功能为重。

从长远来看，经重建后的湿地应与类似的自然湿地功能相近。对俄勒冈（Kentula and others, 1992）、康涅狄格（Confer and Niering, 1992）和佛罗里达（Brown, 1991）的重建湿地的植物多样性对照调查显示，虽然每一工程的植物多样性水准不一，但都高于各自作对照的自然湿地。经调查的湿地类型是有淡水沼泽围绕的一片池塘，如果恢复湿地按照计划发展的话，则2~5年后可能会出现与类似的自然湿地相同或比其更高的植物多样性；一旦恢复湿地内植被完全覆盖整个场所，对空间和资源的争夺就会加剧，那么植物多样性通常会减小，而植物群落会逐渐展现出成熟湿地的特点。

3.6.2 生物管理和维护

蚊虫控制
在湿地区域，尤其是城市边缘的湿地区域，蚊虫问题是一大隐患。有很多案例显示，食蚊鱼（Gambusia sp.）和捕食孑孓的动物（如蜻蜓幼虫）能够有效控制蚊虫数量，其中最有名的一个案例是在加利福尼亚Arcata的恢复湿地（Gearhartt, 1983）。大

3.6 Post Construction Monitoring and Evaluation

3.6.1 Evaluation of Success

One of the most vexing aspects of wetland restoration and creation projects is defining success, primarily because there are generally no definitive criteria or a consensus that can be applied to different restoration projects. This is true for many reasons: lack of clearly stated objectives, lack of long-term monitoring (Kusler and Kentula, 1990), and the subjective point of view of the authorities and professionals involved (Roberts, 1993). However, it is now generally accepted that to evaluate the success of habitat restoration, a project must have a set of clearly defined goals and objectives, and must include a monitoring program capable of demonstrating unequivocally whether or not these goals and objectives have been achieved.

The vast majority of project wetlands are ecologically young, roughly 10 years of age or less. The lack of information on mature projects limits the ability to predict whether or not the functions of project wetlands can replace the functions of natural wetlands. Nevertheless, ongoing research, current results, and good professional judgment can be used to provide insights into the selection of projects that have a high probability of success.

Various attempts have been made to define criteria and a framework of success for wetland projects. Some of the earliest criteria assumed that if conditions were correct for the establishment of wetland vegetation, than other ecological functions would either be present or would develop over time. It is now known that a site "green" with vegetation does not necessarily imply success, and the standards by which projects are judged are more likely to be tied to a wetland's function.

Over time, successful restoration projects can be expected to show healthy characteristics that are comparable to natural wetlands. A comparison of plant diversity on project wetlands and similar natural wetlands in Oregon (Kentula et al, 1992), Connecticut (Confer and Niering, 1992), and Florida (Brown, 1991) showed that, although the level of diversity differed with each project, diversity tended to be higher at each project wetland than on its natural counterpart. In this case, the type of wetland studied was a pond with a fringe of freshwater marsh. If a project wetland develops as hoped and expected, after two to five years it will probably show plant diversity greater than or equal to that of similar natural wetlands. As competition for space and resources increases, and the

多数捕食孑孓的鱼类和无脊椎动物都需要富含氧气的开阔水域，如果植被过于繁茂的话，这些捕食动物控制蚊虫的效果就大打折扣。其他控制蚊虫的方法有采用引入一种微生物 Bacillus thuringus或引入捕食成年蚊虫的动物如蝙蝠等。

对于蚊虫控制，主要关心的是水中氨的浓度，即使是较低的浓度对鱼也是有害的（见表3-8）。没有天敌的话，蚊子就有机会成功繁殖，尤其在那些流速较慢的湿地边缘地带。

氨有两种形式：NH_3和离子态的NH_4^+，在水中达到平衡，通常测量的氨都是两种形式的总和。与离子态的NH_4^+相比，非离子态的NH_3更具有毒性。美国环保署推荐NH_3限制浓度为 0.02mg/L（USEPA，1986），防止鱼对氨的慢性中毒。总的氨氮水准变化区间浓度为160mg/L（温度5℃，pH6.0）到 0.06mg/L（温度25℃，pH9.0）。通过设置低温和低pH的环境条件能降低氨的毒性。表3-9说明非离子态氨NH_3占总氨的百分比随温度和酸度变化是如何变化的。

植被管理
采用机械或人工的方法收割成熟的植被，是防止植物凋落物的过度堆积，导致湿地向陆地演替的有效方法（见图3-49）。此外，对于那些用于水质净化的人工湿地，收割挺水植物（如芦苇）或打捞漂浮植物（如浮萍）能有效移除营养物质，防止凋落物成为湿地的内源性污染源。但是，打捞或收割后的植物残体应该得到妥善的解决，否则可能成为堆放地的污染源，以及导致物种的随意扩散。最常见的处理方式为资源化利用或焚烧，前者是一个更好的方法，因为它能使收割或打捞计划得以长期开展，典型的有收割芦苇用于造纸原料等。

防止物种过度扩散
如果在湿地生态恢复后的监测中发现入侵行为，必须尽快采取机械或人工的方法去除已经显示入侵特征的外来物种，防止它们过度生长。入侵物种不仅侵占空间，并且消耗大量养分或分泌次生物质，抑制其他物种的生长。针对该物种的限制性生态因子的方法会比较有效，比如，对于不耐淹的物种就应调节水文，延长水淹时间。引入食性专一的天

plants more completely cover the site, diversity usually decreases and the plant community tends to become more like that of a mature site.

3.6.2 Maintenance Techniques

Mosquito Control
Mosquitoes are a common concern for wetlands, particularly near urban areas. Mosquitoes can be successfully controlled through the use of mosquito fish (commonly gambusia sp.) and other mosquito larvae predators such as dragon fly larvae. Both predators have been shown to be effective at reducing mosquito populations in numerous case studies, including a well known study on a constructed wetland in Arcata, California (Gearhartt, 1983). Most predators of mosquito larvae, including fish and invertebrates, require oxygenated water with open space. If vegetation becomes too dense, predators may not be as successful at controlling mosquitoes. Other means of mosquito control include introducing the micro-organism Bacillus thuringus, and attracting predators such as bats, which prey on adult mosquitoes.

A primary concerns regarding mosquito control is the presence of ammonia, which is toxic to fish at relatively low concentrations (see Table 3-8). Without fish, mosquitoes may have the opportunity to breed successfully, particularly at the slower flowing edge zones of a wetland.

The term ammonia refers to two chemical species of ammonia which are in equilibrium in water (NH_3, un-ionized and NH_4^+, ionized). Tests for ammonia usually measure total ammonia (NH_3) plus NH_4^+. The toxicity of ammonia is primarily attributable to the un-ionized form (NH_3), as opposed to the ionized form (NH_4^+). The USEPA recommended allowable limit for NH_3 is 0.02 mg/L NH3 (USEPA, 1986) to prevent chronic fish toxicity. Total ammonia ($NH_3 + NH_4^+$) levels at this limit range from 160 mg/L at pH 6.0 and a temperature of 5 degrees C to 0.06 mg/L at pH 9.0 and 25 degrees C. Low water temperature and pH are desired conditions for attaining low ammonia toxicity. Table 3-9 shows how un-ionized ammonia concentrations, as a percentage of total ammonia, increases with temperature and pH.

Vegetation management
Harvesting mature vegetation by mechanical or manual means helps reduce the excessive accumulation of plant litter and is an effective way to prevent the wetland from evolving into a successional upland ecosystem. In artificial wetlands constructed for water purification, harvesting

▼ 图 3 - 49：湿地植被需要仔细的管理，通过经常性的收割来促使陆生植被的发育（摄影：EDAW）
Figure 3-49: Wetland vegetation requires careful management, and often has to be harvested to maintain development of upland vegetation communities (Photograph by EDAW)

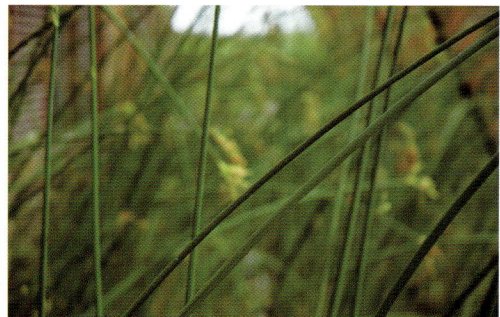

表 3-8: 25°C 下 氨 对 淡 水 鱼 类 的 致 死 浓 度 (来源：USEPA, 1986)
Table 3-8: Lethal Ammonia Concentrations to Freshwater Fish at 25 Degrees Celsius. Note that toxicity increases at higher pH values due to higher concentrations of NH_3. (Source: USEPA, 1986)

pH		时间 Duration	致死浓度(mg/l) Lethal* Ammonia Concentration (mg/l)	
			总量 Total	NH_3
6.5		1-hr	14.3	0.036
		4-day	0.73	0.002
7.0		1-hr	11.6	0.093
		4-day	0.74	0.006
7.5		1-hr	7.3	0.181
		4-day	0.74	0.019
8.0		1-hr	3.5	0.260
		4-day	0.47	0.035
8.5		1-hr	1.3	0.260
		4-day	0.17	0.035

表 3-9: 非 离 子 态 NH_3 随 温 度 和 pH的 变 化 在 总 氨 中 的 百 分 比
Table 3-9: Un-ionized NH_3 as a Percent of Total Ammonia (by temperature and pH) (Source: USEPA, 1986)

温度 Temperature		非离子态NH_3在总氨中的百分比 Percent NH_3 of total ammonia				
		pH				
		6.5	7.0	7.5	8.0	8.5
20		0.13	0.4	1.24	8.82	11.2
25		0.18	0.57	1.77	5.38	15.3
28		0.22	0.7	2.17	6.56	18.2
30		0.26	0.8	2.48	7.46	20.3

敌也是一个有效方式，但是应该对天敌种群的生态风险进行事前的预测。只有在上述方法无效的前提下，才可使用杀虫剂或除锈剂，因为这些化学制剂除了会对生态系统产生长期持续的影响外，也会通过湿地水体扩散到地下水中，并影响周边区域。

3.6.3 监测

完成重建湿地的规划、建设和种植等诸项工作，只是任何恢复、创建和改善工程的第一步。在很多情况下，如果建成后放任自流，让重建湿地完全自然发展，会引发很多不良后果，最后只能采取种种矫正措施，如重新种植幼苗、挖掘新沟渠、重新修筑堤坝、清除野草及其它非本地植被、调整水位和水流量等。以上管理措施应纳入一个适应性管理方案中，该方案系通过适当的监测计划来获得湿地信息。

湿地监测为长期和短期的决策制定提供了参考，比如某种类型的湿地或重建场所中的某块区域特别能够吸引鸟类，通过监测了解这一情况后，湿地管理者就可以在今后的改进或"翻新"工程中加入这些吸引鸟类的因素。湿地监测还可让湿地管理者能够对非本土植被、水管等设备故障、穴居动物情况等简单问题进行日常观察和维修。

监测计划旨在对湿地中水文、水质、生物和土壤的长期和短期变化进行记录（见表3-10）。在计划中应确定测量参数、监测方式和监测间隔；根据场所不同，应测量的参数和监测间隔也有不同。接受工业废水或其他废水排放的湿地，应就生物参数进行大量监测，以防止有毒物质扩散并威胁有机生物；易道公司就是这样为上海化学工业园规划废水处理湿地的。

除了常规检测外，监测计划还包括了流出毒性总量（WET）检测、土壤毒性检测、生物积累量检测和含60多个有机、无机化学物参数的水质测量，以充分确保该生态系统不受有毒物质的潜在侵害。总体来说，对于无需进行水处理的自然恢复湿地，监测计划就不必包含过多测量参数；不过，在制定监测计划前一定要对湿地内的具体危害和不确定因

emergent plants like reeds or drifting plants like duckweed may effectively remove nutrients even as it prevents litter from becoming an internal pollution source. However, plant residues that result from harvesting or clearing must be properly disposed. If left onsite, they may continue to act as a pollution source or cause the proliferation of unwanted species. Disposal methods include recycling and incineration. Recycling options, such as using harvested reeds as biomass for paper-making, is a better option in that it allows for a long-term, self-sustainable harvesting or clearing plan, and includes the potential to generate revenues.

Species proliferation control

Following ecological restoration, the proliferation of exotic species may be identified through a monitoring program, which could assess any need for immediate action regarding the removal or curbing the growth of exotic species by mechanical or manual means. Invasive species may occupy large areas and suppress the growth of other species by consuming excessive nutrients or secreting growth-prohibiting chemicals. Countering invasive species through prohibitive ecological means may be effective. For instance, regulating wetland hydrology to increase submersion times may be useful in deterring invasive species that are not resistant to submersion. The introduction of natural enemies including insects with specific diets can also be an effective control means, but the ecological risks of introducing natural enemies must be assessed beforehand. Pesticides or defoliants must only be applied as a final resort when all of the aforementioned biological methods fail, since these chemicals will bring long-term damage to the sites ecosystem, and to its adjacent areas through transport in surface and groundwater infiltration.

3.6.3 Monitoring

Completing the design, construction, and planting of a restored wetland is only the first half of a successful restoration, creation or enhancement project. Leaving a restored site to allow nature to run its course after initial construction will often bring on undesirable results that require corrective actions, such as replanting of seedlings, digging additional channels, re-contouring levees, removing weeds and non-native vegetation, and finessing water levels and flow rates. These management actions should instead be part of an ongoing adaptive management program that is informed by data collected through appropriate monitoring.

Monitoring informs both short and long term decision

可供监测的湿地特征 Wetland Characteristics for Monitoring	
特征 Characteristic	定性方式 Qualitative Method
水质 Water Quality	
水质（pH值、盐度、营养度、污染物、重金属等） Water Quality (pH, salinity, nutrients, pollutants, heavy metals, etc.)	野外考察工具和仪表 Field kits and meters
沉积物 Sediment	观察清澈度或使用塞齐盘 Observe clarity or use secchi disk
毒性（仅用于受到此种危害的场所） Toxicity (for at risk sites only)	流出毒性总量 (WET) 检测 WET (Whole Effluent Toxicity) test
水文学 Hydrology	
水深 Water Depth	人员测量 Staff gauge
水流模式 Flow Patterns	直接观察，将主要水流途径绘成地图 Directly observe and map major flow pathways
流量 Flow Rates	根据湿地大小和地区降雨规律来估计水流量 Estimate flow based on wetland area and regional rainfall patterns
地表水 Groundwater	水位测量仪 Water level gauges
其他间接观察 Other Indirect Observations	最高水位、漂流线等 High water marks, drift lines, etc.
毒性 Toxicity	
如果水质毒性检测反应为阳性，就要进行本项附加监测，并须对水体污染源做出适当处理 If Water Quality Toxicity tests yield positive results, perform additional monitoring and deal appropriately with the source of the water pollution	
鸟类毒性 Avian Toxicity	通过野生雌雄对和野外鸟栏，对鸟类及其胚胎的毒性进行评估 Evaluation of bird and embryo toxicity, using either wild nesting pairs or field pens
贝壳毒性 Bivalve Toxicity	在湿地的可控环境内，监控存活率和健康状况 Monitor survival rates and health in controlled environments within the wetland
土壤 Soil	
土壤深度 Soil Depth	挖开压缩土或挖到至少0.5m深度，观察土壤颜色和结构 Dig to compacted soil or at least 0.5 m and observe changes in soil color and structure.
土壤颜色 Soil Color	无 n.a.
土壤纹理 Soil Texture	根据触觉，用土壤纹理三角进行分类 Use soil texture triangle to classify based on feel
有机物 Characteristic	
土壤沉积 Characteristic	就地测量沉积深度并读取变化 Read changes in sediment depth from staff gauge
植被 Vegetation	
物种多样性 Species Diversity	辨认常见物种，并确定无法识别物种的数量 Identify common species and note number of unidentified species
覆盖率 % Cover	大致估计植被覆盖率，误差在10%以内；为主要种群绘制地图 Estimate coverage to within 10%. Map major communities
存活率 Survivorship	根据肉眼观察确定存活植物的百分比 Visually determine % of plants alive
植物高度 Height	测量具体植物的高度 Measure heights of particular plants
植物结构 Structure	计算具体植物的主干和枝条数量 Count stems and branches of particular plants
繁殖 Reproduction	观察具体植物的开花和结子规律 Observe blooming and seed setting for particular plants
动物 Animals	
观察 Observation	记录对野生动物、鱼类和无脊椎动物的直接、间接观察结果 Record direct and indirect observations of wildlife, fish and invertebrates
栖息地评估 Habitat Evaluation	
物种多样性和数量 Species Diversity and Abundance	每季度计算鸟类种数和大致数量；可向当地观鸟社团取得资料 Count bird species and their abundances on a quarterly basis. Inquire with local bird watching clubs for data
存活率 Survivorship	
繁殖成功率 Breeding Success	记录在当地繁殖的物种及其幼崽数量 Record any species breeding on site and number of offspring
稀有物种 Rare Species	

定量方式 Quantitative Method	自动监测 Automated
野外考察仪表和实验室分析 Filed meters or lab analysis	大多数情况可行 yes for most
野外考察仪表和实验室分析 Use field meter or lab analysis	可行 yes
在经认证的实验室内监测地表水和地下水中的有机和无机污染物；流出毒性总量 (WET) 检测 Monitor for organic and inorganic pollutants in surface and groundwater using a certified laboratory, WET test	不可行 no
自动水位测量仪 Automated water level gauges	可行 yes
通过航空和卫星照片观察并将主要水流途径绘成地图 Observe and map major flow pathways using aerial or satellite imagery	不可行 no
在堰上统计水流输入或输出量，用流量仪统计内部水流速度 Inflow-outflow weir calculations, and flow meters for internal velocities.	可行 yes
自动水位测量仪 Automated water level gauges	可行 yes
无 n.a.	不可行 no
取至少0.5m的土壤核，让专家检查土壤层次和构成 Soil core at least 0.5 m and have expert examine soil horizons and composition	不可行 no
用Munson色表比较填质颜色 Munsen color chart to determine matrix color	不可行 no
在实验室内对每一土壤层次作微粒分析 Laboratory particle size analysis for each soil horizon	不可行 no
对表层土（包括土壤潮湿层）的有机物含量进行实验室分析 Lab analysis for percentage of organic matter in top layer including soil moisture content	不可行 no
每年调查地形或取沉积核以供实验室分析 Survey topography on a yearly basis, or take sediment cores for laboratory analysis on a yearly basis	不可行 no
辨认所有本土和非本土物种 Identify all native and non-native species	不可行 no
在一个横切面上采集地块植被资料，并为每个种群估计覆盖百分比；为主要种群绘制地图 Collect plot data along a transect and estimate % coverage for each community type. Map major communities	不可行 no
在指定地块内计算每一株植物以确定存活率 Count individual plants to determine survivorship within designated plots	不可行 no
随机抽取植物并测量高度，以进行统计对照 Measure heights of randomly selected plants for a statistical comparison	不可行 no
随机抽取植物并计算其主干和枝条数量，以进行统计对照 Count stems and branches of randomly selected plants for a statistical comparison	不可行 no
随机抽取植物并观察其开花和结子规律 Observe blooming and seed setting for randomly selected plants	不可行 no
无 n.a.	不可行 no
使用"栖息地评估方式"(USFWS, 1980) 或其他类似方式 Use Habitat Evaluation Procedures (USFWS, 1980) or comparable procedure	不可行 no
通过设陷阱捕猎、定点计算或其他定量方法，推断物种多样性 Use trapping, point count or other quantitative methods to determine diversity and of indicator spp	不可行 no
通过对动物"标记或重新捕获"进行研究 Mark and recapture study	不可行 no
通过定点计算、勘察和其他方式，推断种群内生育百分比和新生幼崽数量 Use point counts, surveys, or other protocols to determine percent of population breeding and numbers of young produced	不可行 no
在野生动物或自然资源管理机构的法律许可范围内进行研究 Conduct studies as legally permitted by the jurisdictional wildlife or resource agency	不可行 no

素有所估计。

对重建场所的建设情况的详细纪录，是建立监测计划的良好基础。根据规划的精确度要求和管理预算，可以选择定性或定量的信息采集方式。用定性方式不能精确地判定目标是否符合标准，但是可以对湿地变化的类型和情况有一个大致的了解。

定性方式通常包括以下内容：

- 通过航空照片确定地形、沟渠构建和地表植物；

- 通过地面照片对可见的湿地变迁如植物高度、物种变化、水位升降和季节变换等进行观察；

- 对以下因素进行大致观察：水质和清澈度、人类活动痕迹、鸟类物种、植被情况如枯黄和其他不良表现、水土流失、检查各种建筑设备（如水管和水泵）是否需要维修等等。

采用上述方法时，可结合使用定量方式，为湿地的适应性管理计划提供最优化的信息。定量方式是指对测量参数给出具体数值的测量方式，采用这一方式时，不妨参考一些有重要统计学意义的研究成果。采集定量资料的方法有多种多样，从简便的现场测量、实验室分析到长期研究（如对野生动物及其存活率的生物积累评估）。一定要由相关领域如水文学、生物学和毒理学的专家作定量监测，各大学院系由于科研力量强、年轻研究人员多，可邀请其主持监测和研究项目。定量方式虽然颇耗金钱和时间，但能提供有关湿地场所变迁的精确资料。

有些重建工程需要即时监测，或者需要减少中间步骤或资料记录过程，或者因偏僻而难以定期监测，那么可使用自动资料获取技术。

对每一测量参数的监测频度取决于项目目标和成效标准、自然变化程度和变化速度。如果某一重建工程旨在增加野生动物栖息地而不是改善水质，那么对植被、栖息地品质和鸟类物种的监测就比对质量和水文状况的监测更为重要。由

making. For example, one type of wetland or restoration area of a site may be more attractive to birds, or have better water treatment capabilities. Understanding the factors which make one area functionally superior to another will allow the manager to integrate those components into future enhancement or wetland "retrofits". Monitoring will also allow managers to routinely observe and repair common maintenance concerns such as non-native vegetation, broken pipes and structures, and the presence of burrowing animals, to name a few.

Additionally, a monitoring plan is designed to track short and long term changes in hydrology, water quality, biology, and soils at a restoration site. The plan should identify appropriate parameters to be measured, monitoring methods, and monitoring intervals. Different sites will require monitoring of different parameters at different intervals. A site which receives industrial wastewaters or waters polluted in other ways may require heavy monitoring of its biological parameters to safeguard against toxicity and ensure the health of its organisms. This was the case for the treatment wetland designed by EDAW for the treatment of wastewater from the Shanghai Chemical Industrial Park.

In addition to other parameters mentioned here, measurements of whole effluent toxicity (WET), soil toxicity, bioaccumulation, and water quality measurements of over 60 organic and inorganic chemicals were included in the monitoring plan in order to adequately protect the system from potential toxicity. Generally, at natural wetland restoration sites which are not designed for water treatment, fewer parameters are required in a monitoring plan. Nonetheless, planners should closely assess specific site risks and uncertainties before developing the monitoring plan.

Thorough documentation of the built conditions of a site will set a strong baseline for the monitoring program. Qualitative and quantitative methods of data collection may be appropriate depending on the manager's accuracy needs and management budget. Accurate determinations of target criteria cannot be created using qualitative methods, though they do enable general insights into the type and manner of changes occurring in the wetland.

Typically, qualitative methods may include:
- Aerial photography to identify hydrology, channelization, and ground cover;

- Ground-level photography to observe visual transformations in plant height, species changes,

于湿地的物理、化学和生物作用会随季节不同而有很大变化，因此，在不同季节采集资料有助于大多数参数的成功测量。比如，应在成长季节监测植被——在成长季节的早期和末期进行监测时，计算全部植物数量就容易得多；应在繁殖、筑巢或迁徙季节监测野生动物。在汛期和旱期监测水文状况，就能得到有关整个水文系统的全面信息。不管监测计划是怎样制定的，一定要保持其稳定性，这样才能建立一个切实可用的资料库（见图3-50）。

water levels and seasonal changes; and

- General observations of water quality and clarity, evidence of human use, bird species, vegetative conditions such as yellowing and other stress indications, presence of erosion, checking of facilities and structures such as pipes and pumps that may need repair.

These methods can be used in conjunction with quantitative measures to obtain optimal information for the adaptive management plan. Quantitative methods are those that generate real values against measured parameters. These methods, while not essential, are still beneficial as statistically significant research methods. The methodologies used to collect quantitative data are varied, ranging from simple on-site measurements, to analysis in laboratories and long term studies such as those required to assess bioaccumulation in wildlife and wildlife survivorship. It is important too that quantitative monitoring is carried out by experts in particular fields such as hydrology, biology and toxicology.

University departments can often be employed to undertake monitoring and research programs due to their keen interest in scientific research and the abundant availability of young researchers. While quantitative methods can be expensive and time consuming, they provide the most accurate information regarding site changes.

New technologies in automated data collection are also available for projects requiring real time monitoring, low procedural and data logging efforts which are remote and therefore difficult to monitor regularly. Typical parameters to be measured at the initiation and at intervals during monitoring are included in Table 3-10 (previous page).

The monitoring frequency for each parameter is a factor that has several variables, including project goals and success criteria, natural variability, and rate of change. If a project intends to increase a wildlife habitat, but is unconcerned about water quality improvement, than the monitoring of vegetation, habitat quality and bird species may be of far greater importance than water quality and hydrology. Obtaining information at different seasons is helpful for most parameters as wetlands' physical, chemical and biological processes tend to change greatly with the seasons.

Vegetation should be monitored during the growing season (monitoring in both the early and late growing season will make it easier to identify all plants), and wildlife should be monitored during breeding, nesting, and/or migration seasons. Monitoring hydrology during flood and drought periods will give the manager a better picture of the overall hydrologic regime. Whatever the monitoring plan, consistency is paramount for the establishment of a useable database.

▼ 图 3 - 50: 对美国加州米拉码季节性湿地的监测 (摄影 : Dixi Carrillo)
Figure 3-50: Monitoring vernal pool wetlands in Miramar, California, U.S.A. (Photograph by Dixi Carrillo)

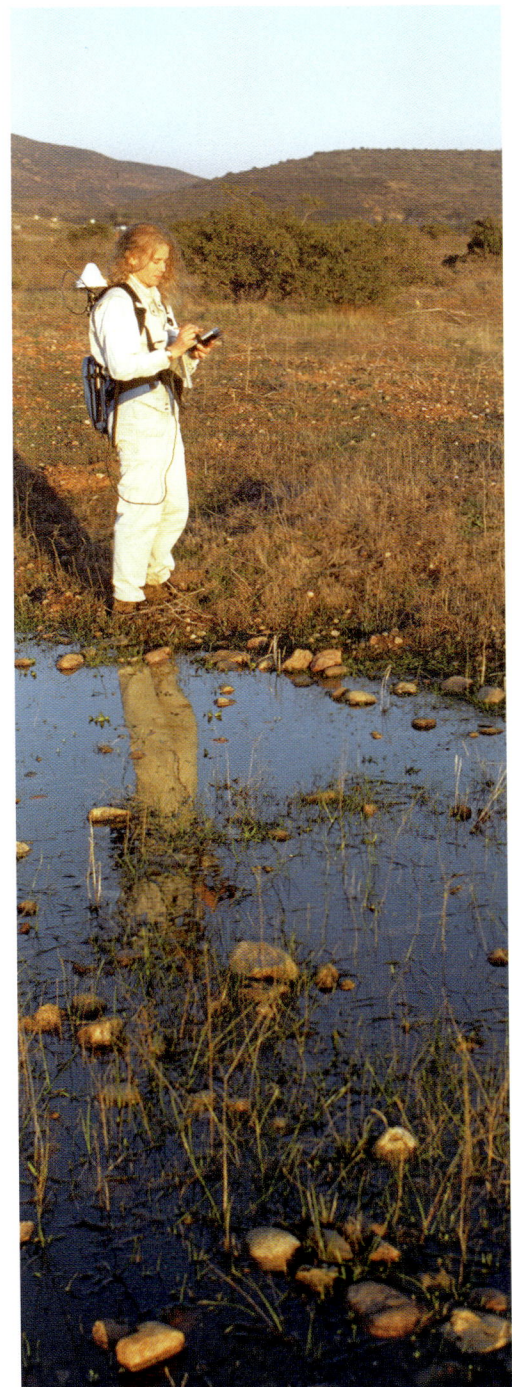

第 四 章
Chapter 4

案例分析
Case Studies

4.1 淡水湿地恢复

4.1.1 建立区域性的湿地恢复框架——佛罗里达大沼泽,美国佛罗里达州

背景

大沼泽(Everglades)位于佛罗里达州南部（见图4-1），与路易斯安那州的海湾沼泽(Bayou)并列为美国最有名的两大湿地。大沼泽一望无际，是很多动物的家园，栖息着广为人知的美国短吻鳄(Alligator mississippiensis)和佛罗里达黑豹(Felis concolor coryi)（见图4-2）等珍稀或濒危物种。大沼泽以北的欧基求碧湖(Lake Okeechobee)是主要水源。欧基求碧湖湖面辽阔，但是湖水不深。每年雨季时，水位上涨，漫出湖堤后在南端缓缓流过大沼泽，最后流向佛罗里达湾。到旱季，留在大沼泽的洪水水位回落，最浅时几乎与地面齐平。因此，大沼泽的泥土常年保持湿润，加上水位涨落不定，形成了大沼泽湿地特征：以锯叶草(Cladium jamaicense)（见图4-3）为典型物种的泥炭沼泽。

在过去的125年里，大沼泽受到了严重的人为破坏。19世纪后期，人们将水从长满锯叶草的大沼泽排出，部分湿地被改造为农田或用于城市开发。但是欧基求碧湖的湖水每年依旧从南岸溢出，因此建在该区域内的农田和开发项目很容易发生洪涝灾害。20世纪20年代末该

▼ 图4-1: 美国佛罗里达州大沼泽位置图(绘制: 易道)
Figure 4-1: The Florida Everglades Location Plan (Graphic: EDAW)

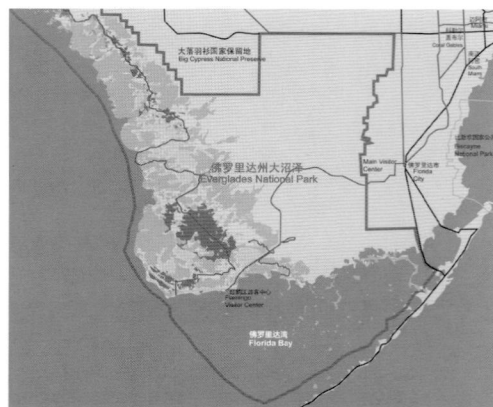

4.1 Freshwater Wetland Restoration

4.1.1 Establishing a Regional Restoration Framework: The Florida Everglades, Florida, U.S.A

Background

Together with the Louisiana Bayou, the Florida Everglades are America's best known wetlands (Figure 4-1). South Florida's "sea of grass" is home to many famous animals such as the American alligator (Alligator mississippiensis, Figure 4-2), and to critically endangered species including the Florida panther (Felis concolor coryi). The wetland habitats that provide a home for these animals are maintained by flood waters from Lake Okeechobee, a large, shallow waterbody lying to the north of the Everglades. During the annual wet season, flood waters spill over the southern margins of the Lake, slowly moving south through the Everglades towards Florida Bay. The floodwaters recede during the dry season, although water levels remain close to the land surface. Water-logged soil conditions and variable water levels have encouraged the formation of peat land soils and marshland dominated by saw-grass (Cladium jamaicense), a characteristic habitat type of the Everglades (Figure 4-3).

The Everglades have come under intense pressure from human activities over the past 125 years. Drainage of the saw-grass marshlands to make way for agriculture and urban development began in the late nineteenth century. Because these new developments were constructed in areas inundated by the annual overflow from Lake Okeechobee, they were highly susceptible to flooding. After a series of severe floods and hurricanes in the late 1920s and again in the late 1940s, Florida citizens and officials petitioned the US Congress to address the flooding problem. Construction of flood control measures was approved in 1948, and over the next few years, a system of levees and canals was built along the southern shore of Lake Okeechobee to divert flood waters from the lake, reducing flood water inputs to the Everglades by over 40%. The diverted waters were used to irrigate farmland and provide drinking water for new developments, but substantial quantities (over six billion liters a day) of freshwater were effectively wasted, discharged directly through canals into the Gulf of Mexico to the west and the Atlantic Ocean to the east (Figure4-5).

Adverse environmental impacts to the Everglades resulting from drainage, development and flood protection measures have been devastating. Approximately half of the original 1.17 million ha of wetland habitats in the Everglades were lost between the 1880s and the 1960s.

◀ 图 4 - 2: 美国短吻鳄
(资料来源: http://images.fws.gov/ 美国渔业
与野生动物保护署服务中心数位图库)
Figure 4-2: American alligator
(Source: http://images.fws.gov/ The U.S. Fish &
Wildlife Service Digital Library System)

◀ 图 4 - 3: 锯叶草的栖息地
(资料来源: http://images.fws.gov/ 美国渔业
与野生动物保护署服务中心数位图库)
Figure 4-3: Saw-grass marshland habitat
(Source: http://images.fws.gov/ The U.S. Fish &
Wildlife Service Digital Library System)

地区发生了一系列的严重洪灾，并受到飓风袭击；40年代末悲剧重演。于是当地居民与政府向美国国会提出请求，要求解决洪水问题。1948年，防洪工程获得批准。随后几年里，沿欧基求碧湖南岸建起了大堤和沟渠，湖中的洪水通过这些沟渠分流，流入大沼泽的水量因此减少了40%以上。分流的湖水用于农田灌溉，并为新开发区的居民提供饮用水。同时每天都有大量的淡水（超过60亿L）无法有效利用，它们或者通过建筑的沟渠向西流入墨西哥湾，或者向东流入大西洋（见图4-4）。

大沼泽上的排水、开发和防洪工程对湿地的环境造成了毁灭性的影响。19世纪80年代到20世纪60年代期间，原117万hm²的湿地生态环境几乎丧失了一半。由于外部欧基求碧湖的洪水发生的频率和时间变化不定，加上大沼泽本身地势下沉、盐水倒灌、有毒动植物入侵、水体污染，剩余的湿地生态环境也逐渐退化，每况愈下。这种退化从湿地的禽鸟种类长期减少可见一斑：根据历史资料统计资料，螺鸢(Rostrhamus sociabilis plumbeus)的数量已经减少了50%；木鹳(Mycteria Americana)的数量降低得更多，从1960年的2500对骤减到20世纪80年代中期的250对（见图4-5）。

为保护和恢复大沼泽剩余的湿地，联邦和州政府部门以及有关方面于20世纪90年代制定出一个大规模的管理计划——《恢复大沼泽全面计划》(CERP, Comprehensive Everglades Restoration Plan)，于2000年获得联邦政府批准。该计划继续为大沼泽内已有设施提供洪灾保护，利用大沼泽提供农业和城市生活用水，同时该计划还包括60多个湿地工程。项目将持续几十年时间，预算成本为80亿美元。

项目目的

CERP计划工程巨大，涉及到南佛罗里达地区450万hm²内很多水管理的问题。通过实施CERP计划，实现以下主要目标：

- 增加湿地及其他自然生态环境的总面积；

- 改善栖息地及其功能和质量；

- 提高生态系统多样性和乡土动植物种类；

The surviving wetlands are being degraded by changes in the degree, timing and duration of flooding from Lake Okeechobee, along with subsidence, saltwater intrusion, the invasion of exotic plants and animals and finally, pollution. The impacts to the ecology of the Everglades have been severe, as indicated by the long-term decline in wetland dependent bird species. For example, snail kite (Rostrhamus sociabilis plumbeus) populations are thought to have declined by 50% based on historical records, and the breeding population of wood storks (Mycteria Americana, Figure 4-4) has declined even more dramatically, from 2,500 pairs in 1960 to 250 pairs in the mid-1980s.

In the 1990s, state and federal agencies along with other concerned parties developed a large scale management plan for the protection and restoration of the remaining wetlands of the Everglades. The plan, known as the Comprehensive Everglades Restoration Plan (CERP), would also provide continued flood protection and safeguard water supplies for agriculture and urban consumption. The CERP was approved by the federal government in 2000, and includes more than 60 wetland projects with construction that will span several decades at an estimated cost of US$ 8 billion.

Goals and Objectives

CERP is a huge undertaking, addressing numerous water management issues across 4.5 million ha of south Florida. Key goals that will be achieved through the plan's implementation include:

- Increasing the total area of wetlands and other natural habitats;

- Improving habitat and functional quality;

- Improving diversity of eco-systems and abundance of native plant and animal species;

- Increasing potable and agricultural water supplies;

- Reducing flooding;

- Providing recreational opportunities and improve navigation; and

- Protecting cultural and archaeological resources and values.

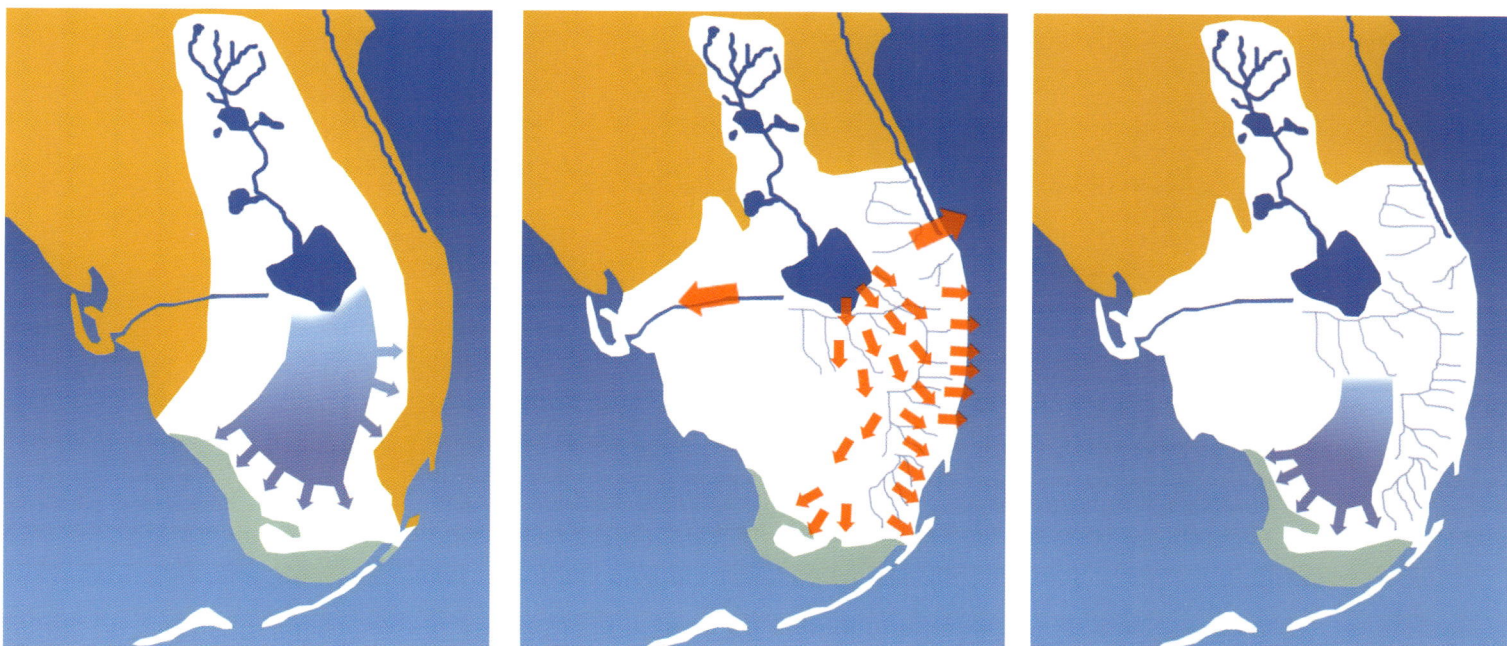

◄ 图 4-5: 木鹳
(资料来源: http://images.fws.gov/ 美国渔和
野生动物保护署服务中心数位图库)
Figure 4-5: Wood stork
(Source: http://images.fws.gov/ The U.S. Fish &
Wildlife Service Digital Library System)

- 增加饮用水和农用水供应；

- 减少洪水水患；

- 提供休闲娱乐功能，改善河道；

- 保护文化资源和遗产资源及其价值。

恢 复 策 略

研发区域性水文及生态系统模型

在CERP区域性框架中，有许多项目是为了恢复大沼泽生态功能和确保生活品质的。这些项目耗资高达数十亿美元，因此准确地评定其效益和成本极其重要。人们对改造南佛罗里达地区的水资源体系提出了各种各样的建议，要评估这些方案对整个水资源系统的影响，采用类比模型成了唯一可行的办法。整个规划过程中广泛使用了两个模型：南佛罗里达水管理模型（SFWMM）和整个食物链系统模型（ATLSS）（见图4-6）。

SFWMM模型是电脑模型，模拟出北起欧基求碧湖，南至佛罗里达湾，面积近200万hm²的整个南佛罗里达地区的水文状况。该模型以南佛罗里达地区30年的气候资料为依据，并运用数百个观测站点的水位和流量资料进行了校准和验证。SFWMM能模拟南佛罗里达地区所有的水文过程，包括降雨、蒸发或散发、下渗、地表和地下径流、河川径流、管道—地下水渗漏、堤防渗水以及地下水抽取。在制定CERP计划的过程中，该模型被广泛使用，分析各种旨在提高南佛罗里达地区水管理的策略所能达到的功效，并选出成本有效性最高的策略。

ATLSS模型主要用于比较各种水文情景对未来系统中生物组成的影响。ATLSS由一系列综合的过程组成，其中包括：

- 低级食物链（包括底栖无脊椎动物、周丛生物和浮游动物）模型；

- 五种鱼类及大型底栖动物功能群的结构种群模型；

- 基于大型消费者个体（木鹳、大蓝鹭、白鹭、美国短吻鳄、白尾鹿、佛罗里达黑豹）的模型。

Restoration Approaches

Development of Regional Hydrology and Ecosystem Models.

CERP provides the regional framework, within which are numerous projects necessary to restore the ecological functioning of the Everglades and to sustain the quality of life. These projects will cost billions of dollars, so an accurate determination of their benefits and costs is of paramount importance. Simulation models have become the only feasible means of assessing system-wide impacts of the various proposed modifications to the water resources system in south Florida. Two models in particular were used extensively in the planning process: the South Florida Water Management Model (SFWMM) and the Across Trophic Level System Simulation (ATLSS) (see Figure 4-6).

The SFWMM is a computer model that simulates the hydrology of virtually all of south Florida: an area of almost two million ha, stretching from Lake Okeechobee in the north to Florida Bay in the south. The model is based on 30 years of climatic data for the south Florida regions, and has been calibrated and verified using water level and discharge measurements at hundreds of locations. SFWMM is capable of simulating all major hydrological components of south Florida including rainfall, evapotranspiration, infiltration, overland and groundwater flow, canal flow, canal-groundwater seepage, levee seepage and groundwater pumping. It has been used extensively in the development of CERP to analyze the efficacy of different strategies to improve water management in south Florida and also to identify the most cost-effective strategies.

The key objective of the ATLSS model is to compare the future effects of alternative hydrologic scenarios on the biotic components of the systems. ATLSS is comprised of a set of integrated programs that include:

- Process models for lower trophic levels (including benthic insects, periphyton and zooplankton)

- Structured population models for five functional groups of fish and macroinvertebrates, and

- Individual-based models for large consumers (wood storks, great blue herons, white ibis, American alligators, white-tailed deer, and Florida panther).

These models are integrated with GIS-based maps, and the whole system is then coupled to a hydrology model that is used to assess the effects of alternative restoration proposals on ecological structure and function.

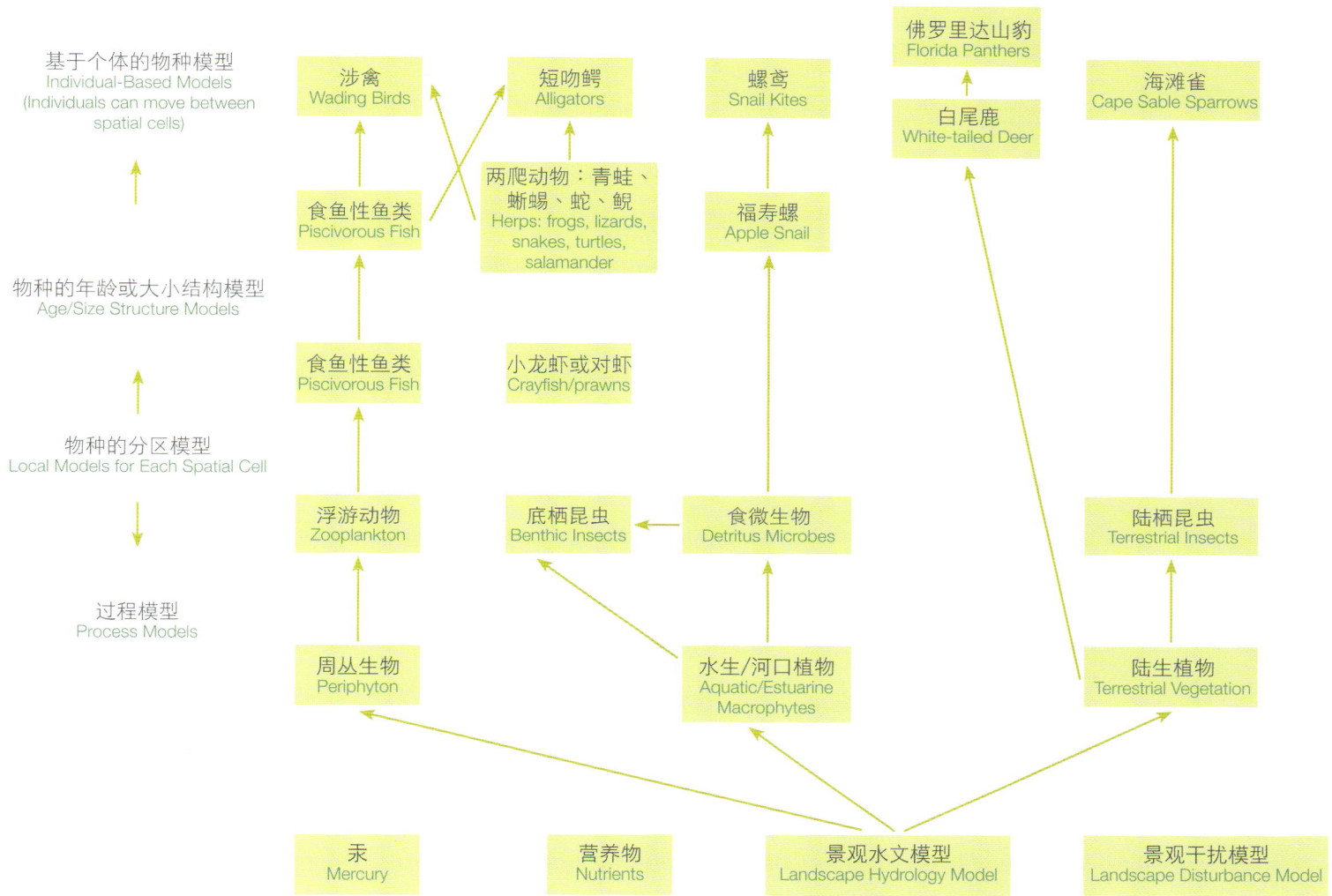

基于个体的物种模型
Individual-Based Models
(Individuals can move between spatial cells)

物种的年龄或大小结构模型
Age/Size Structure Models

物种的分区模型
Local Models for Each Spatial Cell

过程模型
Process Models

涉禽
Wading Birds

短吻鳄
Alligators

螺鸢
Snail Kites

佛罗里达山豹
Florida Panthers

白尾鹿
White-tailed Deer

海滩雀
Cape Sable Sparrows

食鱼性鱼类
Piscivorous Fish

两爬动物：青蛙、蜥蜴、蛇、鲵
Herps: frogs, lizards, snakes, turtles, salamander

福寿螺
Apple Snail

食鱼性鱼类
Piscivorous Fish

小龙虾或对虾
Crayfish/prawns

浮游动物
Zooplankton

底栖昆虫
Benthic Insects

食微生物
Detritus Microbes

陆栖昆虫
Terrestrial Insects

周丛生物
Periphyton

水生/河口植物
Aquatic/Estuarine Macrophytes

陆生植物
Terrestrial Vegetation

汞
Mercury

营养物
Nutrients

景观水文模型
Landscape Hydrology Model

景观干扰模型
Landscape Disturbance Model

上述模型与基于GIS(地理信息系统)的地图相结合后形成一个系统，该系统与一个水文模型耦合，用于评估各种湿地恢复方案对生态结构和功能的影响。

恢复项目的组成

CERP计划中各种不同的工程基本可分为三大类：洪水管理、水质改进和大沼泽恢复/改善工程。

洪水管理：防洪与蓄洪

南佛罗里达地区现有的防洪工程在为城市地区提供饮用水、为农业提供灌溉用水的同时，将大量来自欧基求碧湖的洪水直接引流入海，本可用来养护大沼泽湿地生态环境的淡水资源被白白浪费。因此，CERP计划建造多处蓄洪设施，从而便于高效保存并管理这些水资源。

最重要的防洪工程包括在欧基求碧湖周围建造一些地上蓄水库（见图4-7），总面积约73000hm²，库容达1.85×10¹²L。最大的工程是将建于卡鲁沙哈奇流域的大型蓄水库，它由一个或多个地上蓄水库构成，可以截获并留存暴雨径流以及来自欧基求碧湖的洪水。这不仅降低了洪灾风险，还可以调节入海口处淡水的注入时间，改善注入水量和水质，从而恢复卡鲁沙哈奇河流本身的生态功能。目前，河流下游非自然的淡水洪峰降低了河口的盐度，而在旱季灌溉需求高时，只有少量淡水或根本没有淡水注入河流，致使河口的盐度升高。河口盐度的巨大变化对环境造成不利影响，使海草、牡蛎及其他极其敏感的河口生物渐渐绝种。卡鲁沙哈奇流域水库将于2007年动工，它的建成将使淡水流入大海的时间更符合自然规律，水量更加稳定。

CERP洪水管理计划中第二项重要的工程是在欧基求碧湖周围建设地下水库（ASR）网。ASR将已经处理并达到饮用水标准的淡水压入到地下300m处，蓄水层的咸水受到压力时被排挤出去，从而淡水能够在此储存，形成了一个地下淡水库，在旱季时能够调用。与其他蓄水技术（如地面水库）相比，ASR的优点是因蒸发而损失的水量非常之小。ASR系统具有很多优点：可以多年蓄水，在干旱严重时，不会因蒸发而限制水库的长期蓄水能力。此外，水井可以建在最需要

Restoration Components

The various projects that will be constructed as part of the finalized CERP fall under three main categories: flood water management, water quality improvement, and restoration/enhancement of the Everglades.

Flood Protection and Storage

Existing flood protection works in south Florida divert large amounts of floodwaters from Lake Okeechobee directly into the sea, a process that wastes precious freshwater which could otherwise be used to maintain wetland habitats of the Everglades, as well supply potable water for urban areas and irrigation for agriculture. The CERP would provision for the construction of numerous facilities to store excess flood waters that will allow for more efficient retention and management.

Key flood protection works will include the construction of numerous above-ground storage reservoirs around Lake Okeechobee covering a total area of approximately 73,000 ha, with the capacity to store 1,850 billion liters of water. A typical planned storage reservoir project recently approved under CERP would involve the construction of one or more above ground reservoirs in the Caloosahatchee River Basin. Stormwater runoff, along with floodwaters from Lake Okeechobee, will be captured and stored in these reservoirs. In addition to alleviating flood risks, the reservoirs will restore ecological functioning of the Caloosahatchee River itself by improving the timing, quantity, and quality of freshwater flows to the River estuary. Currently, unnatural, fresh water flood surges downriver reduce estuarine salinity levels. Alternately, during drought periods when irrigation demands are high, little or no fresh water is released to the river, causing estuarine salinity levels to rise. These rapid changes to the estuary's salinity levels have had an adverse impact on the environment, causing die-offs of sea grasses, oysters and other sensitive estuarine species. Construction of the Caloosahatchee River Basin Reservoirs will begin in 2007, and will ensure a more natural, consistent flow of fresh water to the estuary.

A second major component of the CERP flood management program will be the construction of a network of Aquifer Storage Reservoirs (ASR's) around Okeechobee Lake (Figure 4-7). ASR involves pumping freshwater that has been treated to drinking water standards approximately 300 meters underground where it is stored. The pumped freshwater displaces brackish water in the aquifer, resulting in an underground reservoir of freshwater that can be retrieved during periods of drought. The key advantage of ASR when compared to other storage technologies such as above ground storage reservoirs is that very little water is lost to evaporation. ASR systems allow for multi-year storage,

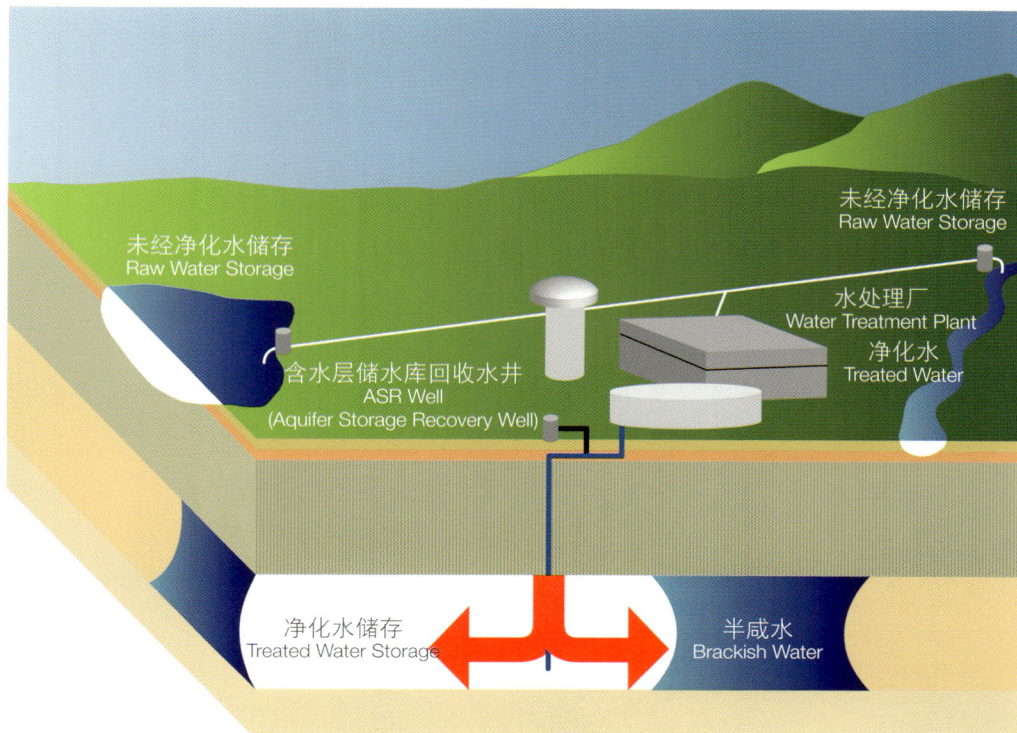

未经净化水储存
Raw Water Storage

未经净化水储存
Raw Water Storage

含水层储水库回收水井
ASR Well
(Aquifer Storage Recovery Well)

水处理厂
Water Treatment Plant

净化水
Treated Water

净化水储存
Treated Water Storage

半咸水
Brackish Water

◄ 图 4-7: 欧 基 求 碧 湖 周 围 拟 建 的 储 水 系 统
(绘 制 : 易 道) (资 料 来 源 : USACE)
Figure 4-7: Aquifer Storage Reservoir
(Redrawn from USACE, 2002)

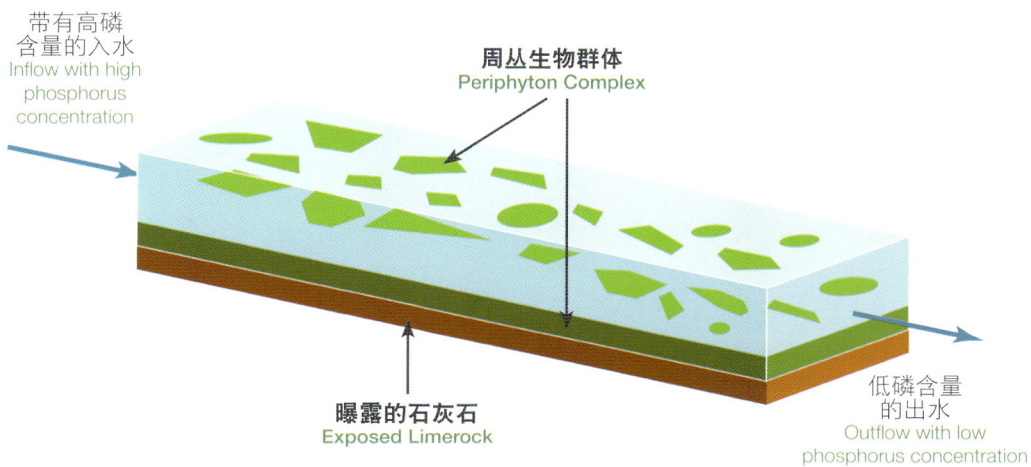

带有高磷
含量的入水
Inflow with high
phosphorus
concentration

周丛生物群体
Periphyton Complex

曝露的石灰石
Exposed Limerock

低磷含量
的出水
Outflow with low
phosphorus concentration

◄ 图 4 - 8 : 暴 雨 径 流 处 理 系 统 (PSTA)
(绘 制 : 易 道) (资 料 来 源 : SFWMD, 2004)
Figure 4-8: Periphyton-Based Stormwater Treatment
Area (PSTA) (Redrawn from SFWMD, 2004)

用水的地方，而且水井用地很小，可以节约征用大块土地的费用。

为保证充足供应饮用水，为农业提供灌溉并避免咸水倒灌进沿岸水井，佛罗里达州已经在小范围内应用ASR技术。但是该技术还存在一些不确定性，在CERP这样的大项目下大规模应用更是如此。特别令人关注的是大型地下水库对土壤含水层的潜在影响。此外，ASR技术的运作与维护成本可能高于其他技术。考虑到这些因素，目前进行了一项ASR区域性研究项目和三项ASR试点工程，目的是收集资料，降低ASR运用过程中的不确定性。

在ASR地域性研究项目中使用了一个地域性水文模型，研究大规模应用ASR对含水层的水位、水质、地表水及动植物的潜在影响。ASR区域性研究包括以下目标：

- 确定用于储存的水源、水质和水量；

- 研究当在咸水层中储存地表水时可能发生的地质化学反应；

- 描述区域性佛罗里达上层蓄水层的水文地球特征，评估CERP计划ASR系统建设中因压力可能产生的变化和水流走向模式；

- 理解调用量、补给量和蓄水时长之间的关系；

- 建立并校验佛罗里达区域蓄水层系统的理论框架并开发出模型；

- 评估释放蓄水层蓄水对大沼泽承受水体、植物群落、鱼类和野生动物的生态影响，包括调查汞在生物中蓄积的可能性。

ASR区域性研究项目和三项试点工程将分别于2009年和2010年完成。按照CERP计划，届时将有300多口水井，每天将60亿L水泵入佛罗里达南部地下蓄水层。

改进水质
南佛罗里达地区的水资源受到农业及城市暴雨径流的污染，水质下降。根据CERP计划在污染源（即相关城市、农村

mitigating evaporation during severe droughts that limits the ability of reservoirs to provide long-term storage. An additional advantage is that wells can be located where most needed and because the wells require little land, the costs of large land acquisitions are avoided.

Florida already uses ASR technology on a small-scale to maintain adequate supplies of drinking water and to provide irrigation for agriculture and protect wells located near the coast from saltwater intrusion. Nevertheless, there are a few uncertainties associated with the technology, particularly when implemented on the large scale required under CERP. Of particular concern are the potential impacts of large-scale underground storage on the aquifer. Operation and maintenance costs may also be higher for ASR than for other means of water storage. To address these issues, an ASR Regional Study and three additional ASR Pilot Projects are underway in order to collect data that will help reduce uncertainties over the use of ASR.

The ASR Regional Study involves the use of a regional hydrological model to address the potential effects of large scale ASR on water levels, on water quality within the aquifer, on surface waters, and on plants and animals. The ASR Regional Study Goals include:

- Identifying sources, quantity and quality of water to be stored;

- Studying potential geo-chemical reactions that may occur when storing surface water in a brackish aquifer;

- Describing hydro-geologic characteristics of the regional Upper Floridan Aquifer, and assessing potential pressure induced changes and water-flow patterns due to the proposed CERP ASR system;

- Understanding the relationship between recovery rates, recharge volumes, and length of time that water is stored;

- Developing and verifying a conceptual framework for modeling the regional Floridan Aquifer System;

- Assessing ecological impacts of releasing aquifer-stored water on Everglades receiving waters, plant communities, fish and wildlife, including investigating the potential for mercury bioaccumulation.

The ASR Regional Study and three ASR Pilot Studies will be completed in 2009 and 2010. CERP eventually calls

▲ 图 4-9: 夕阳中的佛罗里达大沼泽（摄影：易道）
Figure 4-9: The Florida Everglades at sunset (photograph by EDAW)

地区）和敏感的湿地栖息地之间将建起一个缓冲区，用来存放被污染的径流。同时，将兴建约 15000hm² 的暴雨径流处理湿地（见图 4-8），受污染的径流在这里经过处理后再流入湿地。这样水质可望得到有效的改善。

大沼泽的生产力和组成主要取决于磷可利用性。佛罗里达国际大学曾经进行一项历时五年的详细研究，研究表明，磷的浓度只要比周围环境浓度高出 5μg/L，就会损害周围湿地的功能。最近，佛罗里达州环境管理委员会规定了一个长期标准，流入大沼泽的径流磷浓度不能高于 10μg/L。因此，暴雨水处理项目的主要目标就是降低从农业和城市地区流入大沼泽的磷。为此，科学家们提出了多种处理暴雨径流的方案，包括化学添加、固体分离、微滤、沉水植物（SAV）和周丛生物的暴雨径流处理区（PSTA）。在这些方案中，沉水植物和 PSTA 无论从作用、成本，还是可行性来看，都是最有前途的。在 PSTA 中起核心作用的周丛生物是由藻类、藻青菌、异养菌以及碎食生物组成的复合体，这种复合体附着在几乎各种水生生态系统的水底。

for the construction of over 300 wells that will be used to pump six billion liters of water a day down to the aquifer underlying south Florida.

Water Quality Improvement

CERP also addresses water quality issues in south Florida, where water resources are affected by pollution from agriculture and urban storm-water run-off. Polluted run-off will be routed through buffer areas between pollution sources (urban and agricultural areas) and sensitive wetland habitats. Approximately 15,000 ha of storm water treatment wetlands will be constructed to remove pollutants from the runoff before it is discharged into the wetlands.

The productivity and composition of the Florida Everglades wetlands is largely controlled by phosphorus availability. A detailed five-year study conducted by the Florida International University found that phosphorous levels as low as five parts per billion (ppb) above ambient concentrations could cause detrimental impacts to wetland functioning. The Florida Environmental Regulatory Commission recently set the long-term standard for phosphorus entering the Everglades at 10 ppb. Thus, a key objective of the stormwater treatment programme is

在CERP计划中使用PSTA技术处理雨水的建议于1996年首次提出，试点处理工程于2002年竣工。周丛生物群落建成后（这一过程一般需要六至八周）的研究表明，这种处理方法可以在7天内将30m范围内的磷的浓度从80μg/L降至10μg/L以下，比用其他技术低20%~30%。

此外，还采取了其他措施改善水质，其中包括疏浚欧基求碧湖底富含营养的沉积物；建造高级污水处理厂净化现有设施处理过的污水，使之可以沿比斯坎湾安全地排入湿地。

大沼泽恢复或改善工程：生态环境和生物效益

洪水管理和改进的水质将有助于保护和恢复大沼泽。根据CERP计划，排入大沼泽关键区域的水量将比现在增长约四分之一，而且大沼泽内近400km的运河和人工堤岸将被夷平，届时，水将流淌在这片平坦广阔的湿地上，恢复成湿地生态环境的自然水流。

全面恢复大沼泽计划的经济分析

由于CERP规模大、跨度广，且所在地区的淡水、农产品和旅游资源丰富，因此对佛罗里达州具有极高的经济价值。来自公共机构、独立的科学家以及当地高等学府的学者创建了多种经济模式来研究大沼泽对区域经济的影响，"区域综合经济模式（REMI）"就是其中一例。一些由公共机构开发的模式显示，1995年，大沼泽生态服务的功能价值为316.6亿美元，加上水域产生的经济价值，生态服务的价值总共达到587亿，占佛罗里达州正常经济增加值的34.8%，约三分之一。至于CERP的影响本身，迈阿密大学的Richard Weiskoff带领团队运用了广为公众认可的IMPLAN模型进行了计算，结果表明，CERP计划的影响预计占该地区全年总产量、就业数量及收入的0.31%。

南佛罗里达每天都要接待大量乘坐汽车或飞机来此观光的游客（见图4-9）。2001年，游客数量达到近2500万人，生态旅游收入超过50亿美元。大多数游客来到大沼泽，游览开阔的淡水空间，并在其中休闲、娱乐。从长远预测来看，游客数量将不断增长。在这种形势下，如果旅游规划出现偏差，将造成大沼泽的生态和环境

to reduce phosphorus inputs from agricultural and urban areas into the Everglades. Various options have been considered to treat stormwaters, including chemical additions and solids separation, microfiltration, submerged aquatic vegetation (SAV) and periphyton-based stormwater treatment areas (PSTA). Of these options, SAV and PSTA (Figure 4-8) are considered the most promising in terms of performance, cost and feasibility. The periphyton that plays a central role in PSTA is a complex matrix of algae, cyanobacteria, heterotrophic microbes, and detritus that is attached to submerged substrata in almost all aquatic ecosystems. The use of PSTA technology to treat stormwaters under CERP was first proposed in 1996, and a pilot treatment project was completed by 2002. After the establishment of periphyton communities (a process that took six to eight weeks), studies showed the treatment could reduce phosphorus levels from 80 ppb to less than 10 ppb within a flow distance of 30m and a retention time of seven days: results with 20% to 30% lower levels of phosphorous than those achieved by other technologies. This promising small-scale test has led to approval of funding for larger scale field tests. Additional water quality improvement measures will include dredging of nutrient enriched sediments from Lake Okeechobee, and the construction of advanced wastewater treatment plants to polish effluent from existing treatment facilities, allowing for the safe discharge of waste-waters into wetlands along Biscayne Bay.

Habitat and Biological Benefits

Improved water quality and management of flood waters will contribute to the preservation and restoration of the Everglades. Under the CERP, water flow to key areas of the Everglades will be increased by about a quarter compared to current conditions, and almost 400km of canals and artificial levees within the Everglades will be removed to reestablish natural sheet-flows of water through the wetland habitats.

Economic Aspects of CERP

Given its size and breadth, the CERP has significant economic value to the state as a source for freshwater, agricultural produce and tourism. Numerous public and independent scientists (as well as academics from local universities) have created several economic models to study the impact of the Everglades on the regional economy – such as the Regional Economic Model, Inc. (REMI). Certain models developed by public agencies have estimated that the economic contribution of the ecosystem services from Everglades land was US$ 31.66 billion in 1995, and that by adding the value created by the

退化。因此，CERP计划要围绕保护大沼泽环境进行旅游休闲的规划，即在通过生态旅游创收的同时，保持大沼泽的环境质量。

CERP计划在经济方面还考虑了农业及农业产值在区域经济中的作用。1997年，南佛罗里达地区的农业产值为29亿，占该州农业总产值的48%、全国农业总产值的1.5%。因此，CERP计划将继续保持农业的比重，使该地区农业得到进一步的可持续的发展。

结论

大沼泽项目率先将新技术应用于区域性水系统管理与湿地生态系统管理的规划中，并应用耦合了食物链生态系统模型的水文模型系统，评估个体工程对水源、水质和生态的影响。

随着CERP计划的实施，大沼泽的发展已经到了一个关键点。多种迹象表明，南佛罗里达地区剩下的湿地生态系统正处于崩溃的边缘：据估计，从20世纪30年代到90年代，"大沼泽"地区涉禽的栖息数量骤减了90%以上。CERP计划的优势在于以科学为本，并得到多数利益相关方的支持。同时，该计划的管理模式适应性强，大大提高了成功的几率。在美国政府及其他利益相关方的持续支持下，CERP计划可望能够帮助大沼泽湿地恢复，促进佛罗里达州长期、可持续发展。

water area of the Everglades, the total value of ecosystem services reaches a staggering US$ 58.7 billion. This amounts to 34.8% or about a third of the total value added of the normal Florida economy. For the CERP impact itself, a team from the University of Miami, led by Richard Weiskoff, used a widely-respected model (IMPLAN) to show that the forecasted impacts of CERP amounted to roughly 0.31% of the total annual output, jobs, and earnings of the region.

A large tourism population arrives in south Florida every day by both automobile and air. In 2001, this amounted to nearly 25 million tourists, with tourism revenues of over US$ 5 billion. The majority of these visitors use the Everglades for freshwater and recreational open space. However, misguided tourism planning may lead to the degradation of the Everglades, particularly as the number of tourists is forecasted to rise in the long-term. As such, CERP aims to maintain an environmentally-friendly destination in order to achieve a tourism and recreation plan that will continue to generate significant revenues as well as maintain the environmental quality of the Everglades.

Final and important economic considerations of the CERP are the agricultural characteristics and production values for the regional economy. The value of south Florida's agricultural produce in 1997 was US$ 2.9 billion, 48% of the state total and 1.5% of the overall US farm output. CERP has thus emphasized maintaining sustainable agricultural practices such that farm output can continue producing for the region.

Summary

The Florida Everglades case study is an exemplary lesson in using innovative approaches to large-scale planning for regional water systems and wetland ecosystem management. Utilization of system-wide hydrologic modeling, coupled with trophic level ecosystem modeling has allowed individual projects to be assessed based on how they would affect water supplies, water quality, and ecological impacts.

The implementation of CERP has come at a critical juncture in the history of the Everglades, where many indications are that the remaining wetland ecosystems of South Florida are on the brink of collapse: from the 1930s to the 1990s, populations of wading birds nesting in the Everglades plummeted over 90% by some estimates. CERP has the advantage of being based on sound science, and in having the support of the majority of stakeholders. It also benefits from a strong adaptive management program that will maximize the chances of success. With continued support from the US Government and other stakeholders, it is hoped that CERP can achieve its goals in restoring the Everglades and contributing to the long-term, sustainable development of Florida itself.

资料提供
易道公司

Information provided by

EDAW | AECOM

4.1.2 实现多用途的平衡：栖息地、农业、洪水控制 —— 约罗野生动物保护区，美国加利福尼亚州

背景

加利福尼亚州渔猎部（DFG）的维克.法奇奥约罗分洪道野生动物保护区（约罗野生动物保护区）是非常独特的资源，可以为加州人民带来巨大的环境、社会和经济利益。约罗野生动物保护区位于加利福尼亚北部萨克拉门托河以西约罗县的约罗分洪道（见图4-10）。约罗行洪道长约75km，是一条地势较低、建有部分堤坝的防洪走廊。每到洪水季节，大流量的洪水被引到这里，萨克拉门托河的水位得到控制，从而保护了萨克拉门托、西萨克拉门托和大卫斯市及其他当地社区和农场免遭洪水侵袭。约罗野生动物保护区占地6,700hm²，其独特之处在于农业、野生动物栖息地和防洪三个目标在这里同时得到实现，互不冲突。此外，保护区还提供了丰富的公众游览、休闲以及环境教育的机会。

约罗盆地曾是一片面积为33000hm²的湿地，野生动物种类非常繁多：沼泽地上游荡着大量加利福尼亚马鹿，冬天成群结队的迁徙水禽在这里觅食栖息（见图4-11）。约罗盆地坐落于加利福尼亚南部中央大河谷萨克拉门托河冲积平原上，是那里为数不多的几个盆地之一。当时，冬春两季汛期时洪水流入那些盆地，形成湿地。候鸟从北方千里迢迢飞到这里，以湿地上的植物种子和无脊椎动物为食；鱼类在这块季节性淹没的洪泛区上产卵、孵育后代；当地居民也依靠约罗盆地上的丰富资源维持冬春两季的生计：季节性出现的水禽和鱼类成为他们的食物，湿地上生长的柳枝和芦苇等植物原料则可用来编篮子、绳索和用作建筑材料。直到今天，约罗盆地的水文周期还决定着人类和野生动物对它的取用方式。

约罗盆地生态系统已经因为人类活动发生了很大变化。变化首先开始于19世纪中期，当时内华达山区上游开始出现水力采金，这使金矿所在地的河流下游出现大量沉积物；同时出现了农用土地开垦，其负面影响便是大规模兴建洪水控制工程，以保护自家的低地。20世纪初期，在约罗盆地两边邻近萨克拉门托河

▼ 图4-10: 约罗野生动物保护区位置图（绘制：易道）
Figure 4-10: Yolo Wildlife Area Location Map (Graphic: EDAW)

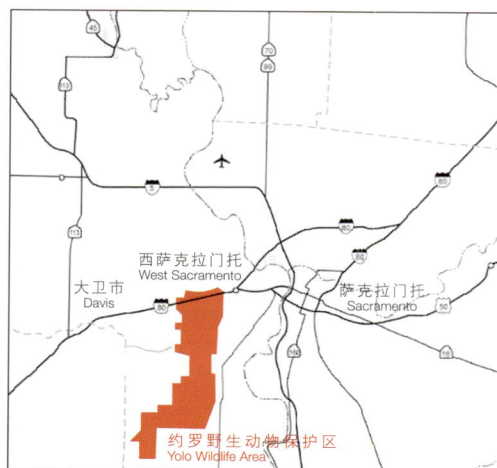

西萨克拉门托
West Sacramento

大卫市
Davis

萨克拉门托
Sacramento

约罗野生动物保护区
Yolo Wildlife Area

4.1.2 Balancing Multiple Uses - Habitat, Agriculture, Flood Control: Yolo Wildlife Area, California, U.S.A

Background

The California Department of Fish and Game's (DFG) Vic Fazio Yolo Bypass Wildlife Area (Yolo Wildlife Area) is a unique resource that provides substantial environmental, social, and economic benefits to the people of the State of California. It is located in the Yolo Bypass in Yolo County, west of the Sacramento River in northern California. The Yolo Bypass is an approximately 75-kilometer, low-lying, partially levied flood protection corridor that conveys seasonal high flows away from the Sacramento River channel. This helps control river levels and protects the cities of Sacramento, West Sacramento, and Davis and other local communities and farms, from flooding.

The 6,700 ha Yolo Wildlife Area is unique in the way agriculture, a wildlife habitat, and flood protection objectives are achieved in a highly compatible manner. The Wildlife Area also provides ample opportunities for public access, recreation, and environmental education.

The Yolo Basin was once a vast, 33,000 ha wetland teeming with wildlife, with herds of tule elk roaming its marshes and dense clouds of migratory waterfowl seeking winter food and shelter. Yolo was one of several basins located within the Sacramento River floodplain in the Great Central Valley of northern California. All of the basins received water during high winter and spring flows as a normal occurrence. Migratory birds came from the far north to feed on seeds and invertebrates present in the wetlands. Several native fish species used the seasonally inundated floodplain as a vital spawning and rearing habitat. The resources found in the Yolo Basin also sustained small communities of Native American populations through the winter and spring months. The seasonal presence of waterfowl and fish provided food, while the wetlands provided plant materials such as willow and reeds for baskets, cordage, and building material. To this day, the seasonal hydrologic cycles of the Yolo Basin determine its use by people and wildlife.

Over time, the Yolo Basin ecosystem has been profoundly altered by human activity. Beginning in the mid-1800s, the adverse effects of hydraulic mining for gold upstream in the Sierra Nevada (which caused a tremendous accumulation of sediment in rivers downstream of the mining) and land reclamation for agriculture led to the construction of large-scale flood control projects to protect private lowlands. In the early 1900s, a system of major levees was constructed along both sides of the Yolo Basin from north to south

▲ 图 4-11：休耕期稻田中的长脚鹬
（提供：该稻田为涉禽栖息地和稻米生产两
用地）（Dave Feliz，加州渔猎局）
Figure 4-11: Stilts on a fallow rice field managed for
shorebirds in rotation with rice production (Photo
courtesy of Dave Feliz, California Department of Fish
and Game)

南北向修建了大型堤坝。建成的长堤和
洪水控制工程就形成了著名的"约罗分
洪道"。

到了近代，分洪道上的大部分土地都被
用于耕牧，而私人所有的地产上，湿地
管理微乎其微，很难为候鸟和本地水禽
提供很好的生存环境。邻近约罗野生动
物保护区的地产上，私人俱乐部猎捕水
禽的历史传统一直延续到今天。事实
上，加利福尼亚州有2／3的湿地仍为私
人所有。

现在的约罗保护区建于2001年，是约罗
盆地基金会（以下称"基金会"）、加州
渔猎部、一些地方、州及联邦防洪或野
生动物保护机构及组织等各方面为恢复
约罗盆地的湿地及栖息地共同努力12年
的成果。约罗盆地基金会属非政府组
织，成立于1990年，代表民众呼吁创建
约罗保护区。基金会的工作人员及志愿
者们付出了多年的努力，终于与相关方
面建立起伙伴关系，正是这种关系使该
事业获得最后的成功。期间，他们始终
保持与各方的对话，共同解决与重要分
洪河道上的湿地恢复相关的复杂问题。
在约罗野生动物保护区的创建过程中，
社会各界人士功不可没，他们通过合作
与创新的伙伴关系恢复了约罗湿地——
太平洋候鸟迁飞路线上的一个关键节
点。保护区的管理工作延续了这合作的
精神，与当地社区持续保持沟通的平
台。

adjacent to the Sacramento River. The construction of
these levees and flood control structures formed what is
now known as the Yolo Bypass.

More recently, the majority of lands within the bypass have
been used for grazing and farming, with limited wetland
management taking place on private land to provide
habitat for migratory and resident waterfowl. The historic
culture of hunting waterfowl on private clubs continues to
this day on properties neighboring the Yolo Wildlife Area.
In fact, two-thirds of all wetlands in California remain in
private ownership.

Establishment of the current Yolo Wildlife Area in 2001
was the result of a 12-year cooperative effort to restore
wetlands and associated habitats in the Yolo Basin that
involved the Yolo Basin Foundation (Foundation), the DFG,
several local, state, and federal flood control and wildlife
agencies, and conservation organizations. The Yolo Basin
Foundation is a nongovernmental organization formed in
1990 representing the grassroots support for creation of
the Yolo Wildlife Area. Foundation staff and volunteers
worked for many years to create partnerships that have
made the effort successful. By doing so, they maintained
the dialogue on solving complex problems associated
with restoring wetlands in an important floodway.

The Yolo Wildlife Area was founded by community
interests that worked to restore a critical link in the Pacific
Flyway through cooperative and innovative partnerships.
The management of the Wildlife Area continues in this
spirit, with frequent interactions with the local community.

目的与目标

建立约罗野生动物保护区的主要目的是恢复约罗分洪道的湿地栖息地（见图4-12），同时保持其农业特征和洪水控制功能。约罗　野生动物保护区的核心使命是在这块土地上同时实现这三种用途（栖息地、农业、洪水控制）。《约罗野生动植物保护区土地使用规划》包括以下明确目标：

- 指导栖息地和物种的恢复和管理，规划公共用途并指导加州渔猎部的其他项目，同时保持行洪道的农业特征和防洪功能；

- 在约罗野生动物保护区的管理工作与区域规划中采用生态系统管理方法；

- 确定并指导约罗保护区内合适的、相容的公共用途；

- 在指导约罗野生动物保护区的管理工作时，要促进与邻近私人地产主之间的合作关系；

- 建立约罗野生动物保护区内场地、野生动植物资源的描述性名录；

- 概述约罗野生动物保护区的运营、维护状况以及为实现管理目标的人力需求，并辅助加州渔猎部制订年度预算。

恢复策略

约罗野生动物保护区的恢复和管理由加州渔猎部负责实施。加州渔猎部的职责是管理加利福尼亚州的各种鱼类、野生动植物资源及它们赖以生存的栖息地，实现它们的生态价值，使鱼类及野生动植物能够繁衍生息。

加州渔猎部管理着这些自然群落，保护它们的自身价值、生态价值及对人类的益处。具体工作包括保护、维持栖息地的数量和质量，以确保所有本地物种及其赖以生存的自然群落得以存续。此外，加州渔猎部还负责通过娱乐、商业、科学和教育等多种方式充分利用鱼类和野生动物资源。

Goals and Objectives

The primary goal of the Yolo Wildlife Area is to reestablish the wetland habitat in the Yolo Bypass, while maintaining the agricultural character and flood control functions. The need to address multiple uses across the landscape lies at the core of the Yolo Wildlife Area's mission. Other stated purposes of the Yolo Wildlife Area Land Management Plan are to:

- Guide the restoration and management of habitats, species, appropriate public use, and programs to achieve the DFG's mission, while simultaneously maintaining the agricultural character and flood control function of the Bypass.

- Direct an ecosystem approach to managing the Yolo Wildlife Area in coordination with regional planning efforts.

- Identify and guide appropriate, compatible public-use opportunities within the Yolo Wildlife Area.

- Direct management of the Yolo Wildlife Area in a manner that promotes cooperative relationships with adjoining private property owners.

- Establish a descriptive inventory of the sites and the wildlife and plant resources that are present in the Yolo Wildlife Area.

- Provide an overview of the Yolo Wildlife Area's operation, maintenance, and personnel requirements to implement management goals, and serve as a planning aid for preparation of the annual DFG budget.

Restoration Approaches

The Role of the California Department of Fish and Game Habitat restoration and management of the Yolo Wildlife Area are conducted by the DFG. The DFG's mission is to manage California's diverse fish, wildlife, and plant resources, and the habitats upon which they depend, to realize their ecological values and to allow for their use and enjoyment.

The DFG manages these natural communities to safeguard both their intrinsic and ecological value and their benefits to people. This includes the goal of habitat protection and maintenance to a sufficient degree and quality, and to ensure the survival of all native species and the natural communities that support them. The DFG is also responsible for the diversified use of fish and wildlife,

▲ 图 4-12：约罗分洪道中稻田附近的湿地恢复
（提供：Dave Feliz，加州渔猎局）
Figure 4-12: Wetland restoration adjacent to rice fields
in the Yolo Bypass (Photo courtesy of Dave Feliz,
California Department of Fish and Game)

湿地管理

湿地管理采用了一个复合系统，包括抽水装置、运河和水控制工程。该系统根据规划的水文状况，并模拟约罗盆地当初的自然给水排水状况，为湿地提供给水和排水。此外，还通过割草、耙地或水管理来调节植被状况，为各种物种提供所需的湿地环境，提高季节性湿地的生态系统生产力。

约罗保护区中的多数湿地是季节性湿地，每年4月初干涸，这为长势缓慢的沼泽梯牧草提供了生长条件。这种草不会影响流经约罗行洪道的水量，同时为水鸟提供充足的食粮。春季干涸的湿地抑制了香蒲和芦苇等挺水植物的生长。有些池塘被用来在夏季储水，为鸭科留鸟及其他动物提供孵卵育雏的环境；香蒲和芦苇是很多种鸟类的营巢地；还有些池塘在春季和夏末是泥滩，为迁徙和营巢的涉禽提供栖息地。

生物资源管理

约罗野生动物保护区地处太平洋候鸟迁徙路线的中心位置，是200多种鸟类的家园，有在北极繁殖后飞到这里过冬的候鸟，也有在这里繁殖但飞往中美和南美等南方热带地区越冬的鸟类。当第一拨阿拉斯加涉禽来到约罗过冬的时候，颜色艳丽的金莺、蓝松雀和食蜂鹟还在哺育雏鸟。游禽的到来晚于涉禽，在秋冬两季大群地来这里觅食栖息。成千上万只北方尖尾鸭（见图4-13）、美洲赤颈鸭、绿头鸭、雪雁和白额雁蜂拥而至，来到约罗野生动物保护区的水淹稻田和季节性湿地上。此外，还可以见到

including recreational, commercial, scientific, and educational uses.

Wetland Management

A complex system of pumps, canals, and water control structures are maintained and used to flood and drain wetlands according to the desired hydrologic regimes and to mimic the natural flooding and drainage that once occurred in the Yolo Basin. Additionally, vegetation is disturbed by mowing, discing, or water management to maximize the habitat value for a diverse array of wetland species. This activity can increase ecosystem productivity in seasonal wetlands.

Most wetlands within the Wildlife Area are seasonal, draining in early April. This encourages the production of swamp timothy, a low-growing grass species that is an ample food source for waterfowl without impacting the flow of floodwaters through the Yolo Bypass. Drainage during the spring also discourages the germination of emergent vegetation such as cattails and bulrush. Some ponds are managed to hold water through the summer months to provide brooding habitats for resident ducks and other species. Cattails and bulrushes provide nesting habitats for a number of bird species. Other ponds are managed as mudflats in the spring and late summer, providing a habitat for migrating and nesting shorebirds.

Biological Resource Management

Located at the heart of the Pacific Flyway migratory route, the Yolo Wildlife Area is home to over 200 species of birds, ranging from migratory arctic breeding birds in search of a more temperate winter home to species that breed locally and then fly south towards tropical climates in Central and South America. The brilliantly colored orioles, blue grosbeaks, and western kingbirds are still feeding their

几种猛禽，如稀有的斯温氏鵟，在新割的苜蓿地里觅食，或在约罗野生动物保护区的水淹地上空盘旋，寻找猎物。

偶尔还可以看到郊狼、浣熊、灰狐和长耳鹿在约罗保护区出没；水道中还有河狸、水貂和河獭等水生哺乳动物在此繁殖。约罗保护区庞大的水系里汇集了大量鱼类、两栖类和无脊椎动物。

保护区里的常栖鱼类包括引进的外来鱼类，如鲶鱼、阔嘴鲈鱼、鲤鱼、身体较小的内陆银汉鱼和美洲西鲱等。随着秋天雨季的到来，本地的大马哈鱼就会逆流而上，进入约罗分洪道，回到它们祖先产卵的地方——蒲塔小溪（Putah Creek）。白色的鲟鱼和条纹鲈鱼也会季节性地来到约罗行洪道。

为继续给各种野生动物提供栖息地，约罗保护区中的栖息地实施了多样化管理，包括季节性湿地、残存的河滨森林和农业休耕地。约罗保护区内地势较高的地区洪涝灾害较少，仍然留有古老而独特的春池。春池是美国西岸特有的潮湿凹地，冬春两季雨水或融雪流入成水池，夏秋多呈干涸，是一种季节性池塘湿地。春池本身极为稀有，滋养着多种珍稀植物、无脊椎动物及许多珍稀物种，如蚂蚱麻雀（Ammodramus savannarum）、紫云英（Asragalus tener var. ferrisiae）和保护神仙虾（Branchinecta conservatio）等（见图4-14，4-15）。保护现有栖息地与创建新栖息地具有同等重要的意义。

分洪

树木和大量的挺水植物可以减缓约罗行洪道中洪水的速度。由于防洪是约罗行洪道的主要用途，因此野生动物保护也不能影响这项功能。保护区与洪水控制机构达成协定，限制挺水植物和湿生植物在保护区的种植数量，并运用二维水文模型来确定高大植物的种植量（见图4-16）。

农业

农业不仅有助于约罗保护区栖息地的管理，也能为保护区的管理提供重要的收入来源。种植水稻是这里的农业活动之一，水稻收获后在市场上售出。洪水随后淹没收割完的稻田，为水禽提供了湿

young when the first Alaskan shorebirds arrive on their Yolo wintering grounds. Following on the heels of the shorebirds are waterfowl, arriving in tremendous waves through the fall and winter in search of food and shelter. Thousands of northern pintail, American widgeon, mallard, snow geese, and white-fronted geese swarm onto the flooded rice and seasonal wetlands of the Yolo Wildlife Area. Several species of raptors, including the rare Swainson's hawk, can also be found foraging in fresh-cut alfalfa fields or soaring over flooded fields in search of prey in the Yolo Wildlife Area.

Coyotes, raccoons, gray fox, and mule deer may occasionally be spotted at the Yolo Wildlife Area. Waterways are also home to resident aquatic mammals, such as beaver, mink, and river otters. The extensive water system maintained on the Yolo Wildlife Area harbors large numbers of fish, amphibians, and invertebrates.

Resident fish include non-native species introduced to the area, such as catfish, largemouth bass, carp, and smaller species, such as inland silversides and threadfin shad. With the arrival of fall flows, native Chinook salmon travel upstream into the Yolo Bypass, returning to their ancestral spawning grounds in Putah Creek. White sturgeon and striped bass also move into the Yolo Bypass on a seasonal basis.

In order to continue providing habitats for this broad swath of wildlife, habitats in the Yolo Wildlife Area are managed for diversity, from managed seasonal wetlands to remnant riparian forests and fallow agricultural fields. In the higher parts of Yolo, flood inundation is less common and a unique and ancient vernal pool community continues to exist or even thrive in the presence of many years of cattle grazing. Vernal pools – seasonally ponded wetlands that support many rare species of plants and invertebrates – are rare. Likewise, rare species inhabit these vernal pool areas, including grasshopper sparrows, Ferris' alkali milk vetch, and conservancy fairy shrimp. Conserving the existing habitats is as important as creating new ones in the Yolo Wildlife Area.

Trees and excess emergent vegetation can slow the movement of flood water through the Yolo Bypass. Because flood control is the primary purpose of the bypass, it is critical that wildlife management does not compromise this function. Agreements with flood control agencies place limits on the amount of both emergent (plants that grow up through the wetland surface) and riparian (streamside) vegetation allowed in the Wildlife Area. These limitations on tall vegetation were determined

图 4-14: 约罗野生动物保护区中的白鹭（提供：Dave Feliz，加州渔猎局）
Figure 4-14: The Yolo Wildlife Area supports various wading birds such as the Great egret *(Ardea alba)* (Photo courtesy of Dave Feliz, California Department of Fish and Game)

图 4 - 15: 在该野生动物保护区栖息的联邦级保护动物巨型乌梢蛇（提供：Dave Feliz，加州渔猎局）
Figure 4-15: Federally protected Giant Garter Snakes *(Thamnophis gigas)* have been discovered residing within the Wildlife Area (Photo courtesy of Dave Feliz, California Department of Fish and Game)

图 4 - 16: 洪水期的约罗野生动物保护区（提供：Dave Feliz，加州渔猎局）
Figure 4-16: The Yolo Wildlife Area is designed to convey stormwater flow during high flood events (Photo courtesy of Dave Feliz, California Department of Fish and Game)

图 4 - 17: 约罗野生动物保护区的校车（提供：Dave Feliz，加州渔猎局）
Figure 4-17: School bus at the Yolo Wildlife Area (Photo courtesy of Dave Feliz, California Department of Fish and Game)

地栖息地。由于野鹅和鹤类通常沿着太平洋候鸟迁徙路线上收割过的玉米地觅食，所以保护区与当地农民共同提供高粱地、玉米地和苏丹草地，以便候鸟觅食。　播种和收割红花草等庄稼可为如环颈雉和哀鸽等山地鸟类提供草籽。这些收割后的庄稼地常可提供丰富的捕猎机会。

保护区南部大部分地区都是牧场，春季时野花盛开，蔚然成景。可是，如果外来一年生草本植物（如一年生黑麦草）过多，会抑制本土植物群落的生长，其中包括几种珍稀濒危植物。牛以这种黑麦草为食，使土壤表面暴露在阳光下，促使各种本地植物生长。历史上，叉角羚羊和加利福尼亚马鹿具有这种作用，但这些动物早在19世纪中期已经灭绝。随着欧亚禾草的引进，当地生态系统的动态平衡被打破，需要大群食草牲畜才可以控制住一年生黑麦草的增生。因此，与许多敏感性湿地不同的是，牛群在约罗保护区景观中是非常受欢迎的功能要素。

公共场地、休闲娱乐和环境教育

自约罗保护区创建以来，该地区成千上万的游人到这里散步、远足、观看野生动植物、打猎、捕鱼并参加各种环境教育活动（见图4-17、4-18）。教育活动的主题是"探索候鸟迁徙路线"，每年吸引4000多名学龄儿童来到这里。约罗保护区设有步道和公路网，方便这些活动的开展。

昆虫或蚊子控制与管理

为了减少湿地中的蚊虫繁殖，萨克拉门托—约罗蚊虫控制区（SYMVCD）参加了约罗保护区湿地的设计、建造和管理。近年来，在这些湿地上大规模修建了多种地貌，有较深的水洼、低矮的高地以及小岛。这些改进不仅提供了深浅不一的水塘栖息地，还进一步提高了这些湿地的旱涝调节能力，这对控制子孓非常必要。早在项目的设计阶段就开始考虑SYMVCD以帮助确保改善后的湿地不会对生活在附近的居民带来健康危害。

using a two-dimensional hydraulic model.

Agriculture

The Yolo Wildlife Area uses agriculture to sustain habitats while providing important income for the management of these same areas. Agricultural activities occurring on the Wildlife Area include rice production that is harvested for the market. Harvested fields are then flooded to provide wetland habitats for waterfowl. Geese and cranes commonly forage in harvested corn fields throughout the Pacific Flyway. Working with local farmers, the Wildlife Area provides fields of milo, corn and sudan for this purpose. Crops such as safflower are cultivated and mowed to provide seed for upland species such as ring-necked pheasant and mourning dove. These harvested fields often provide hunting opportunities for scores of people.

Much of the southern portion of the Wildlife Area is grazed by cattle, resulting in spectacular blooms of wildflowers during the spring months. The predominance of nonnative annual grasses (such as annual rye grass) in this area can inhibit the production of native plant communities that include several rare and endangered species. Cattle eat the rye grass, exposing the soil surface to sunlight, which promotes the growth of a diverse number of native plant species. Historically, pronghorn antelope and tule elk provided this function, but these grazers were extirpated from the area by the mid-1800s. The introduction of Eurasian grasses radically changed the dynamics of the ecosystem and only the intense grazing pressure of livestock can reduce the proliferation of annual rye grass. Unlike many sensitive wetland areas, cattle are welcome as a functional component of in Yolo Wildlife Area landscape

Since the inception of the Yolo Wildlife Area, tens of thousands of visitors from throughout the region have used the area for walking, hiking, viewing wildlife, nature photography, hunting, fishing, and a broad range of environmental education activities. At the core of the educational program is the "Discover the Flyway" program, which annually hosts over 4,000 school children per year. A trail and road network in the Yolo Wildlife Area supports these activities.

Vector/Mosquito Control and Management

Wetlands in the Yolo Wildlife Area were designed, constructed, and managed with the involvement of the Sacramento-Yolo Mosquito and Vector Control District (SYMVCD) to minimize mosquito production. In recent years, extensive topography has been built into these wetlands, providing deeper swales, shallow rises, and islands. Not only will these improvements provide a variety of water depths and habitats, they will further improve the flooding and draining capabilities of these managed wetlands, which is essential for controlling mosquito larvae. The SYMVCD was brought into this project during the design phase to ensure that the improved wetlands would not create unwanted health hazards for communities nearby.

结论

约罗野生动物保护区项目获得了越来越多的认可，人们认同它是美国具有生态价值的综合资源有效管理的先例，也是多个机构及利益相关者间组成合作联盟的典范。它提出了一种崭新的思路，在土地使用中，农业、自然资源、防洪、娱乐和教育等多种功能得到和谐的平衡。

Summary

The Yolo Wildlife Area is increasingly recognized as a national model for effective, ecologically-based integrated resource management and for its exemplary partnerships among many agencies and stakeholders. It is providing a new understanding of how lands may be used for compatible agricultural, natural resource, flood protection, recreation and educational opportunities.

▼ 图4-18: 保护区南部牧牛场的景观，同时也吸引游客来此观光 (提供：Dave Feliz，加州渔猎局)
Figure 4-18: Management strategies in the Wildlife Area encourage native wildflower communities (Photo courtesy of Dave Feliz, California Department of Fish and Game)

▼ 图4-19:上多奇河及湿地位置图(绘制:易道)
Figure 4-19: Upper Truckee River and wetland location map (Graphic: EDAW)

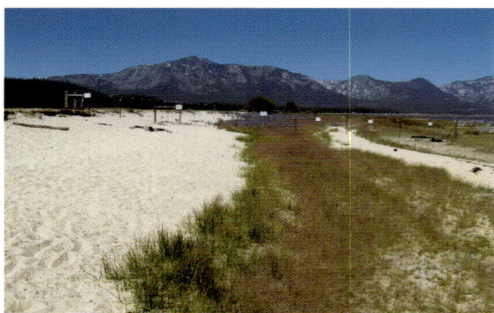

▲ 图4-20:已被提名为濒危物种的太浩湖藓菜及栖息地(提供:Alison Stanton, Steve Matson)
Figure 4-20: Lake Tahoe beach habitat of Rorippa subumbellata (Photo courtesy of Alison Stanton and Steve Matson)

4.1.3 河流形态及功能的恢复 —— 上多奇河河流及湿地修复项目,美国加利福尼亚州太浩湖

背景

太浩湖位于加利福尼亚州和内华达州之间,西接内华达山脉,东临卡森山系的两座山脉(见图4-19)。太浩湖是世界上最大最深的高山湖泊之一,又名"高山上的明珠"。湖水面积495km²,最深处超过500m,其深度名列美国第二、世界第十。

太浩湖的水质和清澈度在很大程度上取决于多奇河。在流入太浩湖的诸多河流之中,上多奇河的流域面积最大,达90多km²。河水主要来自融雪、降雨、地表径流和地下水,每年为太浩湖提供的径流量平均达1.1亿m³。河流源自花岗岩密布的高山,但是在注入太浩湖之前,一路流经森林、牧场、高尔夫球场和城区,在此过程中挟带了大量沉积物和营养物,成为太浩湖的主要污染物。

太浩湖清澈的湖水极负盛名,在近几十年中清澈度曾一度达到最深的30.5m。但是近年来,由于地表径流挟带的细小泥沙和营养物质过多,加上空中悬浮物的沉淀,清澈度已降低至22.8m。

此外,太浩湖是太浩湖藓菜(Rorippa subumbellata)的惟一产地,目前这种植物已申报美国《联邦濒危物种法》规定的濒危物种。这种低矮植物紧贴地面,每到花开季节,黄色的花遍布太浩湖滨及附近星星点点的沙地上(见图4-20)。

太浩湖南岸的上多奇湿地面积最初为526hm²,比现在要大得多。那时,河水泛滥后流入湿地,水中携带的营养物和沉积物在湿地内沉淀下来(见图4-21)。正因为湿地的这一关键性作用,太浩湖才得以有闻名遐迩的碧绿澄清的湖水。但是,美国西部的开发完全改变了这片沼泽地,使这片湿地受到逐渐的毁坏。最初的破坏来自19世纪的淘金热,该流域的树木被大肆采伐;接着是河流下游的渠化,目的是让原木从上游漂流到湖中,再运往康斯托克金矿区。后来,下游成片的土地被用于养牛,农牧业使这片沼泽地进一步退化(见图4-22)。

20世纪50年代后期至70年代,沼泽中心

4.1.3 Restoring a River's Function and Form: Upper Truckee River and Wetland Restoration Project, Lake Tahoe, California, U.S.A.

Background

The Upper Truckee River is critically important to the clarity and water quality of Lake Tahoe. It is one of the deepest and largest alpine lakes in the world, and is otherwise known as the "Jewel of the Sierra". At over 500 meters in depth, Tahoe is the second deepest lake in the United States, and the tenth deepest in the world. Its surface area is 495 square kilometers. Lake Tahoe is located on the borders of the states of California and Nevada, bound on the west by the Sierra Nevada and in the east by two mountain chains that are part of the Carson Range.

The Upper Truckee River drains the largest watershed of Lake Tahoe, covering over 90 square kilometers, providing the largest volume of runoff to the lake, averaging 110 million cubic meters annually. River flow originates from snowmelt, direct rainfall, runoff, and groundwater sources. Hence, the river conveys sediment and nutrients from the single largest watershed contributor of these pollutants. With headwaters in the granite-dominated high Sierra, the river flows through forests, grazing land, a golf course, and urban areas of the communities of Meyers and the City of South Lake Tahoe before reaching the lake.

The famous clarity of the lake reached a peak depth of up to 30.5 meters, but has since deteriorated to 22.8 meters in recent years, attributable to fine sediment and excessive nutrient discharges from surface sources and deposits from other airborne sources.

Lake Tahoe is also the only location for its endemic plant, the Tahoe yellow cress (Rorippa subumbellata), which is presently a candidate for listing under the US Endangered Species Act. The diminutive ground cover plant, with yellow flowers is located on Tahoe's beaches and other nearby sandy soil areas that dot the lake's margins.

Historically, the Upper Truckee Marsh occupied a much larger area along the Lake Tahoe's southern shore. Originally spanning 526 ha, the marsh served to capture waterborne nutrients and sediment when river flows overtopped the banks and wetted the marsh. This critical function helped maintain Lake Tahoe's famous clarity. The marsh has been drastically altered since the settlement of the American West, with initial disturbances taking place during the Gold Rush era of the 1800s, when the watershed was heavily logged. The lower reaches of the river were subsequently channelized, allowing for the

约202hm²的土地被挖开，用来兴建名为"太浩岛"的湖滨社区。大型的挖掘工程营造出运河和码头，挖出的泥沙被泵入沼泽作为堆填物料，成为太浩岛开发区的建筑地基。此工程将沼泽地一分为二，西部是"蒲柏沼泽"（Pope Marsh），东部则是"上多奇湿地"。除了东侧面积约9.3hm²的"东部小湾"（Cove East)外，太浩岛社区已基本建成。

通常，春季时由于冰雪融化，上多奇河的流量为5.66~14.1m³/s，秋季为低流量期，流量一般降至0.7m³/s。当春雨降在积雪上造成洪水时，流量最大可高达85m³/s。太浩湖的水位对河流下游的影响很大。湖面的自然最高水位为1,897m；有了大坝的拦护，实际可允许最高水位达1899m。当湖面达到或接近其上限时，湖水将向河流延伸近2km。

项目目的

自20世纪90年代以来，加利福尼亚州太浩自然资源保护局（本案例中称"保护局"）、加利福尼亚总务部房地产服务处(RESD)和太浩区域规划局(TRPA)共同参与了上多奇河下游及其周边湿地的恢复。

为了满足各利益相关者的不同需要，综合湿地的各种功能，项目制定了以下指导规划过程的12个项目目标：

1. 恢复河流及其冲积平原的自然流程和功能

2. 保护、改善和恢复天然栖息地的功能

3. 恢复和改善鱼类和野生动物的栖息地

4. 通过改善自然地理和生物流程，使水质得到提高

5. 保护太浩湖藓菜(其濒危植物地位现正待审批)；如果可行，扩大其种植面积

6. 对公众开放，提供前往景观眺望点的通道，并提供自然资源方面的科普教育

7. 防止对附近私人物业造成洪灾危险

8. 保护自然景观品质

9. 保护当地的历史和文化遗产

floating of logs from the upper watershed to the lake for shipment to the Comstock gold mining area. Agriculture further altered the marsh, as large land portions in the lower watershed were used for cattle grazing.

Beginning in the late 1950s and continuing through the 1970s, approximately 202 ha in the center of the marsh were excavated to create a marina community known as the Tahoe Keys. Substantial excavation created canals and a marina, and fill material was pumped into the marsh to create building pads for the Tahoe Keys development. Its construction included bisecting the marsh into what is now known as Pope Marsh on the west and Upper Truckee Marsh on the east. The fill area of the Keys was fully built out, with the exception of a nine-ha area on the east side of the community, known as Cove East.

Typical spring snowmelt flows range from 5.66 to 14.1cubic meters per second (cms) and the low flows in the autumn can typically drop to 0.71 cms. Maximum flood flows, often associated with warm spring rains on snowpack, can reach up to over 85 cms. Water levels of Lake Tahoe have a substantial influence on the river's lower reaches. The natural rim elevation of the lake is 1,897 meters and approximately 1.8 meters of additional lake stand is controlled by a dam at the outlet. With operation of the dam, the maximum legal limit is approximately 1,899 meters. When the lake is at or near its upper limit, lake waters will extend nearly two kilometers up the river.

Objectives

The California Tahoe Conservancy (Conservancy), California Department of General Services' Real Estate Services Division (RESD), and Tahoe Regional Planning Agency (TRPA) have been involved in the restoration of the lower reach of the Upper Truckee River and surrounding Upper Truckee Marsh since the 1990s. Restoration of the Upper Truckee River and its adjacent wetlands is guided by several underlying principles established in the Conservancy's mission, the reasons for purchasing the restoration area properties as adopted by the Conservancy Board, the Lake Tahoe Basin-wide carrying capacity thresholds as adopted by the TRPA, and the Plan Area Statements, which serve as zoning regulations in the Basin.

Ultimately, the restoration plan must be consistent with these guiding rules and regulations, within which the project's objectives have been developed. The restoration area falls under two Plan Area Statements that differ in

▲ 图4-21: 太浩岛及其附近的沉积情形
（提供：加州太浩湖保护委员会）
Figure 4-21: Tahoe Keys and sediment plume (Photo courtesy of California Tahoe Conservancy)

▲ 图4-22: 在1930太浩湖南岸的上多奇湿地，当时已经因为滥伐、滥掘水道和过牧造成湿地流失的情形
（提供：美国加州太浩湖保护委员会）
Figure 4-22: The Upper Truckee Marsh circa 1930. The Marsh had already been degraded through deforestation, channelisation and overgrazing. (Photo courtesy of California Tahoe Conservancy)

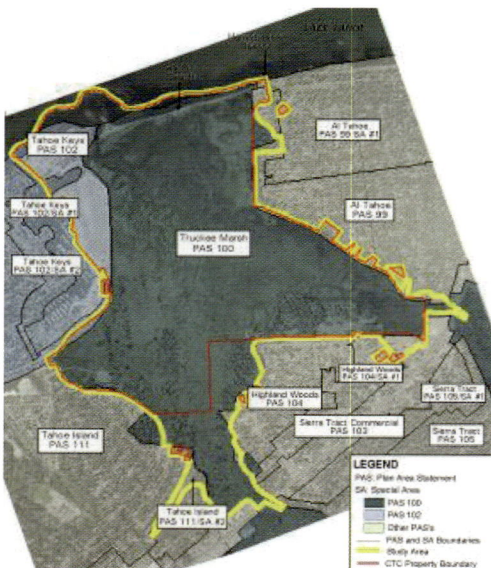

▲ 图 4 - 2 3：总体规划（图片来源：易道）
Figure 4-23: Plan area statement (Source: EDAW)

10. 认真设计湿地与城市的交界处，保护资源

11. 采取的措施不能与上游的状况和活动相冲突

12. 通过适应性管理，认识并规划将来可能发生的变化

湿地恢复方法

易道与保护局、房地产服务处（RESD）签订了合同，易道负责组织由规划师和科学家组成的跨学科团队，制定并执行河流与湿地恢复、野生动物保护与管理、珍稀植物保护的规划方案（见图4-23），并根据该地区宝贵的自然环境，设计通道和解说系统。规划过程包括咨询公民顾问团和政府顾问团，确保大众的参与和机构间的协调。此外，保护局还成立了一个科学顾问团，负责重要专业问题的调查工作，公众与政府的意见可以反映给设计创作室和其他小组会议。

第 一 阶 段 ：西 侧 下 游 湿 地 恢 复 工 程

在"东部小湾"区，有一区域因开发"太浩岛"项目而填平却一直未动工，其中有9.23hm²属于过去的上多奇沼泽地。第一阶段工程将此堆填区域还原成冲积平原和湿地，工程名为"西侧下游湿地恢复工程"（以下简称为"LWS工程"）（见图4-24，4-25）。项目要求清除约67800m³的填料，重新种植乡土植物群落，从而修复当地的天然湿地及水文功能。场地共清除61000多立方米的沙质填料，这些填料被用于修复附近的瓦肖草原国家公园内一个已关闭的采石场。对河流东部破坏较小地区的调查研究表明，恢复工程开展后，该地区建起了三个目标植物群落和类似地貌：长有柳树的湿草地、过渡草地和狗尾草湿地。

LWS工程于2000年获保护局批准，2001年动工，2003年完成。施工期间，在场地通往上多奇河的交界处安设了很多橡胶水囊，这些水囊是为了让湿地植物充分扎根，这既固定了土壤，而且在河水涌入时能够避免沉淀物激増。水囊于2004年移除，河流自此能够接纳速度较快的水流，高山湖泊的水也开始能够进入这片建好的人工湿地（见图4-26）。

建成后的监测结果表明，参照附近地区，西侧下游湿地的填料去除以及植被修复方

their requirements. The project's objectives expressed below may thus be developed differently in their respective areas.
Twelve project objectives have been identified to help guide planning for restoration of the Upper Truckee River and marsh:

1. Restore natural and self-sustaining river and floodplain processes and functions

2. Protect, enhance, and restore naturally functioning habitats

3. Restore and enhance fish and wildlife habitat quality

4. Improve water quality through enhancement of natural physical and biological processes

5. Protect and, where feasible, expand Tahoe yellow cress (under consideration for status as endangered plant species) populations

6. Enhance the quality of public access, access to vistas, and environmental education consistent with natural resources needs

7. Avoid increased flood hazards on adjacent private property

8. Protect natural scenic qualities

9. Design with sensitivity to the site's history and cultural heritage

10. Design the wetland/urban interface to benefit resource needs

11. Take actions that are compatible with upper watershed conditions and activities

12. Recognize and plan, through adaptive management, for changes over time

Restoration Description

Under contract to the Conservancy and RESD, EDAW is leading an interdisciplinary team of planners and scientists that are developing and implementing plans for river and wetland restoration, wildlife protection and management, rare plant protection, and creating trail and interpretive facilities compatible with the area's valuable natural resources. The planning process has included both an extensive program for the public's involvement and an

湿地恢复手册 原则・技术与案例分析

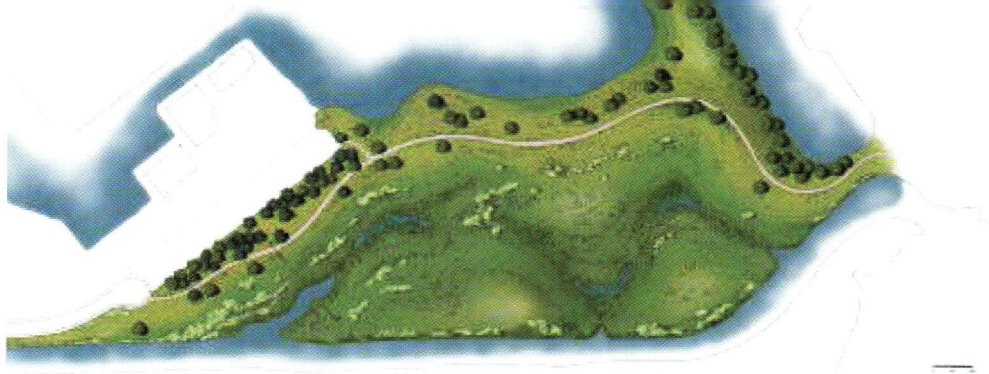

◀ 图 4-24: 西侧下游湿地修复规划（图片来源：易道）
Figure 4-24: Lower West Side plan (Source: EDAW)

▲ 图 4-25: 西侧下游湿地修复工程
（提供：美国加州太浩湖保护委员会）
Figure 4-25: Lower West Side construction
(Photo courtesy of California Tahoe Conservancy)

▲ 图 4-26: 西侧下游地区的灌溉
（提供：美国加州太浩湖保护委员会）
Figure 4-26: Lower West Side irrigation
(Photo courtesy of California Tahoe Conservancy)

法取得了成功。如果挖掘过程中遇到有机土壤，则表明已达到被覆盖的湿地表面时，可以把参照高度（基地必须整平到的高度）调整至更精确的水准。恢复后的高度使景观与地下水深度、河流周期以及湖水水位等比例恰当，关系和谐，从而恢复了湿地生态环境的功能。植被恢复的速度已超过项目开始前的期望。

上多奇河与沼泽的恢复

修复工程的第二阶段规划面积近242hm²，包括上多奇沼泽、上多奇河以及美国第50号高速公路与太浩湖之间的鳟鱼溪。19世纪时，上多奇河的最下游处被裁弯取直，变成一条直线形的人工河道（见图4-27）。结果，河水不能够定期漫出河岸流到邻近的沼泽地里，失去了改善沼泽地的湿地生态功能。此外，该区有许多无组织的休闲活动，野生动植物和当地景观都遭到了破坏。

目前，保护局正考虑进行多项河流和湿地改善工程，包括在适当的地方恢复河流自然形态，增加冲积平原草场的泛滥频率。第一个方案是缩小河道宽度并通过控制坡度造成沉积物下沉，抬高河床；第二个方案是参照历史河道形态和土壤状况评估，恢复河床提高后的河道走向；第三个方案是引导河流注入沼泽地的中心段，使河流在还留有旧河道的地方恢复从前的走向。最后一个方案是在当前河流的两旁挖掘出较低的梯形，使之形成一个河流冲积平原。

规划中的其他修复工程包括调整河口处的河床高度、修复沙丘及林地的植被、重整地形、重建湖滨障蔽沙滩后与河流相连的泻湖。这些实际修复措施将根据包括维护、监测和适应性管理的自然资源管理计划开展、进行。

公众通道以及基于野生动植物保护的使用管理

目前，规划区内有一条主要通道，能够让公众前往东部小湾的海滩及一些资源敏感度较高的地方。因此，修复规划要求改进通道，建立解说系统和观景点，维护并改善通往东湾海滩的公众通道。此外，规划中还有关于在其他地方建立野生动植物保护设施的建议，对资源使用实施管理并指导公众远离敏感资源，这样就不必绝对禁止公众访问和使用。这些设施可将通道的

interagency coordination process, involving a Citizens Advisory Group and Agency Advisory Group. In addition, the Conservancy has assembled a Science Advisory Team to assist in investigating key technical issues. Public and agency input has been sought through design charettes and other small group meetings and workshops.

Lower West Side Wetland Restoration Project

The first phase of the project involved the restoration of a filled area into a functioning floodplain and wetland at the Cove East site, where land had been filled for the Tahoe Keys development, but never built. Known as the Lower West Side (LWS) Wetland Restoration Project, it involved a 9.23 ha portion of the historic Upper Truckee Marsh (see Figures 4-24 to 4-26). The project required removal of approximately 67,800 cubic meters of soil, re-contouring and re-vegetation using native plant communities to restore natural wetland and hydrologic functions on the site. Over 61,000 cubic meters of sandy fill was removed from the site and used to restore a second, former quarry site in the nearby Washoe Meadows State Park. Studies of less disturbed areas east of the river produced three target plant community and topographic regime analogs to work in conjunction with restoration actions: willow-wet meadow, transitional meadow, and wiregrass wetland.

The LWS project was approved in 2000 by the Conservancy. Construction began in 2001 and was completed in 2003. Rubber water bladders, set across the openings of the site to the Upper Truckee River during construction, were kept in place until wetland vegetation had been sufficiently reestablished to stabilize site soils and to avoid a surge of sedimentation once river flow entered the site. The bladders were removed in 2004, allowing the river to overbank higher flows, and allowing high lake stand waters to enter the restored wetland.

Post-construction monitoring has demonstrated the success of the Lower West Side fill removal and re-vegetation approach, using nearby analog sites as guidance. The lowering of the site to the analog elevation was also refined in the field when organic soils were encountered during excavation, indicating that the surface of the covered wetland had been reached. The restored elevation placed the landscape in proper relationship with the depth of groundwater, its frequency of river flows, and the extent of lake water backup to reestablish a functioning wetland habitat. The rapid rate of re-vegetation has exceeded pre-project expectations.

▲ 图4-27: 笔直的河道
（提供：加州太浩湖保护委员会）
Figure 4-27: This straightened river channel will require restoration (Photo courtesy of California Tahoe Conservancy)

终点和观察点设在非敏感区，避免人们从布设的小道进入野生动物栖息地或敏感植物物种区。保护设施还将用来防止游客进入种有大面积太浩藓菜的沙滩区。规划保护性设施能够实现双重性目标：一方面在敏感度不高的地区为游客提供愉快的游览体验，另一方面降低人们进入重要资源区的欲望，避免旅游带来的破坏。

结 论

上多奇河的恢复是把退化的河道与其所属的冲积平原重新结合的最佳案例。项目更同时解决了一些相关的问题，如水质与公众使用方法等。保护局的使命是保存、保护、恢复、改善并发展太浩湖盆地上有特色的重要自然资源和休闲娱乐资源。保护局制定并实施了多项工程，改善太浩湖水质，保护该区的自然美景、野生动植物栖息地和休闲资源，并为公众提供使用的机会。上多奇河及湿地全面恢复工程已成为加利福尼亚州太浩自然资源保护局最卓著的恢复工程之一（见图4-28）。

Upper Truckee River and Marsh Restoration

The second phase of restoration is planned for a nearly 243 ha site comprising the Upper Truckee Marsh, Upper Truckee River and Trout Creek, running from US Highway 50 to the lake. This lowest reach of the river was straightened and moved in the 1800s from its historic meandering channel and placed in a linear, constructed channel. As a result, the river is incised and no longer overtops its banks at a frequency that would enhance the wetland habitat functions of the adjacent marsh. Additionally, the area has experienced considerable undirected public access and recreation, which has disturbed wildlife and damaged the landscape. A multifaceted restoration and natural resources management plan is being developed to address project objectives.The Conservancy is considering a number of river and wetland improvements, including the reestablishment of the natural river morphology and increased wetting of the floodplain meadow in appropriate locations. Narrowing the channel and elevating the bed through channel aggradation is one approach under consideration. Grade control structures would be used to establish sediment aggradation. "Re-hanging" (i.e., reconstructing) a new channel alignment at the proper elevation using historic channel patterns and the evaluation of soil conditions as guides is also being considered. Diverting the channelized river into the central portion of the marsh without reconstruction of a complete channel is a third option that would allow the river to find its own path in an area where traces of the old channel still exist. The final option involves the excavation of a lowered terrace around the current river alignment for construction of an inset floodplain.

Other restoration features in the plan include adjustment of the bed elevation at the river's mouth, the re-vegetation and re-contouring of the site's dune and forest areas, and the reestablishment of a lagoon connected to the river behind the barrier beach on the lakeshore. These physical restoration actions are being supplemented by a natural resources management plan that includes maintenance, monitoring, and adaptive management responses.

Public Access and Use Management for Wildlife Protection

The restoration plan area includes an important and existing public access channel to Lake Tahoe at Cove East Beach and other portions of property that contain significant sensitive resources where public access can be detrimental. The plan calls for improved trails, interpretive facilities, and observation viewpoints to maintain and enhance public access to Cove East Beach. Elsewhere, wildlife protection

facilities have been proposed to manage and direct public use away from sensitive resources, without simply attempting to exclude all public access and use. These facilities could include a trail-end and viewpoint destination in a non-sensitive location while discouraging the establishment of volunteer trails into wildlife habitats or sensitive plant species locations. These protection features are also being used to discourage public access to portions of the project beach that support the largest populations of the Tahoe yellow cress. The twin goal is to provide a pleasing user experience in less sensitive areas and to diminish the public's desire to access areas where important resources can be harmed.

Summary

The Conservancy's mission is to preserve, protect, restore, enhance, and sustain unique and significant natural resources and recreational opportunities provided by the Lake Tahoe Basin. The Conservancy is pursuing this mission by developing and implementing broad programs through acquisitions and site improvements that enhance water quality in Lake Tahoe, preserves both the scenic beauty and recreational opportunities of the region, provides public access, and preserves wildlife habitat areas. By managing and restoring land to protect the natural environment, the comprehensive Upper Truckee River and Wetland Restoration Project has become one of California's most prominent restoration projects.

▲ 图 4-28: 沼泽地鸟瞰 (Copyright EDAW)
Figure 4-28: Upper Truckee Marsh aerial view (Copyright EDAW)

The Wetland Restoration Handbook
Guiding Principles and Case Studies

资料提供
湖南省林业厅
保护处

Information provided by
Hunan Province Forest-
ry Conservation Bureau

4.1.4 退田还湖与社区共管 —— 西洞庭湖青山垸湿地恢复，中国湖南省汉寿县

背景

西洞庭湖自然保护区地处湖南省洞庭湖西部的汉寿县境内，东抵南洞庭湖，北与安乡、南县接壤，西邻常德鼎城区，南紧连汉寿县南部低山丘陵区，总面积356.80hm²。其中湿地面积269.6hm²，核心区在西洞庭湖的大连障山、目平湖、中尾洲一带，为洞庭湖的西南部分，不足洞庭湖总面积的13%，也是洞庭湖三个湖区中面积最小、淤积最为严重的湖区，承纳长江、澧水、沅水和汉寿县南部低山丘陵区的沧水、浪水、烟包山河等8条支流，其南端通过狭窄水道排入南洞庭湖。

本区域位于中亚热带向北亚热带过渡气候区，气候温和，光照充足，雨量充沛，四季分明，年平均气温为16.6~16.8℃，年平均降雨量1200~1350mm，全年无霜期274日。暴雨是区域内最主要的灾害性天气，每年都有发生，年均3~4次，主要集中在5~8月。西洞庭湖湿地主要特征表现为洲滩密布、江湖交错、水域辽阔，水涨为湖，水落为洲。在低水位时，既有明水，又有芦苇沼泽、苔草沼泽、泥炭沼泽、沙滩等地貌，是湿地生物良好的繁衍生息场所。

西洞庭湖是整个洞庭湖湿地不可分割的重要组成部分，优越的自然环境和地理条件培育了丰富的生物资源。经野外考察，记

▼ 图 4-29: 西洞庭湖青山垸湿地位置图
Figure 4-29: The Restoration of the Qingshanyuan Embankment Wetland Location Plan

4.1.4 Lake Restoration on Farmlands and Community Co-management: The Restoration of the Qingshanyuan Embankment Wetland, Western Dongting Lake, Hunan Province, China

Background

The Dongting Lake Nature Reserve lies in Hanshou County at the western edge of Dongting Lake, in Hunan Province. The reserve borders Southern Dongting Lake in the east, the towns of Anxiang and Nanxian in the north, and Changde City in the west, where it leads to the highlands of southern Hanshou County. Within the 357 ha Nature Reserve, 270 ha are wetlands, with a core wetland area that stretches 70 ha along the Dalianzhang Hill, Muping Lake and the Zhongwei Islet. The wetland that makes up the southwest part of the site is the smallest of the lake's three water bodies, covering less than 13% of the total lake area. Silting is the most severe in this section, as the wetland takes in water from the Yangtze, Li, and Yuan Rivers and eight tributaries, while water only exits out at the south end into Southern Dongting Lake through a narrow channel.

The Qingshanyuan Embankment is located in the Western Dongting Lake Nature Reserve, between the towns of Jiangjiazui and Yangtaohu, in Hanshou County. The total area before the lake's recovery was 1,107 ha. Approximately 5,821 rural residents lived on the Embankment. In 1998, local officials decided to raze the Qingshanyuan Embankment and return it to Dongting Lake. Residents were relocated to Jiangjiazui or Yangtaohu. In 1999, in cooperation with the World Wildlife Fund for Nature (WWF) and the government of Hanshou County, the Western Dongting Lake Nature Reserve was approved by the Hunan Provincial People's Government as a provincial Nature Reserve.

Western Dongting Lake today is an indispensable component of the Dongting Lake wetland system. Its favorable natural environment and geological conditions have are evident in the lake area's abundant biological resources. Field investigations have recorded 168 species of aquatic plants, 207 species of birds (seven of which are under national protection), and 114 species of fish, including rare species and related fish resources that are also under national protection.

In February 2002, Western Dongting Lake was also designated as a wetland of international importance. Measures have since been taken to protect the waterfowl habitat, the wetland ecological system and water birds

录到区内水生植物168种；鸟类207种，属国家一级保护的有7种；鱼类114种，包含国家珍稀保护和珍贵的鱼类资源。

青山垸位于西洞庭湖保护区境内，地处汉寿县蒋家嘴与洋淘湖两镇之间，退田还湖前面积共1,107hm²，垸内共有村民5,821人（见图4-30）。1998年，青山垸被列为洞庭湖区退田还湖堤垸，并交由西洞庭湖自然保护区管理，垸内居民全部搬迁到蒋家嘴或洋淘湖镇。1999年，世界自然基金会（WWF）与湖南省汉寿县人民政府合作，使西洞庭湖自然保护区得到湖南省人民政府批准，正式成为湖南省省级自然保护区。但是，在退田还湖初期，产生了农民的替代生计问题，一些渔民采用非法捕捞等方式营生，以致湖区生态环境遭到极大破坏，成为影响保护区管理的重要隐患。2002年2月，西洞庭湖被中国政府指定申报列入"国际重要湿地"名录。保护区以越冬水禽、水禽栖息地及湿地生态系统为主要保护目标，通过科学管理，阻止洞庭湿地进一步退化，实现湿地资源可持续利用。

目的和目标

1. 通过保护区与社区的共同管理，使退田还湖后的青山湖垸能够得到全面的保护，使湿地资源得到持续合理利用，使社区民众的环境保护意识得到提高；让青山湖垸周边社区的居民在共同管理中受益。

2. 实现社区共管后，这里不仅要逐步成为西洞庭湖区内一块最为理想的水禽栖息地和野生鱼类繁育地，还要建成一个垂钓休闲区、农渔业观光区和生态旅游区；让社区居民在共管中找到合适的替代产业，以顺利推广湿地资源的可持续发展模式。

3. 将青山垸建成一个具有国内外领先水准的退田还湖、湿地保护、社区共管、合理利用、可持续发展的多功能综合示范区。

退田还湖的恢复过程

(1) 早期强制管理阶段
1999年建立保护区后，曾采取一系列手段和措施进行强制管理，包括在保护区树立执法宣传牌，强制撤网行动和执法队拘捕

that inhabit the site during winter months. Scientific and technical management has helped prevent the environment of Dongting Lake from further deterioration, while it has achieved a sustainable utilization of wetland resources.

Part of the restoration project has included reverting farmlands into wetland resources. During the early stages of this process, residents were concerned with their ability to find alternative means that would ensure their livelihoods. During the restoration process, those fears were great enough to prompt some to continue to rely illicitly on fishing, which further degraded the lake's ecology. This was a major issue that for some time marked and affected the management of the Nature Reserve. Part of the solution has been educating residents for community co-management of the reserve.

Project Objectives

The project aims to achieve the following through co-management of both the Nature Reserve management office and community participation:

1. To protect the Qingshanyuan Embankment and Qingshanyuan Lake and achieve sustainable utilization of the wetland resources; To enhance awareness of environmental protection, and to provide economic benefits for community residents through co-management.

2. To build the wetland into an ideal waterfowl habitat, a wild fish breeding ground, and a place for recreational fishing, sightseeing, as well as agriculture, commercial fishing and eco-tourism. The aim is to balance the goals of allowing residents to find suitable alternative industries while facilitating the sustainable development of wetland resources.

3. To build Qingshanyuan Lake into a showcase example of a successful lake recovery, a protected wetland, and a community co-management and reasonable utilization success.

The Restoration Process

(1) Constructive Measures – the Early Stages
Since the establishment of the Nature Reserve in 1999, a series of comprehensive methods and measures have been adopted, including the placement of boardwalks, the methodical removal of fishing nets, and the restriction of unlawful commercial fishing. The initial stages of the project were the most difficult, as there was a need to manage a strong disagreement between the Nature Reserve's management office and the community over

▲ 图4-30：青山垸曾是5,821人的家园（摄影：张翼飞）
Figure 4-30: At one time, 5,821 residents had houses on the embankment (Photograph by Zhang Yifei)

非法捕捞人员等活动，保护区与社区居民形成了强烈的对抗。

(2) 养殖业承包

2002年渔业捕捞队强烈要求承包青山垸，于2003年5月通过转让得到了青山垸的承包权，并与汉寿县签订了承包合同，后由于国家林业局、湖南省林业厅、世界自然基金会以及当地社区群众，为了避免承包后的集约经营导致水体的富营养化，造成湿地生物多样性退化，以及不能满足当地社区的群众利益等主要原因，强烈反对而终止。保护区的恢复行动继续进行。

(3) 实施共管

在世界自然基金会的努力下，2004年1月经保护区、汉寿县林业局、青山垸社区（现蒋家嘴和洋淘湖社区）代表的协商达成了一致意见，就青山垸社区共管事宜制定了共管方案和协议，包括共同达成的目标、共管组织机构、共管方式、监督机制、进行利于生物多样性恢复和保护区划分（见图4-31，4-32）。

2004年3月，从青山垸迁出的180多户渔民和西洞庭湖自然保护区正式签订社区共管经营合同，并取得了湖南省林业厅的认可，保护区与社区在利益上最终达成一致。

2005年3月社区自发地募集资金90万元用于鱼苗投放，常德市政府在沅水流域城区河段将2万kg的鱼苗成功放流，是建国以来该市举行的最大规模人工增殖放流活动。仅2005年，青山垸就收获成鱼120多万kg，帮助退田还湖的农民找到了新的生计。生态恢复后，青山垸的水质也变好了，地方管理者计画在世界自然基金会的帮助下申报有机渔产品生产基地。

渔民已经开始自发协助保护区巡湖，及时报告非法事件，西洞庭湖自然保护区的管理人员兴奋地说："自从青山垸实施社区共管后，这些渔民从湖区生态的破坏者变成了维护者，我们再也不用倾巢出动为保护湖面而疲于奔命了"。

结论

生态恢复了，水质变优了，青山垸在世界自然基金会的帮助下，有望成为有机鱼生产基地。该项目使社区共管的生态效益、社会效益和经济效益齐头并进，为社区共管、湿地恢复提供了一个非常成功的模式。

how to utilize the lake's resources.

(2) Aquaculture

In 2002, a group of commercial fishers petitioned the Qingshanyuan Embankment authorities to be allowed to conduct commercial aquaculture. The team signed a contract with Hanshou County in May of 2003. The contract was subsequently cancelled, because it was deemed that intensive cultivation would lead to eutrophication of the water body, and would ultimately damage the wetland's biodiversity and compromise the lake's benefits to the community. The contracted activity was objected to by the Ministry of Forestry, the Department of Forestry of Hunan Province, the WWF and even local communities. With the cancellation of the aquaculture contract, the restoration process resumed.

(3) Co-management

In January of 2004, the WWF and the Nature Reserve management office organized an agreement between the Bureau of Forestry of Hanshou County and representatives from the Qingshanyuan Embankment community in regard to the Qingshanyuan community's participation in the co-management of the Nature Reserve. A participation scenario was presented, including mapping out achievable goals, implementing bodies, co-management methods, monitoring mechanisms and plans for creating a favorable biodiversity (see Figures 4-31, 4-32).

That March, over 180 households moving out of the Qingshanyuan Embankment signed the co-management agreement with the management office of Western Dongting Lake Nature Reserve. The agreement was approved by the Hunan Department of Forestry, as the management office and the community reached a consensus on the mutual benefits of the rehabilitation and co-management plan.

In May of 2002, the community raised RMB 900,000 for 20,000 kg of fish fry, which were released into the Changde City section of the Yuan River. The activity was the largest of its kind in Changde City since 1949. In 2005, 1.2 million kg of adult fish were harvested for the Qingshanyuan Embankment community. This was an important moment and proof-point that the project provided an effective alternative livelihood for farmers and fishers affected by the lake recovery project. An additional benefit of the ecological restoration has been the improvement of water quality. The local management office, with assistance from the WWF, is planning to apply for qualification as a base for organic fish production and products, which will further increase the economic potential of the site along sustainable guidelines.

◀ 图 4 - 31: 社区共管前，青山垸到处是美丽的陷阱－绵密的捕鱼网（摄影：周怀宽）
Figure 4-31: The Qingshanyuan Embankment, a complex maze before co-management (Photographed by Zhou Huaikuan)

▼ 图 4-32: 社区共管后的青山垸（摄影：周怀宽）
Figure 4-32: The Qingshanyuan Embankment after co-management (Photographed by Zhou Huaikuan)

具体成效包括,社区民众的环境保护意识得到提高,他们自行撤出各种非法捕捞网具并停止非法养殖活动,自发组织巡护小组,制定了冬季禁捕方案,旨在保护冬候鸟栖息和鱼类产卵场,各种鱼类得以在青山垸天然繁殖,能维持较高的渔产品质量。社区居民收入提高的同时,生态环境也得到明显改善,栖息的鸟类在种类、数量和停留时间上,都发生了很大的变化。数量增加到接近60种3万只鸟类,青山垸里30多年来第一次出现了从未有过的万鸟齐飞的壮观景象(见图4-33,4-34)。

2004年10月下旬上万只水禽回归青山垸,这不仅证明了恢复计划的成功,更有效带动了西洞庭湖的生态旅游,以观鸟和摄影为主要目的的生态游客增加了,在保护湿地的同时,创造了比过去更好的经济效益,有利于长期稳定的持续发展(见图4-35,4-36)。

▼ 图 4-33: 青山垸的野雁(摄影: 周怀宽)
Figure 4-33: Wild geese fly along the Qingshanyuan Embankment (Photograph by Zhou Huaikuan)

Local fishers have actively started patrolling the Reserve area, reporting unlawful fishing activities. According to the the management office of the Nature Reserve, since the implementation of co-management, commercial fishers have changed from destroyers to guardians of the reserve.

Conclusion

The obvious effects of change can be seen in the enhanced community awareness on environmental protection. Many residents have voluntarily removed nets and stopped illegal fishing activities, have reported unlawful aquaculture activities, and teamed with patrolling groups to establish a winter fishing ban that aims to protect the habitat of migratory birds and preserve a stable breeding ground for fish eggs. Fish are now reproducing in the natural environment of the embankment, which has contributed to the better quality of the area's fish products.

Importantly, the income of residents is rising in tandem with marked ecological improvements. Positive changes are also occurring for resident and migratory birds, with a greater diversity and quantity of species and an increased length of stay. There are now 30,000 birds of 60 species in the wetland and for the first time in 30 years, the Qingshanyuan Embankment community has witnessed over 10,000 birds utilizing the site at once. In October of 2004, approximately 10,000 water birds returned to the Qingshanyuan Embankment. This was not only another indication of a successful restoration, but has subsequently bolstered eco-tourism around Western Dongting Lake. Increasing numbers of eco-tourists visit the Qingshanyuan Embankment for bird-watching or photography activities. The economic efficiency of the program has been stable, which indicates that the site will be sustained over time.

The restored ecology, an improved water quality and the plan to build the Qingshanyuan Embankment into a production base for organic fisheries reflects – after some hard negotiations – the synthesis of ecological, social and economic efficiency brought by community co-management. The project today serves as a successful case study for community co-management and wetland restoration.

图 4-35: 2004年，经过水质改善后的西洞庭湖，在湖南汉寿县蒋家嘴安东湖举行首届西洞庭湖湿地龙舟竞赛（2004），吸引了近10万人参加盛会，继而更成为湖南省旅游节的主要活动（提供：WWF China）
Figure 4-35: A dragon boat race on the improved Western Dongting Lake in 2004, which drew 100,000 participants (Photo courtesy of WWF China)

图 4-36: 推广和宣传可持续性湿地的概念（提供：WWF China）
Figure 4-36: Promoting the sustainable utilization of wetland resources (Photo courtesy of WWF China)

图 4-34: 青山垸的红嘴鸥（摄影：韦宝玉）
Figure 4-34: Black-headed gulls on the Qingshanyuan Embankment (Photograph by Wei Baoyu)

资料提供
杭州西溪国家湿地
公园管理办公室

Information provided by
Administration Committe
of Hangzhou Xixi National
Wetland Park

4.1.5 再现千年渔耕文化的江南湿地——西溪国家湿地公园，中国浙江省杭州市

背景

自然概况

西溪湿地位于杭州市西部，距市中心武林门约6km，主体隶属于西湖区的蒋村乡，另有小部分属于杭州市区的五常乡。西溪湿地从水系上归属于杭州市区运河水系的运西片，处于低山丘陵与平原的过渡地带，是典型的江南水乡（见图4-37）。

西溪湿地的自然地形为低洼的水网平原，周围村庄、桑田等高出原地面1~1.5m，地面标高在2~5.5m之间，地势略呈南高北低状。农田以池塘、柿林、桑地、茭白田等为主。在长期的江南渔桑农业发展下，该地区已经形成了以鱼塘为主、辅以面积较大的洲渚的湿地类型，当地人的交通运输则依赖大小港汊及狭窄的塘基，形成自然与人工的平原湿地景观（见图4-38）。塘基上一般种有老柿树、柳树、桑树等乔木，渚上除了上述树木和农田鱼塘外，还有不少竹林（见图4-39）。一些废弃的农田也正在自然恢复中，其中有少量典型的湿地环境，如芦苇荡和荻群落。调查统计表明，在西溪湿地共分布着维管束植物85科182属221种，鸟类已发现12目26科89种。此外还有大量的浮游植物、浮游动物以及底栖动物。

西溪湿地土壤以水稻土类、脱潜水稻土亚

▼ 图4-37: 西溪国家湿地公园位置图
Figure 4-37: The Xixi National Wetland Park Location Plan

4.1.5 A Rehabilitated Wetland in Southeastern China that Supports Traditional Ways of Fishing and Farming: The Xixi National Wetland Park, Hangzhou, Zhejiang Province,China

Background

Natural conditions

The Xixi (West Creek) Wetland lies on the western outskirts of Hangzhou city, in Zhejiang Province. It covers most of Jiangcun Village, the famous Xihu (West Lake) District, and a small part of Wuchang Village in the Yuhang District. It lies approximately six kilometers from downtown Hangzhou. The wetland belongs to the district that is west of the Hangzhou Canal, a transitional zone between low hills and floodplains, with the typical wetland features and topography of farming and fishing lands south of the Yangtze River. The natural topography of the site is characterized by low, river-traced floodplains. The surrounding villages and mulberry fields are 1 to 1.5 meters above the original ground level, which has an elevation between 2 and 5.5 meters. The overall site is a bit higher in the south than in the north. The agricultural lands are used mainly as fishery ponds, persimmon woods, mulberry fields and wild rice fields. After decades of intense agriculture, the site has developed into a wetland type with an abundance of fishery ponds and large-sized islets. Locals commute around the site on streams of varying size and the narrow levees of the fishery ponds.

The floodplain's landscape has been shaped by both human activity and natural processes. Persimmon trees, willows and mulberries are planted on the pond's boundary levees. In addition to these trees, fishery ponds and bamboo groves are also present on the islets. A few abandoned farmlands have tended towards natural restoration. There are signs of a burgeoning habitat typical of wetlands, such as reed marshes (Phragmites communis and Miscanthus sacchariflorus). Field studies have identified 85 families, 182 genera and 221 species of vascular plants, and 12 orders, 26 families and 89 species of birds in Xixi Wetland. Moreover, there are an abundance of zooplanktons, phytoplanktons and benthons at the wetland site.

The Xixi Wetland is primarily made up of paddy and humid soil. The site features a distinct alternating monsoon climate due to its position on the northern fringe of a subtropical zone. The yearly mean annual temperature is 16.2° Celsius. The site has a yearly annual precipitation of 140cm, due mainly from spring rains, mould rain and

湿地恢复手册 原则·技术与案例分析

类青紫泥田土属和潮土类、潮土亚类堆叠土属为主，所处地区属亚热带北缘季风气候，季风交替规律明显。年平均气温为16.2℃。年平均降雨量约1400mm，以春雨、梅雨、台风降水为主，降水主要集中在3~9月：3~6月是第一个雨季，其中5~6月为梅雨期；第二个雨季是在8~9月。历史上洪涝灾害较为严重。

西溪湿地的生态和环境问题

西溪湿地地势宽广低平、树木茂盛，空气十分新鲜。据记载，水质曾经非常良好。但近二三十年来，由于当地企业排污管理不善以及当地居民产业结构和生活方式的改变，污染排放已经超出了湿地的水质净化能力，因此已经出现了富营养化的水质问题，渔产品也因此受到严重影响，生物栖息地质量逐步下降。

为了制止西溪湿地环境质量的进一步恶化，治理西溪、恢复湿地环境势在必行，这也是营建国家级湿地公园必须开展的前期工作。2003年9月，西溪湿地保护与恢复工程开始实施，2005年5月1日西溪湿地被国家林业局正式列入"国家湿地公园"，并正式对外开放，是我国目前惟一的集城市、农耕、文化为一体的国家湿地公园。

目的与目标

西溪湿地的生态恢复和保护工程坚持"生态优先、最少干预、修旧如旧、注重文化、以民为本、可持续发展"的六大基本原则，全面加强湿地及其生物多样性的保护，维护湿地生态系统的生态特性和基本功能，保护和发挥湿地生态系统的各种功能与效益的最大值，实现湿地资源的可持续发展。同时，成立西溪生态展览馆，展示西溪湿地生态研究和科教的重要成果（见图4-40，4-41）。

湿地的恢复方法

(i)水体和水生态保护与恢复

根据西溪湿地的实际情况，从水体特征角度分析，由于其本身和周边区域的各类污染物排放，造成的水质降低是西溪湿地生态系统退化的重要原因之一，这也导致了湿地水体自净能力的下降。所以西溪湿地水体修复的重点就是通过各种有效手段，减少和控制进入西溪湿地的污染物量，进而改善水质，并结合其他恢复规划，逐步改善现有湿地生态系统的结构，恢复其自我净化水质的功能，达到水体和生态恢复的目标，促进整个西溪湿地景观的恢复。

typhoons. The rainy season is generally from March through September, with March through June as the initial rainy season (with a mould rain period from May to June) and again from August through September. Not surprisingly in this climate, the site has suffered heavily in the past from floods.

Environmental problems in Xixi Wetland

Xixi Wetland enjoys the natural conditions of low, spacious flatlands, lush vegetation and fresh air. Historically, the site has been known for its high-level water quality. In recent decades, however, due to negligent pollution control by local enterprises, industrial development and lifestyle changes of local residents, pollution in the site has exceeded the wetland's water purification capacities. As a result, the water is eutrophic, fishery production has decreased markedly, and biological habitats are in decline.

In order to curb continued environmental degeneration of the Xixi Wetland and repair its ecosystem, conservation and restoration projects have been undertaken since September of 2003. The Xixi Wetland is now open to the public as a "National Wetland Park", officially designated by the National Forestry Bureau on May 1, 2005, which features a unique combination of city life, agriculture and traditional culture.

Goals and Objectives

There are six principle restoration and conservation goals at Xixi: a priority on minimal ecological interference; preservation of the historic site configuration; recognition of traditional culture; consideration of human needs; and sustainable development. Similarly, project objectives include: the overall protection of the wetland and its biodiversity; preservation of the ecological character and functions of the wetland ecosystem; maximizing the benefits of a wetland ecosystem; and the sustainable development of wetland resources. Supporting these efforts is an exhibition center which has been built to showcase ecological research and provide scientific education.

Restoration approaches

(i) Conservation and restoration of the water body and ecology

A main issue for the Xixi Wetlands has been pollutant discharge from inner sources and adjacent areas that have compromised water quality, which has subsequently

▲ 图4-38：百年传承的人工渔塘，成为湿地恢复的基石(摄影：易道)
Figure 4-38: The Xixi fishery ponds are a hundred years old and have become the basis for wetland restoration (photograph by EDAW)

▲ 图4-39：渚上可见一些柿树、柳树、桑树，形成自然景观(摄影：易道)
Figure 4-39: The islets boast natural scenery of persimmon trees, willows and mulberries (photograph by EDAW)

▶ 图4-40: 西溪湿地的渔农耕文化已经成为生态系统的一部分(摄影:易道)
Figure 4-40: The old ways of fishing and farming have become part of the ecosystem of Xixi Wetland (photograph by EDAW)

▲ 图4-41: 西溪拥有丰厚的人文基础, 在恢复建设中也得到了完整保留(摄影:易道)
Figure 4-41: The cultural heritage of the Xixi Wetland has been preserved as part of the restoration effort (Photograph by EDAW)

▲ 图4-42: 河岸两侧采用松木桩护岸, 将湿地的泥浆翻起平铺岸边, 保持区内水岸的自然弯曲状态(摄影:易道)
Figure 4-42: Pine stakes may be planted as banks, with dredged mud piled onto the banks (Photograph by EDAW)

1.保障西溪湿地水资源总量

(1)保障上游来水
(2)控制地下水的开采
(3)保持湿地的行洪、蓄洪功能
(4)保持一定的水面面积
(5)适当控制水位

2.恢复水体自净能力

(1)恢复和保持水体的生态属性
利用地形和通过适当的人工干预手段,恢复和保持西溪湿地水体的流动性和高低水位的周期性变化规律及洪水处理的过程,保持水体原有的水文特征;建立水体中的各种水生生物群落,形成与西溪湿地水体生境适应、结构完整、功能健全、稳定的湿地生态系统,进而恢复水体的生态属性。

(2)恢复和保持湖塘、河道水陆边界的生态属性

拆除原有的非天然材料护岸,在河岸两侧采用松木桩护岸,将湿地的泥浆翻起平铺岸边,插上柳枝;保持区内水岸的自然弯曲状态;在水面开阔的水岸设置一定的斜坡和水深0.5~0.8m之间的浅水区域,人工种植沉水、挺水和浮水植物,恢复湖塘、河道水陆边界的生态属性(见图4-42)。

(3)加强池塘水质的生态修复

加强内陆池塘和外荡水体之间的循环。对水质较差但有一定换水条件的池塘,进行换水、清淤、放养螺、蚌等底栖动物和各种野生鱼类,采取人工曝气的手段恢复池塘的自净能力(见图4-43)。对水质极差又缺乏恢复可能的池塘,采用微生物处理池的方式,并种植水生高等植物,培育浅水湿地生态景观。

(4)适当配置污水处理设施,处理当地的生活污水

(5)完善西溪湿地生态系统,尤其是水生及陆生植被的培育和保护,使其成为结构完整、功能完善、抗逆性强的湿地生态系统(见图4-44)

(6)水质监测计划,每个季度监测一次水质,选择有代表性的点源,进行半年一次

contributed to the degradation of the wetland ecosystem and a decline in its water purification capacities. Given this, an emphasis on the water body's restoration has and should continue to focus on water quality improvements through the reduction and control of pollutants at the site. In addition, supplemental projects must be conducted to improve the overall structure of the wetland ecosystem and restore water purification functions. Improvements in both the water body and ecology will help to rehabilitate the landscape of Xixi Wetland. The proposed methodology for this is as follows:

1. *Ensuring overall water resources*

(1) Ensuring inflows from upstream
(2) Controlling the excavation of groundwater
(3) Maintaining the floodplain and flood storage functions of the wetland
(4) Maintaining sufficient water space
(5) Controlling water levels and appropriate range

2. *Restoring water purification capacities*

(1) Restoring and maintaining the ecological character of the water body
The mobility of the water body, the interchange between high and low water levels and the wetland's flood management may be restored and maintained through re-contouring and proper interventions. In this way, the water body may be restored to its original hydrological character. Aquatic bio-communities may be established in the water body to form a structurally complete, fully functional and stable wetland ecosystem that is compatible to the Xixi Wetland habitat. This could help restore the ecological character of the water body.

(2) Restoring and maintaining the ecological character of the boundaries of lakes, ponds and waterways
Actions may consist of replacing unnatural materials with pine stakes to buttress water banks, piling dredged mud onto the banks, and planting willow branches in the bottom mud. The bank's natural curves should be retained. Properly graded slopes and shallow areas with a depth of 0.5-.0.8m, accompanied by the growth of sunken, emergent and floating plants, may be created on banks towards open water space. These methods are designed to restore the ecological character of the water-land boundaries of lakes, ponds and waterways.

(3) Enhancing the ecology of fishery ponds
This involves increasing circulation between ponds onsite and outside water bodies. For ponds with poor water

的底泥质量监测工作

3.截污控制
根据西溪湿地污染源的调查和分析，造成水质恶化的主要原因是西溪湿地范围内，以及周边区域的各类污染源排放的结果，因此为了达到水体修复的目的，提高湿地水体自净能力，必须对区域内及周边地区的各类污染源进行截污控制。

(1)控制地区居户数量
对西溪湿地一期范围内624户原住民，除少量因生态保护和旅游服务需要保留外，其他逐村、逐户进行拆迁。

(2)改善土地利用和农业经营方式
按照西溪湿地总体规划的要求，严格控制建筑用地。合理调整鱼塘、水田和旱地的结构和数量，并禁用化肥和农药，提倡使用有机复合肥；水产养殖以自然放养为主，禁止在区内进行一切畜禽养殖，严格控制点源污染。

4.限制通航，减少航运污染
限制通行船只的行船速度、数量、形式、目的。湿地内严格控制机动船的数量，当地特有的摇橹船已经成为游客进入湿地观光的主要交通工具（见图4-45）。当地原有的传统手划木船也得到了保留。这样，不仅能够控制船舶对水质的影响，更能降低行船对河岸产生的冲刷风险。

(ii)植被的保护与修复
柿林培育通过对原有果园进行整理，或重新植种柿树形成景观林，具有采摘和观光价值。

水生植物区在烟水鱼庄配置水生植物园，栽植各类本地水生植物，同时培育沿岸湿生植物，并恢复自然群落（见图4-46）。

西溪梅园-集中种植各类梅花，规划赏梅区、梅节活动和科普教育基地（见图4-47）。

竹溪林径-在西溪梅竹山庄以西，补植竹林。

保护秋雪庵芦苇荡和荻荡-保护芦苇和荻的生长，形成独特的芦苇、荻景观，重现"秋雪蒙钓船"的意境。

quality but the possibility of circulation with outside water bodies, water replacement, dredging, the artificial breeding of wild fish and mollusks, such as snails and clams, and air-pumping are all recommended ways of improving the water's self-purification functions. Ponds with poor water quality and little possibility of restoration may be converted into microbial treatment pools with higher aquatic plants to form a shallow-water wetland landscape.

(4) Constructing wastewater treatment facilities to process wastewater of local residents.

(5) Augmenting the wetland ecosystem, particularly in terms of the growth and protection of water- and land-based vegetation, to create a desirable ecosystem that is structurally complete, fully-functional and resilient to a host of adversities.

(6) Water quality monitoring scheme
Water quality is to be monitored every three months, while the quality of bottom mud is to be monitored every six months at representative point sources.

3. *Pollution control*

According to the field study and analysis of pollution sources in Xixi Wetland, the main reason for water quality degradation lies in the different pollutant discharges from inside and outside the site. All pollution sources should therefore be placed under strict control in order to repair the water body and to improve the site's water purification capacities.

(1) Controlling the local population
The 624 households of residents inside the Phase One restoration area of Xixi Wetland will be relocated, with the exception of those that would provide services in ecological conservation and tourism.

(2) Improving land use and agricultural exploitation
The general planning of Xixi Wetland and the land use for architectural construction is under the strict control of its planners. The proportion and number of fishery ponds, submerged farmlands and dry farmlands must be wisely regulated. The use of chemical fertilizers and pesticides is to be prohibited, and organic composite fertilizers are recommended. Fishery production comes mainly from breeding in the natural landscape, while livestock and poultry are simply excluded from the site.

4. Restricting navigation
Restrictions must be placed on the speed, number,

▲ 图4-43: 利用换水、清淤、放养等方法，加强内陆池塘和外荡水体之间的回圈（提供：西溪湿地公园管理办公室）
Figure 4-43: The water circulation between the ponds in the site and water body outside the site may be enhanced through water replacement, dredging and artificial breeding of fish and mollusks
(Image courtesy of the Administration Committee of Hangzhou Xixi National Wetland Park)

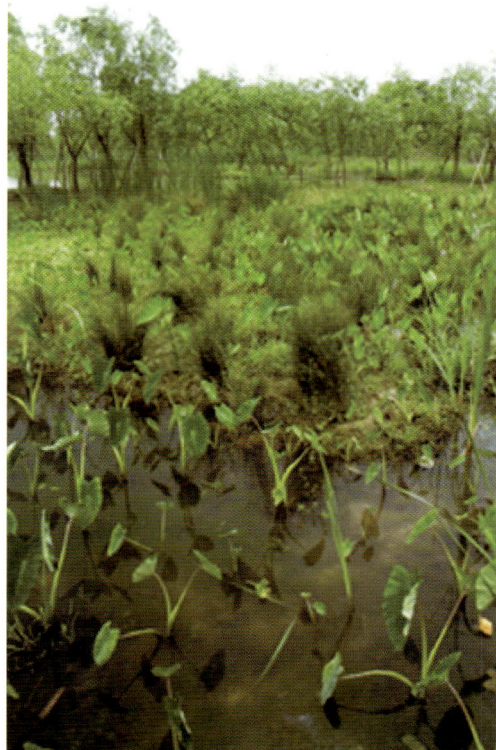

▲ 图4-44: 通过水生及湿生植被的培育，恢复西溪湿地生态系统(摄影：易道)
Figure 4-44: Restoring the wetland ecosystem through the growth of water- and land-based vegetation (Photograph by EDAW)

▶ 图 4-45: 怡然自得地划桨观光能够减少航
运污染 (摄影：易道)
Figure 4-45: Traveling on a rowing canoe provides
a good sightseeing experience. More importantly, it
helps reduce pollution from navigation.
(photograph by EDAW)

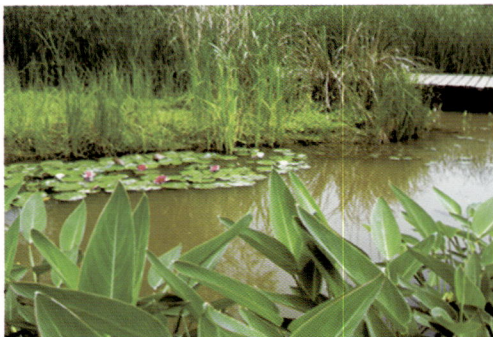

▲ 图 4-46: 烟水鱼庄的水生植物区 (摄影：易道)
Figure 4-46: The garden of aquatic plants in the Manor
of Misty Waters (Photograph by EDAW)

保护杨柳林，防止外来物种入侵-合理
配置草坪面积，配置的草坪具有高的生
物多样性和水土保持能力，另一方面具
有耐践踏，低费用的特性。如：双穗雀
稗、结缕草、狗牙根等。

(iii)景观保护与修复（见图4-48,4-49）
地形、地貌的保护与修复
(1)保持低洼地形，培育低洼湿地景观
(2)加强河网沟通，恢复平原河网景观
(3)扩展局部水面，形成局部湖沼景观
(4)保留现有岛渚和岛渚上的自然植被
(5)保留水质和生态良好的池塘

建立湿地生态景观
(1)建立湿地景观培育区：采用封育为主的
方式
(2)保护和恢复塘－基－渚湿地景观
(3)恢复平原河网的湿地景观

结论
一期工程占地3.46km²，94%是虾龙滩、朝
天暮漾、费家塘3个原生态保护区。在实施
过程中，原先隔开的小面积水体被连通，
这有利于物种在较大的空间自由交流，吸
引水鸟栖息。鱼塘的垂直塘基被改造为缓
坡，较有利于湿生植物生长，为过往鸟类
提供更丰富的食物。现在湿地中保留和恢
复作为次生湿地标志的鱼塘多达383个，原
有的2802棵柿树也都得到保留。

由于实施限建，留出了大片空地，农居拆
迁后也留下不少裸露地表。采用了苦楝、
枫杨、大叶柳、竹等乡土树种进行植被恢
复。补种的约16.67hm²芦苇不仅恢复了湿地
景观，也净化了湿地水体。在现有的大块
湿地、林地、草地、农田、养殖场附近的
树木、院落等地设置了人工鸟巢，吸引更
多鸟类栖息。

西溪湿地的文化也与自然生态同步得到恢
复。湿地公园将散落在民间的农耕用具以
及原居民生活用品收集起来，重现西溪
湿地原貌，不负"文化西溪"的美誉。
（见图4-50）

▼ 图 4-47: 西溪梅园为赏梅区、梅花节活动和
科普教育基地(提供：西溪湿地公园管理办
公室)
Figure 4-47: The Plum Garden for sightseeing, plum
festivals and scientific education
(Image courtesy of the Administration Committee of
Hangzhou Xixi National Wetland Park)

destination and kinds of vessels that use the site as a transportation route. The number of motorized boats must be put under control. Currently, rowing canoes used by locals have been the main vessel for sightseers to the Xixi Wetland. These efforts would help mitigate the negative impacts of navigation on water quality and reduce the risk of riverbanks being washed thin by constant boat-induced surges.

(ii) Conservation and restoration of the vegetation

Persimmon woods: The existing orchards may be reorganized to house persimmon woods, which have high economic yields and add sightseeing value.
Garden of aquatic plants in the Manor of Misty Waters: Local hydrophytes are planted in garden water areas, while phreatophytes are planted on the banks. Native communities have also been restored in the garden.

The Plum Garden: Different species of plum flowers have been planted for sightseeing, plum festivals and scientific education.

The Creek Path: Located in the Bamboo Grove, it offers a tour in artificially planted bamboo in the west of Xixi Wetland.

Reed marshes (*Phragmites communis* and *Miscanthus sacchariflorus*): Grown to improve aesthetic value near the Nunnery of Autumn Snow.

Willow tree groves: Provides cover in wetland areas and improves aesthetics.

Prevention of foreign species invasion: This will increase the survival rate of planted and native species.

Strategic configuration of meadow size: Well configured meadows provide good service in biodiversity enhancement and erosion control. They are cost-effective and resilient to trampling. Some applicable meadow species include knotgrass (*Paspalum distichum*), Chinese lawngrass (*Zoysia sinica*) and Bermuda grass (*Cynodon dactylon*).

(iii) Landscape conservation and restoration

Means of conserving and restoring topography and terrain features include:
(1) Retaining low-land topography and establishing a low-land wetland;
(2) Increasing water circulation between river networks and restoring the river networks of floodplains;
(3) Partially expanding water space to form new lakes and ponds;
(4) Preserving existing islets and the natural vegetation on the islets; and
(5) Preserving the ponds with good water quality and a well-developed ecosystem.

Principles for establishing an ecological wetland landscape include:
(1) Establishing landscape development zones which are fenced off from human activities;
(2) Preserving and restoring the pond islet formation; and
(3) Restoring the floodplain's river network wetland features.

Conclusion

Three natural reserves account for 3.46 square kilometers or 94% of the Phase One construction of Xixi Wetland. During the Phase One restoration, previously isolated water patches were integrated in order to allow the free association of different species in a larger space and to provide desirable food sources for birds attracted to the site. Previously horizontal levees of fishery ponds have also been reconstructed into sloped ponds in order to facilitate the growth of phreatophytes, and to provide sufficient food for passing birds. Three hundred and eighty-three fishery ponds have been retained or restored, and all 2,802 persimmon trees have been kept intact.

As a result of the successful limitation of human activity and development along with the relocation of residential abodes, large patches of land on the site were left vacant. These lands have been covered with local plants like bead trees, Chinese ash trees, large-leaf willows and bamboo. Approximately 16.67 square kilometers of reeds have been planted to restore the wetland landscape and purify the water body. Birds are now attracted to the large wetland areas, woods and grasslands, and to the artificial bird nests in the trees and courtyards near farmlands and aquatic farms.

Additionally, the Xixi Wetland has undergone a restoration of not only its natural ecology, but of its cultural heritage. Local farming tools and implements have been collected for display in order to add to the historical and cultural sensibility of the site. The conservation and restoration measures implemented for the Xixi National Wetland Park have helped restore the site's ecosystem and have maintained the site's unique combination of city life, agriculture and traditional culture. The Xixi wetland example shows how site restoration can occur while maintaining much of the historic use of the site through wise land uses that are advantageous to the ecosystem.

▲ 图 4-48: 对于民居采用修旧如旧的翻新方式
（提供：西溪湿地公园管理办公室）
Figure 4-48: Renovated residential houses that retain their traditional look and form (Image courtesy of the Administration Committee of Hangzhou Xixi National Wetland Park)

▲ 图 4-49: 拆除、改善建筑物以恢复人文景观的协调性（提供：西溪湿地公园管理办公室）
Figure 4-49: The removal or restoration of existing architecture may help restore the harmony of human structures (Image courtesy of the Administration Committee of Hangzhou Xixi National Wetland Park)

Peter Bridgewater
Sec. General Ramsar
Convention : A wonderful
Wetland, an example for
the cities of the world!

▲ 图 4-50: 2006年湿地公约局秘书长 Peter Bridgewater 先生参观西溪湿地公园并题字
（提供：西溪湿地公园管理办公室）
Figure 4-50: The Secretary General of the Ramsar Convention, Dr. Peter Bridgewater, visiting Xixi the wetland in 2006 (Image courtesy of the Administration Committee of Hangzhou Xixi National Wetland Park)

资料提供
优斯公司
PANYA 顾问公司

Information provided by

URS

PANYA CONSULTANTS

4.2 滨海湿地恢复

4.2.1 湿地恢复与水产养殖一体化工程——巴帕南河区域开发项目，泰国南部

背景

区域环境

巴帕南位于泰国南部，被视为洛坤府（见图4-51）的中心。巴帕南河及其邻近区域受到许多人为因素影响，缺乏有效的流域规划和资源管理。该流域面积约3075km²，旱季时，河口边界向上游延伸100km。历史上，该流域大部分地区是红树林和淡水湿地（Boromthanarat, S. et al.）。随着巴帕南的人口增长，重要的红树林和淡水湿地都被开发占用。多数红树林的丧失都归咎于将红树林改造成水产养殖池塘，如图4-52所示土地利用图。

1994年，泰国政府意识到该流域正在退化，于是出台了多个流域规划战略，实施各种结构性控制，促进对水资源的管理。该计划名为"巴帕南河流域开发工程"，该工程的两大重点是：(1)自然资源和环境的修复；(2)针对虾类养殖和稻米种植的管理。该计划的主要目标是提供一个可持续发展的解决方案，使其可以平衡有关各方的需要，包括稻农、虾农、洪水控制区、市政供水机构和环保人士。

▼ 图4-51：巴帕南地区和巴帕南湾位置图（绘制：易道）
Figure 4-51: Pak Phanang Region and Pak Phanang Bay Inset (Graphic: EDAW)

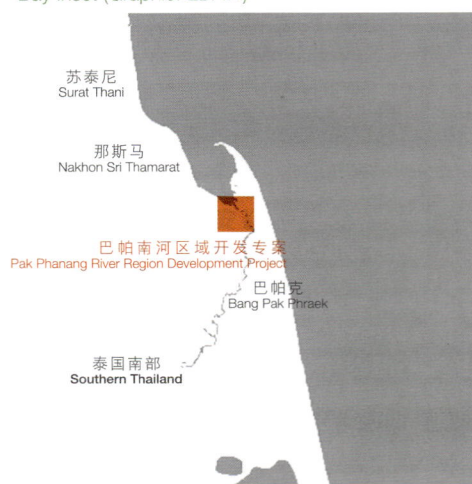

苏泰尼
Surat Thani

那斯马
Nakhon Sri Thamarat

巴帕南河区域开发专案
Pak Phanang River Region Development Project

巴帕宽
Bang Pak Phraek

泰国南部
Southern Thailand

4.2 Coastal Wetland Restoration

4.2.1 Integrated Wetland Restoration & Aquaculture: Pak Phanang River Region Development Project, Southern Thailand

Background

Regional Setting

Pak Phanang is located in southern Thailand, and is viewed as the heart of the Nakhon Si Thammarat Province (see Figure 4-51). The Pak Phanang River and adjacent areas have suffered under the burden of numerous anthropogenic factors and the failure to provide effective watershed planning and resource management. The watershed is approximately 3,075 km² and is characterized by estuarine conditions extending 100 kilometers upstream from the mouth in the dry season. Much of the watershed was previously mangrove forest and freshwater wetland (Boromthanarat, et al, 2005). As the population grew in Pak Phanang, development encroached on vital mangrove and freshwater wetland areas. Much of the mangrove loss can be attributed to conversion of mangrove forest to aquaculture ponds, as is shown by the land use maps in Figure 4-52.

In 1994, the Royal Thai Government realized there was a cycle of watershed deterioration and initiated a number of watershed planning strategies and to implemented various structural controls to aid in the management of water resources. The plan was called the Pak Phanang River Region Development Project. Two of its major focus areas were the rehabilitation of natural resources and the environment in conjunction and the management of shrimp and rice farming needs. The main objective is to provide a sustainable solution that can balance the needs of various stakeholders, including rice farmers, shrimp farmers, flood control districts, municipal water supply agencies, and environmentalists.

There is a growing scientific and professional awareness of the economic, environmental and social importance of mangrove forests (UNEP-WCMC 2006). Likewise, there is a vital need in the Pak Phanang region, and throughout Southeastern Asia, to develop a plan for mangrove habitats to coexist with aquaculture systems to ensure a sustainable economic future for Thailand and its neighbors. The need is based on multiple synergies between mangrove forests and coastal life including minimizing shoreline erosion and generating significant primary productivity that supports estuarine food webs.

人们越来越意识到红树林在经济、环境和社会方面的价值(UNEP-WCMC,2006)。巴帕南地区急需制定红树林与渔农生产的双赢计划，从而确保泰国及其邻国未来经济的可持续发展。这种需要以红树林和滨海生物的多方面协同为基础，包括减小海岸侵蚀、创建支持河口食物网的重要基础生产力。除了泰国之外，在东南亚的许多地区都面临着同样的问题。

2005年，美国贸易和开发署(USTDA)与泰国水资源协会及自然资源和环境政策与规划办公室(ONEP)共同资助了一项研究，调查巴帕南地区滨海和湿地恢复的可行性。优斯公司(URS Corporation)受委托进行可行性研究，并为试点项目场地提供设计方案。

基 地

如图4-53所示，试点项目场地位于巴纳空(Pak Nakhon)区域。场地面积约23km²，北邻巴帕亚(Pak Phaya)运河，南接巴纳空运河，东临巴帕南湾。历史上，整个项目区都被红树林所覆盖(CORIN 1991)，巴帕南湾的整条岸线都布满了茂密的森林和红树林生态系统，从湾边向内陆延伸1000~4000m不等。1974~1989年间，项目区内78%的红树林生态环境被改造成水产养殖池塘，主要用于养虾，其余的则改成农田或鱼塘。

1974年，该地区约592hm²土地被泰国皇家林业部出租，用作虾类养殖的合作试验地。1983年起，合作试验地开始大规模兴建虾塘，1985年起开始养殖和出产虾类。人们在这块土地上修建了纵横交错的进水和出水沟渠。然而，住在这里的人们也面临着不少问题。首先，附近巴帕亚运河和巴纳空运河沿岸的住宅区和工业区排出的废水会进入养虾塘，虾塘排出的废水也进入到两条运河中；其次，该地区在1985、1987和1988年接连发生暴雨，受到特大洪水的灾害；1991年又遭遇了一场大台风的袭击。此外，虾类养殖缺乏水产管理技术。这些因素都严重影响了当地居民的生计。为解决这些问题，渔业部于1994年在这块合作试验地上开始兴建"盐水灌溉工程(SWIP)"，旨在提高水产养殖的水质，并为合作试验地上的居民提供更先进的水产养殖技术。

如今，盐水灌溉工程面临着资金问题，并且该工程所在地区面临着环境继续退化，

In 2005, the United States Trade and Development Agency (USTDA), in cooperation with the Thailand Water Resource Association and Office of Natural Resources and Environmental Policy and Planning (ONEP), funded a study to look at the feasibility of restoring coastal and wetland areas in the Pak Phanang Region. URS Corporation was hired to complete the feasibility study and develop a conceptual design for a pilot project site.

Site Description

The project site selected for the pilot study is located within the Pak Nakhon Sub-district, as shown in Figure 4-53. The project site consists of approximately 23 km² (14,400 rai) situated between Pak Phaya Canal to the north, Pak Nakhon Canal to the south, and Pak Phanang Bay to the east. Historically, the entire project area was covered by mangrove forest (CORIN 1991). In fact, dense forest and mangrove ecosystems extended along the entire shoreline of Pak Phanang Bay, extending from the Bay edge inland, anywhere from 1,000 to 4,000 meters. Between 1974 and 1989, 78% of mangrove forest habitat within the project area was converted to aquaculture ponds, primarily for shrimp farming, and the remaining was converted to integrated farming and fisheries.

Approximately 592 ha (3,700 rai) within the center portion of the project site was leased from the Royal Forestry Department to setup a land cooperative for shrimp farming in 1974. Construction of the shrimp ponds in the area began in 1983, and shrimp aquaculture production began in 1985. A network of interconnected influent and effluent canals was constructed within the land cooperative. However, people within the land cooperative had problems due to wastewater from the communities and industries along Pak Phaya and Pak Nakhon Canals. Wastewater from shrimp farming within the land cooperative also circulated back into the canal system. In addition, major storms caused significant flooding in the area in 1985, 1987, and 1988, in addition to a major typhoon in 1991.

These factors, including the lack of sustainable aquaculture management technology, had a significant impact on the livelihood of farmers in the area. To address these problems, the Department of Fisheries initiated a Saltwater Irrigation Project (SWIP) in 1994 for the land cooperative within the project area to improve the water quality for aquaculture and provide enhanced aquaculture technology for farmers in the land cooperative.

Today, the SWIP faces funding problems and the SWIP areas are confronted by further environmental degradation and water management conflicts among villagers. Many

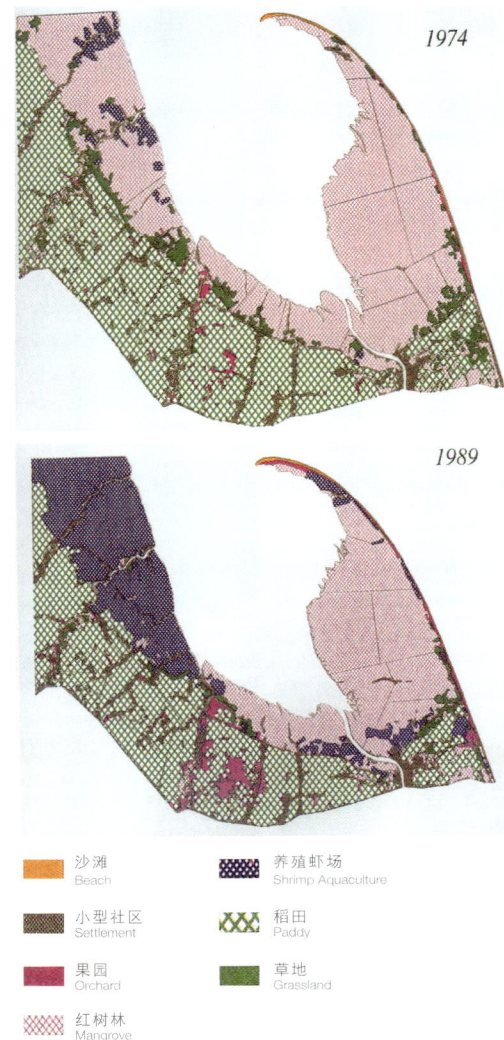

图 4-52：巴帕南地区的土地利用历史资料 (CORIN 1991)(提供：优斯公司)
Figure 4-52: Historic land use data for the Pak Phanang region (CORIN 1991)(Images courtesy of URS)

村民由于水管理而产生纠葛，许多虾塘和鱼塘都基本废弃，在鱼虾或作物发生疾病，或养殖成本升高，使虾塘和鱼塘无利可获时更是如此(Lewis et al. 2003)。当地人主要从事种田、虾类养殖，还有养虾的一般性劳动、工厂和水产业的其他工作。水产业在该区占主导地位，包括生产鱼粉、鱼酱、虾干、仓储、冰块生产以及渔船修理与保养服务等。

项目所在区的主要环境问题是那些大量废弃的虾塘，社会问题则是水产经营失败产生的债务以及行业冲突。这些冲突来自人们对经济区划的不同看法。村民认为政府的政策未能解决他们的经济问题。此外，村民认为红树林退化也是个大问题。他们期望能够有专家给出建议和最好的方法，遏制红树林的退化，恢复红树林。

项目目的

优斯公司的试点研究主要实现以下目标：
(1) 修复或恢复滨海红树林栖息地；
(2) 加强海岸保护，降低海水侵蚀；
(3) 为当地提供经济可持续发展的方案；
(4) 制定监督机制，在项目进行过程中评估项目是否取得成功。

▼ 图4-53：项目基地地图（提供：优斯公司）
Figure 4-53: Project Site (Image courtesy of URS)

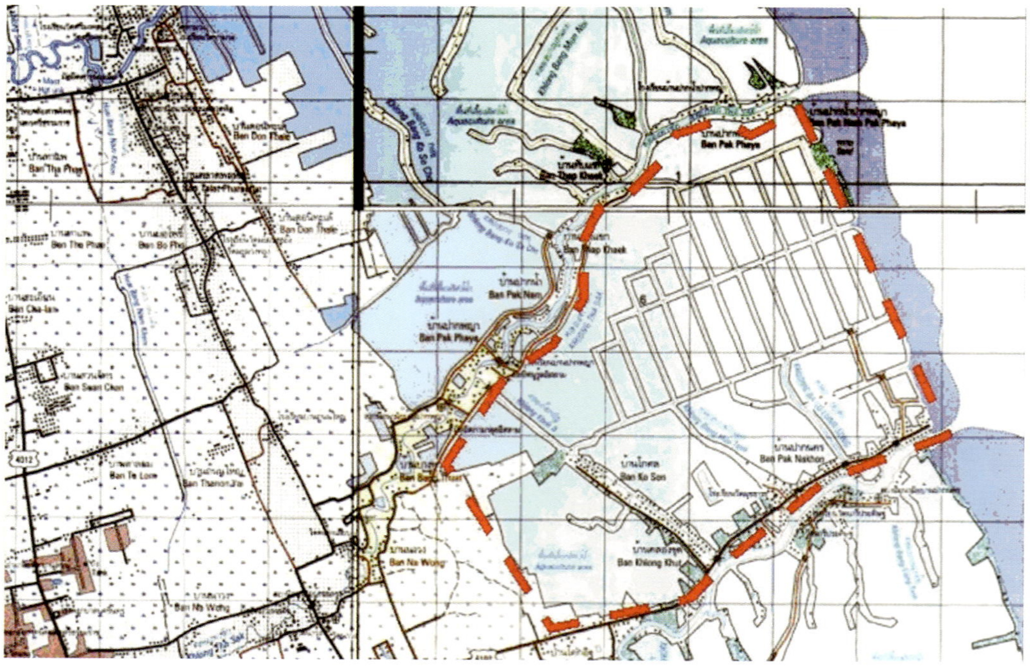

of the shrimp and fishponds have been abandoned or disused. This typically occurs once the ponds become unprofitable, usually because of declining yields or crop failures due to disease or increased production costs (Lewis et al, 2003).

The primary means of income in the community, in addition to agricultural and shrimp farming, are general labor for shrimp farming, industry, and fisheries. Fishery-related industries dominate the area and include production of powdered fish, fish sauce, dried shrimp, storage, rice production, and fishing boat repair and maintenance services.

The main environmental and social problems within the project area are the large number of abandoned shrimp farms, the debts associated with failed aquaculture operations, and business conflicts. These conflicts arise from different perspectives regarding economic zoning. It is the opinion of the villagers that the government's policy does not address their economic needs. In addition, mangrove forest degradation is a major concern. There is a need for immediate rehabilitation of the area. The villagers have expressed a desire to have experts recommend the best way to mitigate forest degradation.

Goals and Objectives

The main objectives driving the URS pilot study were to:

(1) Rehabilitate or restore coastal mangrove habitat
(2) Increase coastal erosion protection
(3) Provide sustainable economic opportunities for the local community
(4) Develop a monitoring strategy to evaluate success over time

Project Components and Restoration Approaches

This section provides a summary of design criteria and system processes associated with the proposed land-use components or opportunities for the project area. These components were developed based on the goals and objectives established for the project, meetings with stakeholders and resource agencies, research on similar projects throughout Southeastern Asia, and the results of the site characterization. Specific opportunities include mangrove forest rehabilitation, sustainable open-water aquaculture, integrated mangrove aquaculture, and wastewater treatment mangrove areas. The integrated mangrove aquaculture component includes various commercial benefits from both aquaculture and silviculture. Project components are listed below.

项目组成与恢复方法

依据本项目的目的、对项目场地的认识和评价，并借鉴东南亚地区同类研究项目的经验，各相关部门和资源机构召开会议，共同确定了本项目的组成部分。主要包括：

- 红树林修复
- 开放水域的可持续水产业
- 红树林与水产一体化系统
- 用于废水处理的红树林区

其中，红树林与水产一体化方案考虑了水产业和造林业的商业利益。

(1) 红树林修复

红树林修复是指部分或完全修复红树林的生态系统结构和功能，红树林恢复是指将生态系统还原成原有状态的行为（Field 1998, Ellison 2000）。生态系统得到恢复是修复工程成功后可能出现的结果，但也存在其他可能性，比如红树林水产一体化系统。本案例中，"修复"一词是指为改善项目区红树林的状况而采取的所有策略和落实措施。

论述修复或恢复红树林的方法的科学文献有很多（Lewis and Marshall 1997; Field 1998; Lewis 2005），例如在废弃不用的虾塘中修复红树林（Steven son et al. 1999）。Lewis & Marshall(1997)认为在成功的红树林恢复工程中，有以下五个重要的步骤：

1. 了解当地红树林物种的生理生态特征（基于物种的生态学研究），特别是繁殖格局、繁殖体分布和种苗定植。

2. 了解、控制红树林分布状况的一般水文形态，目标红树林物种的成功定植和生长。

3. 对阻碍原红树林自然的次级演替的改变进行评估（如对虾塘征税，水控制工程等）。

4. 在设计初期恢复时，先恢复相关水文系统，利用红树林繁殖体自然补充的办法来培育红树林。

5. 如果第1至4步确定了自然补充无法实现恢复工程的幼苗成功定植量、稳定率或幼苗成活率目标，那么只能种植繁殖体、采集的幼苗或培育的幼苗。

Mangrove Forest Rehabilitation

Mangrove forest rehabilitation involves partial or full replacement of mangrove ecosystem structure and function, whereas mangrove forest restoration is the act of returning the ecosystem to its original condition (Field, 1998; Ellison, 2000). Ecosystem restoration is a potential outcome of a successful rehabilitation project, but there are also alternative possibilities such as integrated mangrove-aquaculture systems. As applied to this case study, the term "rehabilitation" refers to the general approach and implementation actions taken to improve the mangrove forest condition in the project area. The term "restoration," is used in a theoretical sense to describe the science of mangrove forest recovery.

There is an abundance of scientific literature that discusses methods to rehabilitate or restore mangrove forests (Lewis and Marshall, 1997; Field, 1998; Lewis, 2005), including mangrove rehabilitation in disused shrimp ponds (Stevenson et al, 1999). Lewis and Marshall (1997) have suggested five critical steps to achieve successful mangrove restoration:

1. Understand the autecology (individual species ecology) of the mangrove species at the site, in particular the patterns of reproduction, propagule distribution, and successful seedling establishment.

2. Understand the normal hydrologic patterns that control the distribution and successful establishment and growth of the targeted mangrove species.

3. Assess the modifications of the previous mangrove environment that prevent natural secondary succession (e.g., levied ponds, water control structures, etc.).

4. Design the restoration program to initially restore the appropriate hydrology and utilize natural volunteer mangrove propagule recruitment for plant establishment.

5. Only utilize actual planting of propagules, collected seedlings or cultivated seedlings after determining through Steps 1-4 that natural recruitment will not provide the required quantity of successfully established seedlings, or a rate of stabilization or growth of saplings that were set as goals for the restoration project.

These five steps were utilized in the design of mangrove rehabilitation areas within the project site.

在本项目区内红树林的修复设计中应用了上述五个步骤。

a. 对滨海红树林的修复

修复滨海红树林对于防止海岸侵蚀至关重要，而且还能够提高红树林的生物多样性。沿海修复区从海岸线向内陆延伸约100m远。在此区域范围内，约有53km²的水产养殖池塘将恢复成红树林。

滨海红树林修复区一般不需要太大的平整工作。有些池塘的情况特殊，可能需要一些平整。将原有沟渠填平，并在塘底建起合适的坡度，促进落潮时的排水状况。水文动态模型得出结论：在已有塘底的高度范围内，塘底坡度最小的状态，是适合红树林生长的水文形态。

沿海岸线的外侧堤岸被保持原样，它们可以保护滨海红树林恢复区内正在生长的红树林。在没有堤岸的地方，则使用灌木篱笆代替。在项目区内的池塘里，可以利用或改造已有的水控建筑，以降低潮汐对红树林的影响，尤其在红树林种下的头1到3年内，这些建筑能够阻止潮汐将红树林幼苗连根拔起。在红树林的根系稳固后，即可拆除外侧堤岸和水控建筑，让红树林接触到自然潮汐。池塘的内岸可以推平或降低，挖出的土壤可以用于塘底平整。最终潮汐的流动更自然。

同时，在沿海水产养殖池塘积极的植树活动可加速修复过程，防止海水侵蚀该海岸，并可尽早带来栖息地的益处。

b. 对内陆红树林的修复

内陆红树林的修复措施与上文论述的滨海红树林的修复相似，但内陆的植树策略稍有不同。

该策略重在通过自然补充实现次生演替，而不是积极植树。最初的三年时间试验了自然补充法。如果这个方法不能达到功能标准，则可换种繁殖体和苗木。在恢复区综合采用自然补充和直接种植的方法将为这里的修复工程建立一个积极的回馈机制。换言之，扎根的树木会长大、成熟、结籽，而这些繁殖体将散播至整个项目区。

在项目区的环境渐趋稳定以后，可在内部地区种植中期或晚期演替物种来提高该地

a. Shoreline Mangrove Forest Rehabilitation

Rehabilitating the mangrove forest along the coastline is critical for erosion protection and has the added benefit of providing mangrove biodiversity. The shoreline rehabilitation zone would extend approximately 100 meters inland from the shoreline. In this zone, approximately 53 hectares (330 rai) of aquaculture ponds would be rehabilitated into mangrove forest.

The area designated for shoreline mangrove forest rehabilitation should not require significant grading activities. Depending on individual pond topography, detailed grading activities may be required to fill in existing ditches and provide a sloped pond bottom to promote draining at low tide. Hydrodynamic modeling concluded that minimal pond grading within the existing pond bottom elevation range would allow for a hydrologic regime appropriate for mangrove growth.

The existing outer levees along the shoreline would remain intact and serve as protection for the developing mangrove trees within the shoreline mangrove forest rehabilitation area. In areas where the existing levee is not available to provide protection, brush fencing may be used for this purpose.

Existing water control structures for ponds in this area would be utilized or modified to provide a muted tidal regime to minimize the uprooting of developing mangroves over the first one to three years of development. After the mangroves are established, control structures and outer levees could be removed to allow for natural tidal hydrology. Internal pond levees in this area could be demolished or lowered to provide soil for pond bottom grading, and eventually to allow for a more natural tidal regime.

Active planting in the aquaculture ponds near the shoreline would expedite the rehabilitation process and provide erosion protection and habitat benefits at the earliest possible time.

b. Interior Mangrove Forest Rehabilitation

Implementation activities are similar to those discussed above for shoreline mangrove forest rehabilitation. However, the planting strategy is altered slightly for these interior areas.

This strategy focuses on natural recruitment for secondary succession rather than direct planting. In the first three years,

区的生物多样性。例如，可以在地势较高的地区种植木果棟属的树木Xylocarpus moluccensis。

执行内陆红树林修复计划是个长期的过程，监控红树林的发展情况是其中一项重要的工作。监控资料可用来优化种植，协同项目区的各个实施战略，而且资料还可应用于其他地方的恢复。

在此项目区内恢复红树林生态环境给当地居民带来很多益处：当地可以出产林业产品，特有的生态环境可以用于水生物种培育，红树林带来了环境效益，同时推动了社会的发展。此外，随着恢复区的发展，该区还拥有很大的生态旅游潜力。

开放水域的水产系统
项目区内有些地方将保留为开放水域养殖池塘，很可能安排在目前的"咸水灌溉工程范围"内。过去的实践表明，在这里运营和管理集约化、半集约化的水产养殖业都会造成"自体污染"。因此，要保证可持续发展，就必须进行改进。改进措施包括改善污水管理；改变过去泵水进池塘的做法，恢复完整的潮汐水文形态，增强冲刷作用和改善水循环。同时，该系统还综合采用不同水生物种的混合养殖、轮换养殖，具有多样化的特征。
实施工作主要包括以下工作：

- 重新设计各池塘的排水渠格局，使其将废水导入红树林废水处理湿地；

- 实施最佳管理方案，实现可持续发展并有效控制疾病；

- 提供潮汐进水渠，不再依靠泵水站进水。

在实际水产养殖中，可将虾类、甲壳类和鱼类混合养殖。养殖多个物种可以使渔民不受市场波动的影响，增加水产系统的弹性。而且单一养殖中，病原体会导致疾病暴发而造成损失。因此，提高品种的多样性是抗击大规模单一养殖带来的市场波动、防止潜在环境问题的关键。

(3) 红树林－水产养殖一体化系统

红树林－水产养殖一体化系统(IMAS)可以在种植本地红树林的同时留出区域继续进行

natural recruitment would be tested. If this method does not meet performance standards, than the site area could be planted with propagules and nursery stock. Natural recruitment and direct planting in rehabilitation areas would establish a positive feedback mechanism for site rehabilitation (trees that establish will mature and produce seed that will be dispersed throughout the site).

As the site conditions begin to stabilize, interior areas would become good locations to increase biodiversity by planting mid-to-late successional species such as the mangrove *Xylocarpus moluccensis* in areas of higher ground.

Monitoring mangrove development is an important part of the long-term project implementation plan for this component since monitoring data would be utilized to optimize planting and other implementation strategies on the site and could be applied to other sites if successful.

The restoration of mangrove habitats in the project area will provide numerous community benefits including the provision of forest products, aquatic species nursery habitat, mangrove environmental benefits, and social incentives. In addition, restoration areas may have significant eco-tourism potential as the area develops in the future.

Open Water Aquaculture Systems
Portions of the site may remain as open water aquaculture ponds. This would most likely occur within the current extent of the Saltwater Irrigation Project. However, historical intensive and semi-intensive aquaculture operation and management has proved to be "self polluting", requiring improvement to maintain sustainability. Improvements may include better wastewater management and the re-introduction of full tidal hydrology, as opposed to pumped influent, to increase flushing and circulation. In addition, incorporation of polyculture, rotation and diversification of aquaculture species are important features of the proposed system.

Implementation activities would consist primarily of the following:

- Reconfiguration of discharge channels for specific ponds to convey wastewater to proposed mangrove treatment wetland areas;

- Implementation of best management practices for sustainability and disease control; and

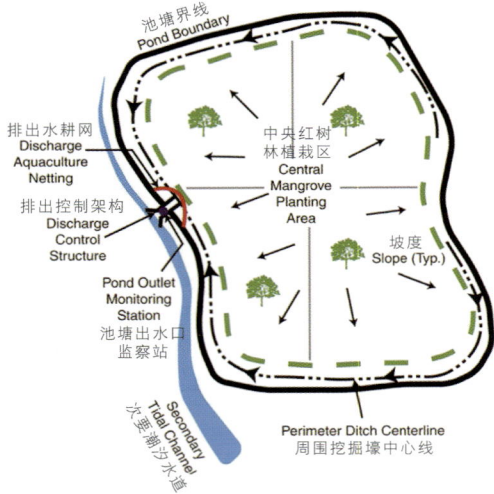

▲ 图4-54：水产养殖一体化系统的典型构造（提供：优斯公司）
Figure 4-54: Typical IMAS Configuration (Image courtesy of URS)

▲ 图4-55：在邻近该区的"绿地毯工程"修复池塘内，红树林枝繁叶茂，呈现出勃勃生机（提供：优斯公司）
Figure 4-55: The *Rhizophora* sp. mangrove species are thriving within the rehabilitated Green Carpet Project pond adjacent to the site (Image courtesy of URS)

水产养殖（见图4-54）。此外，有选择地砍伐红树林后，生产出的商业产品可能带来的效益提高了该系统在经济上的可持续性。IMAS是指全面恢复某些地区的潮汐水文形态，同时通过水控建筑继续进行水控制，从而提高虾塘或其他水产养殖池塘的可持续运营中的灵活性。

印尼的Empang Parit模式、巴帕南的"绿地毯工程"模型（见图4-55）都是典型的林渔系统。这这种系统中，都是将池塘中央部分垫高，种上红树林，周围则是较深的水渠，专门养殖鱼、虾或蟹等。

目前，紧邻项目区，有一个IMAS正在运作。日本资助的"绿地毯工程"是一自然保护团体组织，目前，该组织和巴帕南河开发局与当地农民紧密协作，成功地在恢复了红树林的同时，保留了有赢利的虾蟹养殖系统。

IMAS的实施工作包括土地平整和土木工程、堤坝缺口挖掘、清除水道淤泥和种植红树林物种。

(4) 红树林污水处理湿地

如前所述，红树林可用于处理各种污水，包括养殖池塘产生的污水。据估计，过滤 1km² 池塘产生的氮和磷大约需要 2~22hm² 的红树林 (Robertson 1995)。因此，本项目有 可能利用项目区的部分地区建立

- Providing tidal influent channels independent of pump stations.

Aquaculture practices should include a mixture of shrimp, crustacean and finfish farming. Cultivating multiple species protects farmers from fluctuations in the markets, and adds resilience to the system. Monocultures are more likely to fail as a result of disease because the entire crop is susceptible to certain pathogens. Diversification is thus essential to weathering the market fluctuation and potential environmental problems associated with large-scale monocultures.

(1) Integrated Mangrove Aquaculture Systems

Integrated mangrove aquaculture systems (IMAS) allow for the establishment of native mangrove forests, while reserving area for continued aquaculture. In addition, the potential for marketing commercial products obtained from selective mangrove harvesting increases the overall economic sustainability of the system. An integrated mangrove aquaculture system involves the reintroduction of full tidal hydrology to certain areas, while retaining water control using water management structures to provide flexibility for operation of sustainable shrimp or other aquaculture facilities (see Figure 4-55).

A typical silvo-fisheries system, such as the Empang Parit model from Indonesia, or the Green Carpet Project model in Pak Phanang, features a pond with a raised central platform planted with mangroves, surrounded by a deeper canal that provides a permanent aquaculture area for fish, shrimp, and/or crabs.

Currently, there is a working model of an integrated mangrove aquaculture system directly adjacent to the project area. The Green Carpet Project, a Japanese-funded conservation group, and the Pak Phanang River Development Station are working closely with local landowners and have succeeded in reestablishing mangrove forests while maintaining a profitable shrimp and crab aquaculture system.

Implementation activities for this component would include grading and earthwork operations, levee breach excavation, channel dredging, and mangrove species planting.

(2) Mangrove Treatment Wetlands

Mangrove forests have been shown to provide benefits in the treatment of various types of wastewater, including

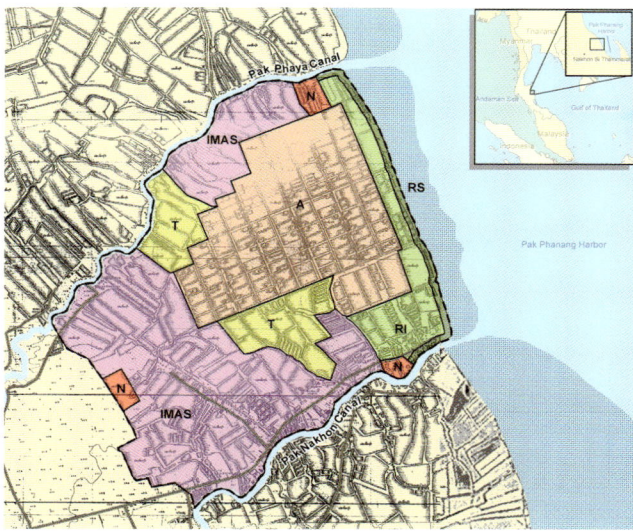

图例
Legend

基地范围
Approxmiate Project Boundary

河道
River

道路
Road

土地利用计划
Land Use Components

A 水产养殖区
Open Water Aquaculture

IMAS 红树林水产养殖一体化系统
Integrated Mangrove Aquaculture System

N 红树林培育区
Mangrove Nursery

RI 内陆红树林恢复区
Interior Mangrove Rehabilitation

RS 滨海红树林恢复区
Shoreline Mangrove Rehabilitation

T 实验性红树林处理湿地
Experimental Mangrove Treatment Wetland

养殖污水湿地处理区。污水在此经过处理再排入巴帕南港，水质可望得到改善。

这一部分的实施工作包括土地平整和土木工程、堤坝缺口挖掘、清除水道淤泥和种植红树林物种。

项目的其他规划
上文所述的各个系统为项目区提供了不同的设计方案。图4-56展示了第一种方案。实施各种方案可能牵涉到重建或推平项目区内的一些池塘。这时，可签订共同所有权或合作协议，分担责任，共享利益。

社区利益
拟建设的工程将利用上述各组成部分，为社区制定一个可持续发展的恢复和经济计划。将给社区带来各方面的利益。

(1) 红树林的用途
系统地、有选择地进行红树林的砍伐工作不仅能够使红树林生产出商品，而且水产养殖的产量会保持相对较高的水准。红树木质耐腐蚀，而且经得起水生无脊椎动物的钻蚀，因此是极好的商业建筑（例如房屋、船舶和防波堤）材料，还可用于生产新闻纸、火柴杆和火柴盒（UNEP 2006）。红树林还出产烧柴、肥料、杀虫剂、木炭、捕鱼和水产养殖材料、纺织品和皮产品、建筑材料、食物和药材。所有这些红树林方面的经济活动都可以用可持续的模式进行。

(2) 水产业持续发展
整个东南亚地区的研究以及本文中"绿地毯工程"表明，IMAS是一个成功的模式。然而，红树林地区并不适合集约或半集约的虾类养殖。除了虾类，混合养殖（虾鱼混养）、轮换养殖（例如泥蟹与虾类）和多样化养殖（例如将虾塘改造成遮目鱼鱼塘）（Lewis et al. 2003）已见成效，因而可在本项目中采用。

(3) 长期的经济可持续性
盐水灌溉工程及邻近地区当前的营运方式不具有可持续性。当一个生产区内汇集了过多的虾塘时，就造成"自身污染"。自体污染不仅会使水质下降，而且会诱发疾病最终暴发，导致经济下滑（Lewis et al. 2003）。1993年以来，本来很丰富的虾苗一直以令人震惊的幅度减少，据在三角洲红树林水渠中放置大型固定

aquaculture wastewater. It is estimated that between approximately two and 22 ha of forest are required to filter nitrogen and phosphorus loads from effluent produced by a one hectare pond (Robertson, 1995). There is the potential to utilize a portion of the existing site area for establishment of a wetland treatment zone for aquaculture wastewater. This area would provide water quality benefits for wastewater prior to release in Pak Phanang Harbor.

Implementation activities for this component would include grading and earthwork operations, levee breach excavation, channel dredging, and mangrove species planting.

Project Alternative Layouts

The components outlined above were utilized to develop alternative site layouts for the project area. Figure 4-56 shows one possible layout. The implementation of various alternatives may involve the reconstruction or loss of certain existing aquaculture ponds within the overall site. Where this situation occurs, communal ownership or cooperative agreements would be developed to share the responsibilities and economic benefits from realigned ponds.

Community Benefits

The proposed project will utilize the components listed above to develop a sustainable restoration and economic plan for the community. The following are specific community benefits associated with this opportunity.

(1) Mangrove Uses
With selective and managed mangrove harvesting, aquaculture yields can remain relatively high while yielding commercial mangrove products. Mangrove wood is resistant to rot and to the boring activities of many marine invertebrates, and is therefore excellent for use as commercial timber for building (for houses, boats and jetty construction) and for making newsprint, matchsticks, and matchboxes (UNEP 2006). Other typical products extractable from mangrove forests include: fuel wood, fertilizer, pesticides, charcoal, fishing/aquaculture materials, textile and leather production, construction materials, food, and drugs. All economic activities from mangrove will be conducted in a sustainable manner.

(2) Continued Aquaculture
Research throughout Southeast Asia, as well as local experience through the Green Carpet Project mentioned above, have shown the successes associated with integrated mangrove aquaculture systems. However, mangrove areas are generally not ideal for intensive or semi-intensive shrimp farming. In addition to shrimp, integrated polyculture (shrimp

网兜的渔民估计，在过去这十年中，他们的收入下降约80%。由此得出的唯一结论就是，红树林遭到毁坏后，造成虾的产量大幅度下降（Lewis et al. 2003）。IMAS是一条可行之路，可以对已有的破坏进行修复，同时，支持当地经济的可持续发展。

(4) 近岸渔业
红树林产生的有机物构成河口和近岸复杂的食物网的基础。食物网中包括细菌、真菌、微型底栖生物（身长小于1mm的动物）和不同营养级等级的大型底栖动物，具有商业价值的鱼虾通常高居食物链的顶端。因此，生产性的红树林系统不仅能够清除污染物、改善水质，而且在提高近岸鱼类丰富多样性方面还具有可观的长期净效益（见图4-57）。

(5) 潜在的市场推动力
水产养殖虾类市场在过去十年间以5%的平均增长率增长，比20世纪70年代（23%）和80年代（25%）的两位数增长相比有所减缓（Lewis et al. 2003）。虾类养殖、食品安全等问题也引起公众舆论越来越关注虾类养殖对环境和社会的影响（Lewis et al. 2003）。虾类产品主要消费国的消费群体已经创建出带

▼ 图4-57：应以更具可持续发展的经营方式来取代集中的虾类养殖（提供：优斯公司）
Figure 4-57: Intensive shrimp farming practices need to be replaced by more sustainable operations (Image courtesy of URS)

with finfish), rotation (mud crab alternated with shrimp) and diversification (shrimp ponds converted to milkfish ponds) (Lewis et al, 2003) have shown successes and could be incorporated into the proposed project.

(3) Long-Term Economic Sustainability
Current practices within the saltwater irrigation project and adjacent areas are not sustainable. When too many shrimp farms are crowded together in a single production area, it is referred to as "self pollution". This practice results in water quality deterioration and the emergence of pathogens, which have led to economically debilitating outbreaks of shrimp disease (Lewis et al., 2003). Since 1993, the abundance of natural shrimp seed has continued to decline alarmingly and shrimp fishermen who operate large, fixed bag nets in the mangrove channels of the delta estimate that their yields have declined by about 80% over the past 10 years. It can only be concluded that mangrove destruction accounts for a large, but unknown, part of the observed decline in the shrimp catch (Lewis et al., 2003). Integrated mangrove aquaculture systems are one potential pathway to repair some of the damage that has been done and to support a sustainable local economy.

(4) Near-Shore Fisheries
Organic materials produced by mangroves are the foundation for complex estuarine and near shore food webs involving bacteria, fungi, meiofauna (animals less then 1mm in size) and macrofauna at multiple trophic levels, often culminating in commercial shrimp and fish species (Lewis et al., 2003). Therefore, along with the fact that productive mangrove systems can be effective at removing pollutants thereby improving water quality, the long-term net effect on near-shore fish abundance and diversity can be very positive.

(5) Potential Market Incentives
Growth in the aquaculture shrimp market has slowed over the past decade (averaging 5% growth) compared to double-digit growth rates observed in the 1970s (23%) and 1980s (25%) (Lewis et al. 2003). In addition, public opinion is being influenced by high profile concerns over the environmental and social impacts of shrimp culture development and food safety issues. (Lewis et al., 2003). Progress has been made by consumer groups in the primary consuming countries to create a sustainably produced shrimp market with a "green" label. This could evolve into a sustainable market to provide shrimp and other aquaculture market stability.

有"绿色标签"的可持续发展的虾类生产市场，可能会演变成一个可持续的市场，以维持虾类及其他水产品市场的稳定性。

(6) 潜在的合作方式

在私营企业，激烈的竞争要求业主深入理解市场、品质标准，采用具有成本效益、能够提高品质的技术，并且决策更迅速正确(Hartwich et al. 2003)。公私合营在水产业中已得到有效应用，在发展中国家更是如此。公私合营能够使从事农业、工业研究的组织与当地农民和私营公司形成很好的协同合作。公私合营能够方便当地生产者获得市场讯息，减小交易和谈判成本，了解产品质量信息及要求，从而更好地满足消费者需要（如产品包装、产品外形和大小等）（NACA 2003）。这种合作方式可以有效地帮助发展中国家实现以消费者为中心、可持续发展的工业生产（见图4-58）。

(7) 红树林的环境效益

红树林的退化加重了暴雨和自然灾害对海岸的侵蚀和破坏，继而影响到滨海居民、旅游业及其他许多经济领域(UNEP 2006)。健康的红树林系统可以减轻洪水灾害，防止海岸侵蚀，承受泥沙沉积，清除有毒物质，补充地表径流，并为鱼类、甲壳类、野生动物提供栖息地，给社区带来娱乐、旅游、教育和研究及野生动物保护等方面的间接效益。

(8) 社会动因

修复红树林、恢复可持续的水产养殖等活动可以建立社区特色，提高社区的凝聚力。人们共同从事种子采集、幼苗培育、具有纪念意义的社区植树日、拆除鱼塘堤坝水控建筑的建设等活动。提高水产业的可持续性可以稳定及提高当地居民的经济发展。

社区代管

该项目的主要目标是从地区开始扭转自然资源被过度损耗和开采的形势。为使项目取得成功，惟一的办法是让当地社区拥有所有权，并接手系统的实施和长期管理工作。此过程的第一步就是成立当地核心工作组，其成员必须在社区中享有较高声誉，并愿意承担责任，向广大民众普及可持续发展的好处，组织初期及后续培训，并建立一个组织机构来进行可持续的红树林水产养殖系统的经营和维护工作。在泰国思琅的工程属于同类工程，那里的实践表明，重视自力更生和村民自己提出的倡

(6) Potential Partnerships

In the private sector, competitiveness requires an in-depth understanding of markets and of quality standards, cost-effective and quality-sensitive technologies, as well as improved decision-making methods (Hartwich et al., 2003). Public Private Partnerships (PPP) have been used effectively in the aquaculture industry, particularly in developing countries, to bring together organizations involved in agro-industry research with local farmers and private companies. Benefits of PPPs may include an increased access for local producers to market information, reduction in transaction and negotiation costs, increased access to product quality information and requirements, and the possibility to better respond to consumer demands through packaging, product form and size, etc. (NACA, 2003). This type of cooperation can effectively support the industry in developing countries to achieve sustainable and consumer focused production.

(7) Mangrove Environmental Benefits

The degradation of mangrove forests has increased coastal erosion and destruction from storms and catastrophic natural events, which in turn affect coastal residents, tourism operations and many other economic sectors (UNEP, 2006). Healthy mangrove systems can provide flooding attenuation, erosion protection, sediment trapping, toxicant removal, groundwater recharge, as well as fish/shellfish/wildlife habitats. Other indirect benefits to the community could include recreation, tourism, education and research opportunities, and wildlife protection.

(8) Social Incentives

Rehabilitation of mangrove forest and sustainable aquaculture activities done in conjunction with a community creates a strong community identity. The activities may include seed collection, nursery establishment, community and ceremonial planting days, breaching levees and constructing and installing water control structures. Improving aquaculture sustainability may stabilize economic opportunities for local residents.

Community Stewardship

The primary goal of the project is to begin, on a local scale, to reverse the depletion and exploitation of Thailand's natural resources. The only way a project like this can succeed is for a local community to take ownership and control over the implementation and long-term management of the proposed systems. The development of a local core workgroup is the first step in this process. The local core workgroup must consist of concerned and respected members of the community

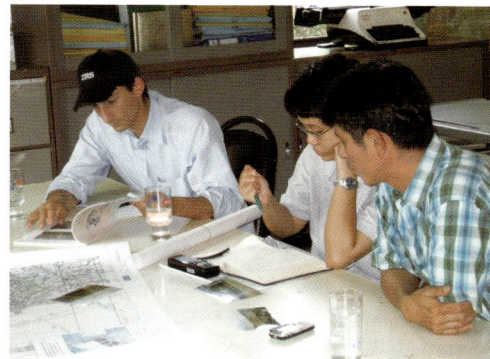

▲ 图4-58：通过合作来增强科学和市场研究可以有助于创造并维持一个可持续的经济系统（提供：优斯公司）

Figure 4-58: Partnerships to enhance scientific and market research could help create and maintain a sustainable economic system (Image courtesy of URS)

议，社区将出现一种团结奋进的意识，进而得到当地及其他政府机构的认可和支持（Charnsnoh 1998）（见图4-59）。

当地工作组成员应代表村民的立场，说出他们的意见。当地工作组必须明确理解项目的目的、方法和益处，并协助向项目区居民做出解释，该恢复工程项目才能得以落实并持续发展。工作组成员从政府和专家那里学习可持续性水产养殖技术后，传播给广大村民，并按照计划监督和评估恢复工程的发展，加强与村民的沟通，回答村民提问，帮助政府管理本区的区域划分。

项目设计与实施团队、当地核心工作组和当地居民之间将进行技术转让，包括在GIS数据库软件应用、监测资料的收集和传输设备、资料分析软件和社会经济监督战略等方面提供培训。

结 论

泰国自然资源的过度损耗与开采由来已久。该项目与东南亚地区同类项目都提倡在农村地区建立起一个不同于以往的模式，完善经济与环境的可持续性，使之代代相传。中国也存在类似情况：大规模的水产养殖基地往往建立在原来的红树林或盐沼生态系统之上。巴帕南项目为中国进行此类湿地恢复和实施自然系统—水产养殖一体化策略提供了可贵的模式，展示出如何应用水文与生物原理和建模工具来设计土地规划策略，并同时实现经济与社区的可持续发展、洪水控制、水质改善和生态环境恢复等多个目标。

willing to take responsibility for educating the community concerning the benefits of sustainability, conduct initial and ongoing training, and in organizing an institutional structure to complete the operation and maintenance tasks required for a sustainable mangrove aquaculture system. A similar project in Trang (Charnsnoh, 1998) showed that by focusing on self-reliance and village-based initiatives, communities developed a sense of unity and morale which was later recognized and supported by the local authorities and others.

Members of the Local Workgroup will understand the unique needs of the community and will act as the voice of the villagers located in the project area. The scope and approach of the restoration from this project will become implementable and sustainable only if the Local Workgroup has a clear understanding of and can help to explain the purposes, methods, and benefits to the communities within the project area. Local Workgroup members will also transfer available technologies for sustainable aquaculture from the Government and consultants to the villagers. The workgroup will further help to monitor and evaluate progress of the planned restoration, gauge villagers' understanding of the opportunities, answer questions, and help government officials regulate zoning in the area.

A technology transfer will thus take place between the project design and implementation teams, the local core workgroup, and the local community. This technology transfer includes training with GIS database software, monitoring collection and use of data transfer equipment, data analysis software and socio-economic monitoring strategies.

Summary

The depletion and exploitation of Thailand's natural resources has been a neglected problem for too long. This project, along with similar projects throughout Southeast Asia, proposes to establish a different model for economic and environmental health in rural communities that can last from generation to generation. Similar situations exist in China, where aquaculture occurs on large scales and often in areas that were historically mangrove or salt marsh ecosystems. Pak Phanang presents a valuable model for using similar integrated wetland restoration and aquaculture approaches and methodologies in China. This study shows how numerous goals including economic and community sustainability, flood control, water quality improvement, and habitat rehabilitation can be solved with a thoughtful landscape approach to planning that utilizes hydrologic and biological principles, and modeling tools.

▲ 图4-59：当地社区的参与对于创建可持续发展的红树林水产养殖模式至关重要（提供：优斯公司）
Figure 4-59: Involvement of the local community is essential for creating a sustainable mangrove aquaculture model (Image courtesy of URS)

资料提供
刘荣成、林竑斌
福建省惠安县林业局

Information provided by
Liu Rongcheng, Lin Hongbin
Fujian Province Forestry Ad-
minstration of Huian County

4.2.2 红树林湿地恢复 —— 洛阳江红树林湿地恢复，中国福建省惠安县

背景

洛阳江位于福建省东南部，北纬24°47`~24°58`，东经118°38`~118°52`之间（见图4-60），是泉州湾河口湿地的主要构成部分。洛阳江流域曾经分布了大片的红树林，由于多年的围垦和养殖业的发展，导致红树林退化严重，面积锐减，广泛分布的红树林在20世纪90年代初已经不足20hm²，红树林湿地濒临消亡。

随着经济和社会的发展，生态环境的建设引起了社会各界的关注。同时，省市领导的高度重视、专家和环保团体的不断呼吁，使得恢复红树林湿地、建设泉州湾湿地保护区的工作逐渐步入正轨。惠安县人民政府1998年在洛阳江屿头设立自然保护小区，该自然保护小区在2002年晋升为省级自然保护区，2003年扩建并更名为"泉州湾河口湿地省级自然保护区"，扩建后的总面积为7039hm²。

洛阳江红树林湿地的恢复工作始于2001年5月，红树林恢复工程规划行动由惠安县林业局实施，其策略是采取先填平补齐，再逐步向周边拓展，恢复红树林面积。规划范围由洛阳古桥至洛阳镇西方村，在该范围内寻找适合种植红树林的潮间带进行恢复工程。到2005年，共恢复以桐花树为主的红树林300hm²，目前该区面积95.6hm²，已成为"泉州湾河口湿地省级自然保护区"的核心区。

▼ 图4-60: 洛阳江红树林湿地位置图（绘制:易道）
Figure 4-60: Location of the mangrove restoration site at Luoyang River, Hui'an County, Fujian Province (Graphic: EDAW)

4.2.2 Mangrove Wetlands Restoration at Luoyang River, Huian County, Fujian Province, China

Background

Located in China's southeastern Fujian Province, the Luoyang River is a principal part of the estuarine wetlands of Quanzhou Bay. Historically, large areas of mangroves were located in the Luoyang River valley, but decades of relentless sea-reclamation and aquaculture have led to the severe degradation of mangroves within these wetlands. The mangrove coverage at Luoyang River has dwindled remarkably, to less than 20 hectares in the early 1990s and was on the verge of extinction.

Popularly known as "marine forests", mangroves are one of the most important wetland resources and constitute typical botanical communities in the tidal mudflats of tropical and sub-tropical zones. Mangroves play an indispensable role in maintaining biodiversity, reinforcing levees, controlling pollution and providing natural resources. Due to their unique ecological character, mangroves have high educational and scientific value. At the same time, they serve as eco-sightseeing destinations for tourists.

In recent years though, public awareness on environmental protection has risen in tandem with social and economic development. The attention from provincial and local governments, environmental experts and ad-hoc groups has finally led to the implementation of mangrove restoration projects and the establishment of a natural reserve for the wetlands of Quanzhou Bay. The local government of Hui'an County set up a minor natural reserve in the estuary of Luoyang River in 1998 and the site was upgraded to a provincial natural reserve in 2002. In 2003, the site was expanded to 7,039 ha and renamed the Provincial Natural Reserve of Estuarine Wetlands of Quanzhou Bay.

The restoration of the mangrove wetlands at Luoyang River started in May 2001 at the discretion of Hui'an County Forestry Bureau. The method employed was to fill bare areas with new mangrove shoots and then expand the mangrove area outwards. Throughout the restoration site, suitable intertidal zones were identified and used for the mangrove restoration projects. Three hundred hectares of mangroves, mainly *Aegiceras corniculata*, were restored in the Luoyang River site by 2005. In its current state, 95.6 ha of the Luoyang River site have experienced notable progress in mangrove restoration and have subsequently become the core zones of the Provincial Natural Reserve of Estuarine Wetlands of Quanzhou Bay.

Goals and Objectives

According to field studies on Luoyang River, there are four

湿地恢复手册 原则·技术与案例分析

目的与目标

红树林有"海上森林"之称，是热带、亚热带海岸滩涂特有的植物群落，也是最重要的湿地资源之一。红树林对维护生物多样性、固堤护岸、防治污染、提供自然资源等有重要作用，其独特的生态特性不仅具有很高的教育和科研意义，而且可以提供丰富的生态旅游资源。调查结果表明惠安县湿地有4个植被类型：红树林、河口沼泽植被、河口沙生植被和浅水植被，包括13个群系和20个群丛。据初步统计，有维管束植物79科221属271种（含变种），其中，红树林有桐花树、秋茄和白骨壤等3科3属3种，桐花树和白骨壤是太平洋西岸自然分布的北界；水生生物种类有172科369种；属国家重点保护对象的野生动物有海鸬、青脚鹬、斑嘴鹈鹕、隼类、鹰类、蟒蛇、虎纹蛙等；被列入中日国际候鸟保护协定的鸟类65种，中澳协定保护的鸟类29种，国家重点保护的珍稀鸟类13种，列入中国濒危动物红皮书的种类有黄嘴白鹭、黑翅鸢、黑嘴鸥等三种。这些珍稀物种都随着洛阳江红树林湿地的恢复而成为潜在的保护对象。

红树林湿地的恢复方法

红树林造林树种选择

惠安县在洛阳江造林时以红树林植物的耐寒性、耐盐性、耐潮力、向海性、生态安全性等特性作为物种选择的原则，最终选择原有的桐花树、秋茄等物种为主要造林树种，白骨壤为次要树种。为提高红树林观赏价值，根据各种植物的形态、花期和花色，计划引种木榄、红海榄等，进行植物景观美学搭配。

造林地选择

造林地选择在潮间带。造林时，遵循"从易到难"的原则，先种植在风浪较平稳的地带，再逐步发展到海水略深、风浪稍大的地带。根据具体地形，因地制宜进行施工设计，选择盐度在5‰~20‰的区域，进行红树林造林工作。

由于历年的围垦和养殖，滩涂已经不平整，高处硬化板结，低处长期水淹，不利于红树林的生长，因此宜在造林前先进行整地。针对不同的造林地，选择不同的品种。在中、高潮位以桐花树混交

vegetation types in its wetlands: mangroves, estuarine marsh vegetation, estuarine psammophytes and shallow-water vegetation, in 13 formations and 20 biotopes. A preliminary estimation indicates that there are 79 families, 221 genera and 271 species of vascular plants in the site, including three each of family, genera and species of mangroves (e. g., *Aegiceras corniculata*, *Kandelia candel* and *Aricennia marina*). The growth of *Aegiceras corniculata* and *Aricennia marina* at the Luoyang River site marks the northern edge of their natural distribution on the west coast of the Pacific Ocean. Of the 172 families and 369 genera of aquatic wildlife on the site, some are under national-level protection such as the sea cormorant (*Phalacrocorax pelagicus*), common greenshank (*Tringa nebularis*), the spot-billed pelican (*Pelecanus philippensis*), as well as falcons, eagles, boas and the tiger frog (*Rana rugulosa*). Sixty-five birds at the site fall under the China-Japan Migratory Birds Protection Agreement, while 29 birds fall under the list of China-Australia Migratory Birds Protection Agreement, and 13 birds are endangered species under national protection. The Chinese egret (*Egretta eulophotes*), black-shouldered kite (*Elanus caeruleus*) and Saunder's gull (*Larus saundersi*) are specifically listed in the IUCN Red Book as being among the most endangered species in China. With the current progress of the restoration projects of the Luoyang River mangrove wetlands, the above endangered species are potential candidates for biological protection.

Restoration approaches of mangrove wetlands

Species selection of mangrove foresting

The selection of mangrove species in the restoration site is based on their resistance to cold, salt, tide, their orientation towards the sea, and their ecological safety. The actual foresting practice has seen the existing species, mainly *Aegiceras corniculata* and *Kandelia candel*, as primary varieties, and *Aricennia marina* as the secondary variety. In order to create aesthetic harmony and enhance the sightseeing appeal of the mangroves, *Acanthus ilicifolius*, *Bruguiera gymnorrhiza* and *Rhizophora stylosa* will be introduced selectively into the site based on their form, flowering season and blossom colors.

Site selection of mangrove foresting

Intertidal zones have the potential to be forested with mangroves. The forestation may begin in shallow areas with milder tides, and proceed to more difficult areas with deeper sea levels and stronger tides. The actual layout of the artificial forests must adapt to the specific topography. The mangrove forestation must be conducted in areas with a salt concentration between 5 and 20%.

Decades of sea-reclamation and aquaculture in Quanzhou

▼ 图4-61: 福建洛阳江红树林湿地的恢复
（摄影：易道）
Figure 4-61: Restoration of mangrove wetlands at Luoyang River, Fujian Province (Photograph by EDAW)

▼ 图4-62: 根据地形地势进行红树林恢复
（摄影：易道）
Figure 4-62: The layout of artificial forests must adapt to the area's topography (Photograph by EDAW)

秋茄种植，在低潮位种植白骨壤，在高潮位种植老鼠簕、在中潮滩中下部种植红海榄，在中潮滩上部至高潮滩下部种植木榄，在大高潮可受到浸淹或浪花飞溅的地方种植黄槿等半红树物种（见图4-61,4-62）。

红树林造林方法

在滩涂地养殖的红树林

宜采取局部整地的方法：在高出滩面的地方按造林株距进行分段切口，每个切口宽约0.3m，降低滩涂高度至能被潮水合理浸淹位置，同时将切下的滩涂土壤按株行距堆放在低洼处，使植株在退潮时能排干积水，不被海水长期浸淹（见图4-63）。

每公顷整地所需土方量从几千立方米减少到三四百立方米，整地费用降至3000~6750元/hm²。这种恢复方法可以将造林点保持在一个受海水合理浸淹的位置，既有利于植株成活，又大幅降低了整地成本。

大米草整治

大米草的过度生长，不仅淤塞潮沟，而且阻抑了红树林幼苗的生长，必须进行清理。采用无毒化学除草剂，整治成本为4500元/hm²。局部试验表明，这种方法具有一定的成效。

设置各种管护牌、标志碑，与当地社区合作进行公众教育（见图4-64）

桐花树（隐胎生植物）造林方法

育苗

在沿岸选择中潮位滩涂，为苗床进行全面平整，苗床与江岸垂直，床宽1~1.2m，床高0.4m，床间距0.4~0.5m，沟深0.3m，待江潮浸淹抚平滩面后使用（约1个月）。

在8~10月利用现有的桐花树资源，选择强壮的优树采种，现采现育，采用三角形或方形法点播，株行距为6~8cm，种子尖头端向上，入土深度为2~3cm，露出床面1cm为宜。次年春季阴雨天，用白僵菌防治病虫害一次，一年后苗高度达到25cm以上时即可以用于造林，

Bay have left the tidal mudflats in the region at uneven elevations, with higher areas suffering from soil hardening and lower areas suffering from long-term submersion. These conditions are detrimental to the growth of mangroves. The reforestation sites must thus first be leveled off before any planting effort is possible. Different mangrove species may be planted at different forestation sites. For example, *Aegiceras corniculata* alternating with *Kandelia candel* may be applied in medium- to high-tide flats, the *Aricennia marina* in low-tide flats, *Acanthus ilicifolius* in high-tide flats, *Rhizophora stylosa* in the middle to lower sections of medium-tide flats, *Bruguiera gymnorrhiza* in upper medium-tide flats to lower high-tide flats, and semi-mangrove species like *Hibiscus tiliaceus* planted in mudflats susceptible to submersion and splash effects during high tides (see Figures 4-61, 4-62).

Forestation approaches

Mangroves in aquaculture mudflats

A partial leveling is recommended for mangroves in aquaculture mudflats. This method involves the following actions: making 0.3 m-wide notches at intervals matching the forest row spaces in flat areas above sea level; reducing the flat area to a specific elevation for reasonable submersion by tides; and returning the flat soil removed during the notching process to the low areas at an interval matching the forest row space. Sea water may thereby be easily drained during ebb flows and the plants will not be submerged for an extended period of time.

This method reduces the amount of processed earth to 300 – 400 cubic meters and therefore the cost of the land leveling to RMB 3,000 – 6,750 per hectare. It allows the forestation site to be constantly and reasonably submerged. This inexpensive method maximizes the survival rates of new plants.

Species control of *Spartina anglica*

The overgrowth of *Spartina anglica* may block flood paths and hamper the growth of mangrove saplings. This weed can be placed under control with the use of non-toxic chemical herbicides at a cost of RMB 4,500 per hectare. Partial tests have proved the effectiveness of this control method.

It is recommended using signage that informs people of the restoration project and provides public education on the significance of wetland protection and the vital role played by local communities.

Planting guide for specific species: How to plant *Aegiceras corniculata* (crypto vivipary)
Culturing the saplings

Select a suitable medium-tide flat, level the land thoroughly, and construct culturing beds perpendicular to the bank. The

但两年苗造林效果更佳。出苗时，用海锄或其他工具挖掘，不得损伤根部，不得用手拔苗，保持根部带土，用塑胶袋包装，可以一袋多装，装载运输。

造林

移植苗用于造林，苗高30cm，入土深为12~15cm，一般选在每年5~8月退潮后进行。整地规格为30cm×30cm×30cm，将苗木扶正放入穴中，回填至根部以上3cm，然后压实填平。造林株行距为60~100cm，每穴1株，采用三角形或方形法种植，每公顷可种植10500~15000株。

8~10月用桐花树种子进行高密度点播补植。营造混交林时，根据其他显胎生红树植物胚轴成熟时间进行株间点播（如秋茄胚轴成熟时间在4~5月）。白骨壤为隐胎生球果，不便育苗，要移取合格苗进行栽植。

容器袋育苗可以提高造林成活率，但育苗成本太高，对要求高密度造林的红树林来说，造林成本太高。容器育苗的方法与上述类似，在整床前选取松软透气的滩涂装袋12cm×18cm为宜，一年后即可用于造林。也可以育大容器苗，用于科研项目和示范工程（见图4-65）。

秋茄（显胎生植物）造林方法

秋茄为显胎生植物，种熟时间约在4~5月，种子较大，无须育苗，随采随造为宜，消毒处理后的秋茄胚轴方可进行点播，株行距为80cm×80cm，胚根朝下插植入土3/5，压紧即可，胚轴轻微损伤，仍可使用，对成效影响不大（见图4-66）。

以本区为例，桐花树年均生长高度为15~20cm，地径年均增长0.7cm，三年生树高为60~100cm，地径为2~3cm，植株粗壮，冠幅在50cm以上。秋茄年均高生长量为20~30cm，地径年均增长0.9cm，分枝较少，三年生树高为80~130cm，地径为3~4cm，冠幅在40cm以上。

幼林管理

种植后，立刻进行补植，3~5年内不间苗，为防止人为破坏，惠安县林业局选派专职护林员10人，完善监管，严禁采伐和破坏红树林的活动，成效显著。

beds must be between 1 – 1.2m wide, with a height of 0.4m, a row space of 0.4 – 0.5m, and a trench depth of 0.3m. The beds are usable approximately a month after initial construction, when the tides have submerged and smoothed the flat.

Culture saplings from the best available plants in the existing resources of *Aegiceras corniculata* during August through October. Plant the sapling tips up into the culturing beds immediately after collecting them from grown plants. Pattern the saplings in triangles or rectangles at a row space of 6 – 8cm. The saplings are best planted 2 – 3cm in the soil and one cm above the bed. During the rainy season of the following year, saplings may be treated with beauvericin for pest control. The saplings usually grow to over 25cm over next year and are ready for forestation, although another year in culturing beds will increase their chances of survival. When they are ready, saplings may be unearthed with sea hoes and other simple tools. It is important to avoid damaging or pulling the roots out with one's bare hands. The saplings, along with the bed soil in the roots, may then be placed in bags for easy transportation.

Foresting the saplings

Use 30cm-high cultured saplings for forestation during the ebb period from May through August. Plant the sapling in an upright position at a depth of 12 – 15cm, and refill the cavity to 3cm above the root. Then press the soil, spread the root, press again and completely fill the cavity. One sapling usually occupies a space of 30×30×30cm. Saplings are usually planted in triangles or rectangles at a row space of 60 – 100cm, with one per cavity. There can be between 10,500 – 15,000 plants per hectare. High-density seed sowing may be advisable from August through October. When building a hybrid forest between *Aegiceras corniculata* and other mangrove species, it is wise to pay attention to the maturing stage of the propagules of other vivipary mangroves (that is, the propagules of *Kandelia candel* mature in April through May) and sow the mature propagules in between rows. *Aricennia marina*, a crypto vivipary plant, has only strobili that are unfit for culturing. Therefore, it has to be transplanted into the hybrid forest.

Culturing saplings with polythene bags is an effective way to increase their survival rates but is also quite expensive. It might be cost prohibitive for a restoration site that requires high-density foresting. Bag-culturing is similar to the aforementioned bed-culturing, with the only difference being that soft, permeable tidal mud is collected into 12×18cm polythene bags before the culturing beds are leveled. The saplings are cultured in these bags and are ready for forestation the following year. Larger culturing tanks are also available for scientific research and demonstration projects.

（摄影：崔丽娟）
(Photograph by Cui Lijuan)

（摄影：易道）
(Photograph by EDAW)

▼ 图4-64: 设立标志碑，与当地社区合作推广湿地保护和恢复的观念（摄影：易道）
Figure 4-64: An example of information signage that communicates the significance of wetland protection and restoration, and the role of local communities
(Photograph by EDAW)

结论

通过湿地恢复工程，现在惠安洛阳江的红树林面积从不足２０ｈｍ²增至３００ｈｍ²，增幅１５倍，有效地保护了本区的红树林资源。

红树林能降低水中的悬浮物、氮、磷和金属等，使水质提高，减少海水赤潮的发生率。２００４年８月时，洛阳江江水曾受到污染，许多养殖锯缘青蟹的渔民反映，出现大量死蟹和病蟹，经济损失惨重，但是在红树林边的养殖区却丝毫未受影响，可见红树林湿地具有超强的净化功能。

在未恢复红树林之前，当地的渔业生产以养蛏和海蛎为主。大规模的滩涂养殖和陆源污染物的不断增加，致使湿地生态遭受严重破坏，养殖业的产量、质量也逐年下降，收益逐年减少，有的甚至无以为继。恢复红树林的工作展开后，湿地环境日渐好转，红树林区水生和底栖动物日渐丰富，社区居民也受益良多。据调查，红树林的高生产力，仅鱼苗一项的年收益就达到２０００多万元，鲻鱼、蟹虎鱼、沙蚕等海产品的产值也大幅提高（见图４-６７）。

公众保护湿地的意识不断增强。通过绿化简报、标语、宣传车等各种形式的广泛宣传，逐渐使群众认识到保护湿地和建设红树林是一项造福子孙的工作。其中不乏附近的各中小学校和青年志愿者加入保护红树林和爱护湿地和野生动物的宣传活动，部分养殖户开始时有抵触情绪，后来也纷纷主动让地造林。

How to plant *Kandelia candel* (vivipary)

Kandelia candel is a vivipary plant, whose propagules are large enough to be planted without a culturing stage. The propagules mature from April through May and must be used immediately after being removed from the host plant. The propagules must also go through a sterilization process before being sowed onto the forestation site. The planting action is simple: insert the hypocotyl with the root end downwards into the soil until 3/5 of the stem is submerged and then press the ground until it is tight. The row space of the plants is usually 80×80cm. Slight bruises on the propagules generally do not affect their chance of survival. The growth rates of artificially forested mangroves have been recorded in Luoyang River wetlands. As for the stout *Aegiceras corniculata*, the height increases by 15 - 20cm every year and the diameter in the ground-level trunk by 0.7cm. A three-year-old plant usually has a height of 60 - 100cm, a diameter at its ground-level trunk of 2 - 3cm and a crown diameter of over 50cm. *Kandelia candel* is slender and has fewer branches. Its height increases by 20 - 30cm every year and the trunk diameter in ground-level increases by 0.9cm every year. A three-year-old plant generally has a height of 80 - 130cm, a trunk diameter at ground-level of 3 - 4cm and a crown diameter of over 40cm (see Figure 4-66).

Managing young forests

After the initial seed sowing, supplements must be planted promptly to replace dead saplings or propagules. No thinning or gapping is allowed in the first three to five years of forestation. The Forestry Bureau of Hui'an County has employed ten full-time rangers to protect mangroves from unauthorized harvesting and other human disturbances. This effort has proved effective in ensuring an adequate survival rate for young mangrove forests.

Conclusion

The restoration projects in the Luoyang River wetlands have expanded mangrove areas from less than 20 hectares in the early 1990s to 300 hectares today. The efforts to preserve the natural resources of the mangroves have had a successful start and have set a precedent for future successes.

These mangroves can effectively reduce floating particles, nitrides, phosphides and metal sediments in the water body, improving water quality and preventing crimson tides. During an outbreak of pollution in the Luoyang River in August 2004, many Samoan crab farmers reported heavy losses because their crabs were killed or diseased by polluted water. However, the sea-farming zones near mangroves were unaffected, sufficient enough proof that mangrove wetlands have remarkable and important purification capacities.

▼ 图 4-65: 桐花树的造林育苗
Figure 4-65: *Aegiceras corniculata* propagates under the mangrove forestation program

（摄影：易道）
(photograph by EDAW)

（摄影：崔丽娟）
(photograph by Cui Lijuan)

Prior to the implementation of mangrove restoration projects, fishery production at Luoyang River focused on clams and oysters. Relentless sea-farming on tidal mudflats and increasing pollution from land sources finally led to the severe degradation of mangrove wetlands. As a result, the output and quality of aquatic yields declined rapidly, and local sea-farmers saw their income levels drop off, so much that some could no longer make a living from fishing. With mangrove restoration projects well underway at Luoyang River, the wetland environment has improved substantially, and mangrove zones have become a rich reservoir for aquatic and benthic creatures. Locals have benefited from the environmental improvement, as mangroves have a high potential for economic production. For example, the annual yield in fry production amounts to over RMB 20 million and the fishery outputs of mullets, gobioids and nereids have increased considerably.

Public awareness on wetland conservation has also been on the rise in Hui'an County, thanks to extensive education campaigns through news bulletins, posters and mobile information-service platforms. Local residents have learned that current and future generations may benefit from wetland conservation and mangrove forestation. School students and young volunteers have actively taken part in education campaigns on the conservation of mangroves, wetlands and wildlife in the County. Sea farmers were at first generally suspicious of the restoration project, but they soon changed their minds and presented their farming zones voluntarily for mangrove forestation. The forestation and management of mangrove areas on the Luoyang River has improved biodiversity, water quality, aesthetics, and its aquatic farming potential.

◀ 图 4-66: 秋茄的造林育苗
Figure 4-66: Propagule culturing and foresting of the *Kandelia candel*

（摄影：易道）
(Photograph by EDAW)

（摄影：崔丽娟）
(Photograph by Cui Lijuan)

◀ 图 4-67: 因为红树林的恢复也使得当地的渔产日渐丰富，带动了当地的经济发展
（摄影：易道）
Figure 4-67: The restoration of the mangrove wetlands has led to the increase in local fishery production and the economic development of the County
(Photograph by EDAW)

深圳
Shenzhen

蛇口
Shekou

后海湾
Deep Bay

米埔自然保护区
The Maipo Nature Reserve

元朗
Yuen Long

4.2.3 滨海湿地生态恢复、栖息地管理与维护——米埔自然保护区，中国香港

背景

后海湾湿地位于中国南部珠江河口东部，北接深圳经济特区，南邻香港特别行政区，是米埔自然保护区的所在地（见图4-68）。落潮时，湾区有2700hm²的潮间带浅滩显露出来。浅滩四周围绕着面积约400hm²的潮间带红树林，红树林后则是传统的虾塘，当地人称之为基围。基围上生长着红树林，还有芦苇湿地以及商业鱼塘。自20世纪70年代城市开发以来，这些海湾湿地不断被填塞，湿地已经退化。

香港特别行政区对后海湾的保护始于1976年，当时米埔的基围和红树林地带被指定为"具有特殊科学价值的地质遗迹"（SSSI）。1983年，世界自然基金会香港分会接管此地，开始米埔的野生动物保护和环境教育工作。1995年，依据《湿地公约》，这块包括米埔自然保护区在内、面积共1500hm²的湿地被列为国际重要湿地，保护工作因此得到进一步的加强。如今，世界自然基金会协同香港政府渔农自然护理署，共同管理米埔自然保护区。

▲ 图4-69：米埔的鸟类数量对地区乃至世界都具有重要意义，图中为反嘴鹬（提供：世界自然基金会香港分会）
Figure 4-69: Mai Po supports globally and regionally significant bird populations, such as the pied avocets (Recurvirostra avosetta) (Photo: Leung Wai Ki, WWF Hong Kong)

4.2.3 Habitat Management and Maintenance in a Restored Coastal Wetland: The Maipo Nature Reserve, Hong Kong Special Administrative Region, China

Background

The Inner Deep Bay wetlands (Deep Bay) are located at the eastern edge of the Pearl River Estuary, in southern China. Deep Bay is bounded to the north by the Shenzhen Special Economic Zone (SEZ) and to the south by the Hong Kong Special Administrative Region (SAR), and is also home to the Mai Po Nature Reserve. During low tide, 2,700 ha of mudflats are exposed in Deep Bay, fringed by some 400 ha of inter-tidal mangrove forests. Behind the mangrove are traditional shrimp ponds (known locally as gei wai) that support stands of mangroves, reedbeds, and commercial fishponds. Since the mid-1970s, the area of these wetlands has declined due to their gradual in-filling for urban developments.

Protection of these wetlands on the Hong Kong SAR side of Deep Bay began in 1976 when the shrimp ponds and mangroves at Mai Po were designated as a Site of Special Scientific Interest (SSSI). Management of Mai Po for wildlife conservation and environmental education began in 1983 when the World Wide Fund for Nature (WWF) Hong Kong started managing the site. Protection was further increased in 1995 when a 1,500 ha area of the wetland, which included the Mai Po Nature Reserve was listed as a Wetland of International Importance under the Ramsar Convention. The WWF now manages the Mai Po Nature Reserve in collaboration with the Hong Kong Government's Agriculture, Fisheries and Conservation Department (AFCD).

Ecological importance

The Inner Deep Bay wetland supports some 54,000 wintering waterbirds, with another 20,000 to 30,000 shorebirds using the site as a staging post during spring and autumn migration. Some 18 species of these waterbirds are considered threatened and 30 species appear in numbers that are greater than one percent of their total estimated population in East Asia (Carey & Young, 2001).

These wetlands also provide an important habitat for other wetland wildlife such as the Eurasian otter (Lutra lutra), and several rare and endangered invertebrate species including the four-spot midget dragonfly (Mortonagrion hirosei) (Young, 1999). The wetland habitats in and around Mai Po are also important, and include mudflats, mangroves, shrimp ponds, reedbeds and commercial fishponds.

◄ 图4-70: 后海湾于20世纪60年代引进了池塘养鱼法，现在仍是米埔附近地区常见的景观（提供：世界自然基金会香港分会）
Figure 4-70: Pond fish-farming was introduced to the Inner Deep Bay area in the 1960s, and still dominates the landscape adjacent to Mai Po (Photo courtesy of WWF Hong Kong)

生态价值

每年冬天，大约54000只水禽来到后海湾湿地过冬；春秋两季，两三万只涉禽把这里当作迁徙途中的驿站。其中，约有18种水鸟属于濒危鸟类，有30种鸟类在这片湿地上的数量高出东亚地区预计总量的百分之一（Carey Young 2001）（见图4-69）。

后海湾湿地还滋养着其他野生动物，其中有欧亚水獭和四斑细螅（广濑妹螅）这样的珍稀濒危无脊椎动物（Young 1999）。米埔及其周围有各种重要的的湿地生态环境，包括潮间带浅滩、红树林、基围、芦苇湿地以及商业鱼塘。

管理历史

后海湾湿地的管理至少可以追溯到13世纪，当时第一批移民在后海湾边定居下来。他们主要靠在当地打鱼为生，在水中捕捞鱼、虾、蟹和牡蛎等。

20世纪初，这里的人口稍有增长，因为一些人从中国大陆南下，来到这片海岸定居。每一拨移民都带来新的开垦和养殖技术，例如19世纪20年代带来了咸水稻田耕作；19世纪40年代开始在基围里养虾；19世纪60年代开始用鱼塘养鱼。后海湾湿地的土地利用方式经历了数次变化，直到今天，仍可以在这片土地上看到各种养殖方法留下的痕迹（Irving & Morton 1988）（见图4-70）。

项目目的

在米埔项目开始时，世界自然基金会香港分会成立了专门委员会来监督保护区的开发与管理。委员会成员来自政府部门、学术团体、绿色环保团体以及其他积极的团体。

根据世界自然基金会香港分会2006年的资料，米埔项目拟定的目标包括：

1. 管理米埔自然保护区，保持并尽量增强、提高湿地生态环境的多样性和本土野生动物种类的丰富性；

2. 作为学生和大众的教育基地（包括为残疾人设计专门设施和游览路线）；

3. 挖掘该保护区作为国际重要湿地的培训潜能，促进东亚或澳大利亚水鸟

Historical Management

People living on the coast of Inner Deep Bay have been managing the area's wetlands since at least the 13th century, when the first settlers established themselves around the Bay. These populations depended mainly on local fisheries for their livelihood, catching everything from fish to shrimps, crabs and oysters.

From the early 20th century, the population experienced sporadic growth, as groups of migrants from mainland China came down to settle along the coastal area. Each time, settlers brought with them new techniques of reclaiming and farming the land, such as;

-   1920s – techniques for brackish rice farming

-   1940s – shrimp farming using local gei wai ponds

-   1960s – pond-fish farming

As a result, the Inner Deep Bay wetlands have been through a series of land-use changes, and the remains of each of these farming practices can still be seen in the landscape today (Irving & Morton, 1988).

Goals

When the WWF Hong Kong began their project at Mai Po, they established a special committee to oversee the Reserve's development and management. Members of the committee include representatives from government departments, academics, green groups and other advocate groups. The goals for the Reserve are (WWF Hong Kong, 2006):

1. To manage the MPNR so as to maintain and, if possible, increase the diversity of habitats appropriate for southern China lowland wetlands and the richness of native wildlife in the area.

2. To promote the use of the area for educational purposes both by students and the general public (including the provision of special facilities and tours for the disabled).

3. To realise the training potential of the Reserve as part of the Ramsar Site so as to promote wetland conservation and wise use in the East Asia/Australasian Flyway, in particular China.

4. To promote scientific research relevant to the management and conservation of wetlands and their biota.

5. To promote and support measures to reduce and minimise external threats to the habitats and wildlife at the Reserve.

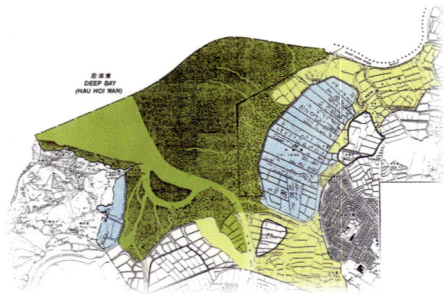

核心区
Core Zone

公众使用区
Public Access Zone

生物多样性区
Biodiversity Management Zone

私人地区
Private Land Zone

受善利用区
Wise Use Zone

▲ 图4-71: 米埔后海湾拉姆萨尔湿地被分为五个生物多样性管理区(提供:世界自然基金会)
Figure 4-71: The Mai Po Inner Deep Bay Ramsar Site is divided into five management zones (Image courtesy of WWF)

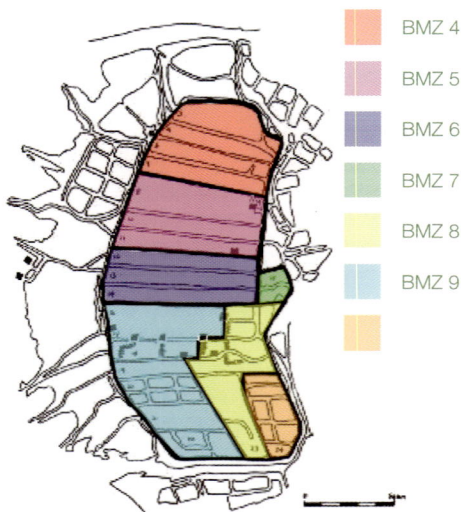

BMZ 4

BMZ 5

BMZ 6

BMZ 7

BMZ 8

BMZ 9

▲ 图4-72: 米埔生物多样性管理区分为七个不同管理体制的小块(提供:世界自然基金会)
Figure 4-72: The Mai Po Biodiversity Management Zone is divided into seven compartments with different management intentions (Image courtesy of WWF)

迁徙路线上的湿地保护和合理利用,尤其位于中国的湿地;

4. 促进湿地及其生物群落管理与保护方面的科学研究;

5. 推行并支持各种措施,将对保护区生态环境和野生动物的外在威胁降至最小。

恢 复 与 管 理 过 程

计 划

1997年,中国香港政府公布了国际重要湿地管理计划(Anon 1997),该计划将湿地分为五个生物多样性管理区(见图4-71):

1. 核心区(CA)。该区域未遭破坏,基本保持原始特色。

2. 生物多样性管理区(BMZ)。该区域将成为水鸟的庇护所。在集中管理、保护生物多样性的情况下,进行教育和训练活动。米埔自然保护区位于此区。

3. 妥善利用区(WUZ)。位于该区域的商业鱼塘应有利于生态可持续发展。

4. 公共参观区(PAZ)。将有组织地向游人开放位于该区的部分国际重要湿地,使游人欣赏其独特价值,并享受与野生动物接触的乐趣。

5. 私人地产区(PLZ)。承认土地当前的法律归属性。

上述各区进一步细分为不同小块,米埔生物多样性管理区共有七个这样的小块(图4-72)。每个小块本身由一组基围组成,每个基围都有各自明确的管理目的:保护某个物种(如黑脸琵鹭)、种群(如岸鸟),或者保护生态环境(如芦苇床、红树林或淡水等)。

植被管理
植物群落对于生态具有重要意义,对其的管理必不可少。一旦一些植物开始入侵、蚕食其他重要的生态环境,就需要控制其长势。例如:

Restoration/Management Approaches
Planning

In 1997, the Hong Kong Government published a management plan for the Ramsar Site (Anon, 1997), and this divided the Site into five Biodiversity Management Zones (Figure 4):

1. Core Area (CA) – to become an undisturbed, largely natural area.

2. Biodiversity Management Zone (BMZ) – to become a refuge for waterbirds and focus for biodiversity conservation, education and training in an intensively managed environment. This zone includes the Mai Po Nature Reserve.

3. Wise Use Zone (WUZ) – to allow ecologically sustainable use of the commercial fishponds.

4. Public Access Zone (PAZ) – to provide managed access to a part of the Ramsar Site in order for people to appreciate its special values and enjoy contact with wildlife.

5. Private Land Zones (PLZ) – to recognize the existing legal status of the land.

Each of these zones is further divided into smaller compartments, with the Mai Po Biodiversity Management Zones being divided into seven compartments (Figure 4-72). Each compartment itself consists of a group of shrimp ponds with its own broad management intention. These intentions are either for conserving particular species (e.g. black-faced Spoonbill, *Platalea minor*), species groups (e.g. shorebirds), or habitats (e.g. reedbeds, mangroves or freshwater).

Vegetation Management

Although the plant communities at Mai Po are ecologically important, they need to be managed and their spread controlled whenever they begin to encroach on other habitats of importance. Examples include:

- Reeds encroaching on open areas of water within shrimp ponds that are used by waterbirds;

- Reeds encroaching on designated mangrove habitats;

- Mangroves encroaching on designated reedbed habitats; and

- 芦苇可能侵入基围中供水鸟栖息的开放水域；

- 芦苇可能侵入指定的红树林栖息地；

- 红树林可能侵入指定的芦苇湿地或水鸟最重要的栖息地——泥滩。

此外，一些植物种类（如薇甘菊、大米草和海桑等蔓生植物）必须清除，因为它们属有毒或入侵物种，这些植物的蔓延会减少湿地生态环境和物种的多样性（见图4-73）。

植被控制

米埔地区的植被控制通过以下方法进行：

- 物理清除（使用或不使用机器）。

- 控制性烧荒。火是防止某些植物（如对热敏感的树和灌木）拓殖并使草场和芦苇床恢复活力的有效管理手段，但使用时要非常小心。这种手段不但危险，而且还给公共关系带来一定风险，因为很多乡民对火有排斥心理。

- 使用除草剂。必须慎重选择除草剂，使其对环境的影响最小。与用火烧荒一样，公众对在自然保护区中使用化学物质一般也都很排斥，因此使用时必须格外谨慎。

- 控制放牧。有时可以用控制放牧家畜的办法来保留需要的植被（如草场）。只要控制得当，还可用家畜来实现植被均衡的效果，为野生动物提供食物或栖息地。

树木管理

20世纪80年代初期以前，米埔地区的土地多为开放空间，渔民总是沿着基围塘边的堤岸走到虾塘靠海一边的水闸捕捞基围虾，在基围边上自然而然走出了小路。每年冬天，渔民都会砍掉或烧毁虾塘堤岸上的植物，防止其过度生长，因此该地区的树木非常有限。

1982年前后，通往海边基围方向修建起一条新公路，自此渔民不再走堤岸小路，而是沿公路驱车前往虾塘。久而久

- Mangroves encroaching on the mudflats which are the most important habitat for waterbirds in the Ramsar Site.

Other plant species may have to be removed because they are exotic or invasive species, and their spread will also reduce the diversity of habitats and species in the wetland. Such species include the exotic invasive climber *Mikania micrantha*, the grass *Spartina* spp., and the mangrove *Sonneratia* spp.

Vegetation control

Vegetation control has been carried out at Mai Po using the following methods:

- Physical clearing with or without machinery.

- Controlled burning. Although fire is a powerful management tool that can be used to prevent colonization by certain plants (e.g., heat sensitive trees and shrubs) and to rejuvenate grasslands and reedbeds, this tool needs to be used carefully. It carries not only safety, but also public relations risks, as many people have a negative impression of fires in the countryside.

- Herbicide Application. The herbicide must be carefully selected so as to have minimal impact on the environment. As with fire control, the public has a generally negative impression over the use of chemicals in a nature reserve, so care must be taken in that regard.

- Controlled grazing by domestic stock can sometimes be used to maintain desired vegetation such as grasslands. If carefully controlled, livestock can also be used to produce a uniform stand of vegetation that can provide feeding or nesting sites for wildlife.

Tree management

Prior to the early 1980s, the landscape within Mai Po was largely open space, with very few trees growing along the shrimp pond bunds. This was because local fishermen regularly used the bunds as footpaths to access sluice gates at the seaward end of the pond where they had homes and where they would harvest shrimp. Each winter, fishermen would cut or burn the vegetation along the bunds in order to prevent overgrowth.

With the completion of a new road along the seaward end of the shrimp ponds around 1982, fishermen stopped using the bunds as footpaths, and would instead drive

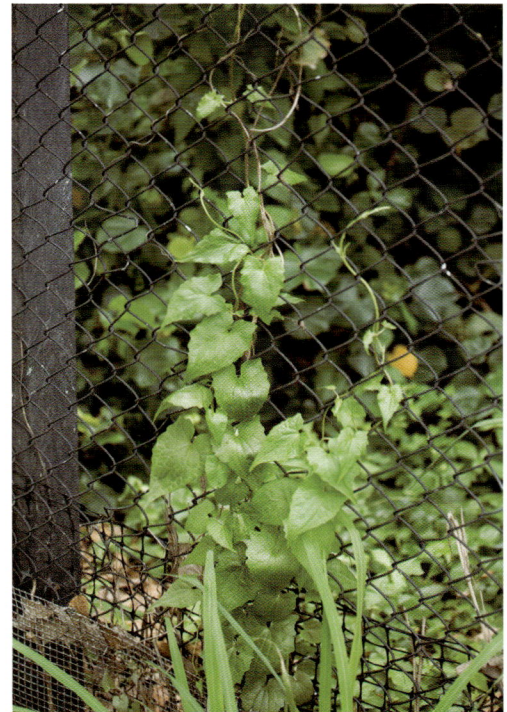

▲ 图4-73: 毒性蔓生植物薇甘菊是米埔需要定期清除的南美洲入侵性植物(摄影：EDAW)
Figure 4-73: The exotic climber *Mikania micrantha* is an invasive South American species that needs periodic removal from Mai Po (Photograph by EDAW)

之，堤岸上人迹罕至，树木开始生长。然而，许多米埔的水鸟（如鸭子和岸鸟）为了更容易发现天敌，及时逃离，更喜欢没有多少高大树木的开放区域。

第16、17号基围塘的实际情况体现出树木管理的好处。在20世纪80年代末和90年代初之间，多达10000只迁徙涉禽将这块基围塘当作春季涨潮时的栖息地，但是自1994年，使用该塘的涉禽数量开始减少，到1997年，已没有涉禽来到这个基围塘。究其因，主要是因为基围塘边的树木过于高大。1997年和1998年清除了这些树木后，涉禽又开始返回这个池塘。

世界自然基金会香港分会专门有一个针对基围塘边高大树木的管理计划，管理重点是那些作为水禽栖息地的池塘。然而，在自然保护区内清除树木会引起公众的关注，因此需谨慎开展此项工作。可以采用修剪枝叶、控制树高的方法，而不一定整棵拔除，从而实现有效的树木管理。

树木种植

在米埔部分地区需要修剪、砍伐树木的同时，还有一些地区需要植树造林，尤其是路旁及保护区通向陆地的一边。在这些地方种树可为游人提供阴凉，还能够作为附近开发区的视觉屏障。所种植的树木都是华南地区低洼湿地的天然物种，如雀榕和乌桕。它们结出的果实能吸引食果鸟类来此觅食。在树木覆盖充足的地区，则选择种植车轮梅和鸭脚木等结果灌木。

芦苇床管理

米埔地区芦苇覆盖了46hm^{2}湿地，是香港同类植物保存面积最大的地区，可能也是广东省最大的芦苇湿地。近年来，由于基围塘淤泥的增加，米埔地区的芦苇生长面积已有所增长。

世界各自然保护区已成功制定出各种策略，将芦苇床作为野生动物栖息地来管理。这些战略包括控制芦苇床的水位，或者以1~15年为周期，在不同时节喷洒除草剂并采用割刈或燃烧的方法控制芦苇的长势（Burgess et al. 1995; Hawke & Jose 1996）。

along the road to access the ponds. As a result, the bunds' vegetation disappeared and trees began to grow along them. However, many of the waterbirds that use Mai Po (e.g., ducks and shorebirds), prefer open areas with a few tall trees, in order to more easily detect predators.

One example of the benefits of tree management is from shrimp pond number 16/17, which between the late 1980s and early 1990s, was used by up to 10,000 migratory shorebirds in spring as a high-tide roosting site. However, the number of shorebirds using this pond started to decrease in 1994, and by 1997, shorebirds had completely abandoned the pond. The main reason for this abandonment was the growth of tall trees along the bunds. When these trees were removed in 1997 and 1998, shorebirds began returning to the pond.

WWF Hong Kong now has a program to manage tall trees along the shrimp pond bunds, especially around those ponds that are managed as waterbird habitats. Yet as there are public concerns about tree removal within a nature reserve, work needs to proceed carefully, by trimming side braches and managing tree height, rather than removing whole trees.

Tree planting

While trees are trimmed in some areas of Mai Po, there is also a program of tree planting in other areas, especially next to footpaths and around the landward edge of the Reserve. Trees planted in these areas provide shade for visitors and act as a visual screen from nearby developments. Tree species planted in the reserve are native to the South China Region, associated with lowland wetlands, and bear fruit attractive to frugivorous birds. These include species such as *Ficus superba* and *Sapium sebiferum*. In areas of Mai Po with sufficient tree coverage, fruit-bearing shrubs such as *Rhaphiolepis indica* and *Schefflera octophylla* have been planted.

Reedbed management

The 46 ha of reed grass *(Phragmites australis)* at Mai Po is the largest area of its kind remaining in Hong Kong, and probably the largest in Guangdong Province. Over the years, the area containing reeds at Mai Po has increased as shrimp ponds have silted up.

Various strategies have been successfully developed by nature reserves around the world to manage reedbeds as a wildlife habitat. This includes controlling water levels in the reedbed, or by managing the reeds directly through a mixture of spraying, cutting or burning at different times of year, on rotation varying from one to 15 years (Burgess et al., 1995; Hawke & Jose, 1996).

为验证米埔的芦苇床是否需要定期割刈，2001年1月开始了一项长期研究，为制定植被管理策略做准备。在该项研究中，有五块面积1hm²的试验地，每年割除四块试验地上达1hm²面积的所有芦苇，并用雾网调查法监控各块地上的鸟类活动情况。第五块芦苇控制地作为参照样地。

米埔的芦苇管理中还采用挖掘淤泥、喷洒获准使用的除草剂（如草甘膦）等方法控制芦苇的蔓延，防止其侵入开放水域（见图4-74）。

基围管理

收获基围虾

20世纪40年代初，沿海红树林被围起来，建成了最初的基围塘。每个基围塘面积约10hm²，不仅给当地居民带来利益，而且对环境的负面影响非常小，是典型的可持续滨海湿地利用模式。在这些基围塘里，基围虾的生产依赖于后海湾的自然生产力。秋天，虾苗从后海湾涌入池塘，这些虾以死的有机物为食，枯死的红树叶沉入塘底，成为基围虾的佳肴。因此，渔民在虾塘中保留了红树林，它们其实是免费的鱼食虾食。

每个基围塘都有一道闸门，潮水穿过滨海红树林的水道通往后海湾，因此潮水可以在基围塘与后海湾之间双向流动。闸门一般宽1.0m~1.5m，有水泥墙体和嵌入墙体沟槽的木板闸。放下或提起木板可调节通过闸门的水流量（见图4-75）。

珠江含沙量大，因此从后海湾涌入基围塘的水里泥沙含量也很大。据估计，基围塘的泥沙沉积率达1.7cm/a（Lee 1988）。为使水道的深度适合虾类生长，每十年要清理一次淤泥。

在基围虾收获期结束时，基围中的水会被完全排干，塘中的鱼也被捕捞起来。这时，每个这样的池塘会吸引多达1600只鸟来到这里过冬，如苍鹭、白鹭和濒危鸟类黑脸琵鹭（Leader pers. comm. 2000）。

To investigate whether the Mai Po reedbeds require regular cutting, a long-term study was initiated in January 2001 to develop a vegetation management strategy. This involved the cutting and removal of reeds in four one-hectare experimental blocks on an annual rotation, and monitoring site use by birds of each block through mist netting. A fifth block of reeds was maintained as a control block.

Management of the reeds at Mai Po also involves controlling their spread into open areas of water by either dredging or spraying with an approved herbicide (e.g. glyphosate).

Shrimp pond management

Gei Wai shrimp harvesting

The gei wai at Mai Po were created in the early 1940s by impounding coastal mangroves. Each pond is approximately 10 ha in size, and is an example of a sustainable coastal wetland because they benefit local communities with a minimal adverse impact on the environment. This is because gei wai shrimp production relies on the natural productivity in Deep Bay. The ponds are stocked with young shrimp from the Bay in autumn, and the shrimp grow and mature by feeding on dead organic matter, mostly dead mangrove leaves on the bottom of the pond. Fishermen therefore maintain the mangrove stands inside the ponds as a free source of food for shrimp and fish.

Each shrimp pond has a sluice gate that allows water exchange with Deep Bay via a channel through the coastal mangroves. The sluice gate is 1.0 – 1.5 meters wide, has concrete walls and wooden sluice boards which are slotted into grooves in the walls. Placing or removing the boards modulates water flow through the gate.

Due to the high sediment load of the Pearl River, the water flushed into the shrimp ponds from Deep Bay carries a high silt load. The sedimentation rate in the shrimp ponds has been estimated to be 1.7 cm. yr^{-1} (Lee 1988). In order to maintain the channel at a suitable depth for shrimp production, dredging is conducted every ten years.

At the end of the shrimp harvesting season, the ponds are completely drained in turn for harvesting the fish inside. Up to 1,600 wintering birds, such as herons, egrets and the endangered Black-faced Spoonbill *(Platalea minor)*, are attracted to a single draining ponds (Leader pers. comm. 2000) to feed on small fish and shrimp trapped at the bottom of the pond.

Water level management

▲ 图4-74: 米埔地区的芦苇(*Phragmites australis*)。为保证开放水域面积，需通过清除淤泥或用除草剂对芦苇进行控制(摄影：EDAW)
Figure 4-74: Mai Po. The reedbeds require management to maintain open areas of water through dredging or herbicide application. (Photograph by EDAW)

▲ 图4-75: 通过与海洋相连的水闸来调节基围塘的水位 (摄影：EDAW)
Figure 4-75: Water levels in shrimp ponds are controlled via sluice-gates connected to the Deep Bay ocean (Photograph by EDAW)

水位管理

根据《米埔后海湾国际重要湿地管理计划》（Anon 1997）和《米埔管理计划》（Young 1999），只有12~14号基围保留为传统的基围虾塘，其他基围都为野生动物栖息地和生态环境（如芦苇床）服务。

如何管理基围塘的水位取决于保护此塘的目的。例如，第11号和16、17号基围塘用于为春秋过境的岸鸟提供涨潮时的栖息地。因而，要在这两季把基围的水位降下来。夏天，此地见不到涉禽，基围水位被抬高，防止植被入侵蔓延到露出水面的泥滩上。

淡水池塘管理

20世纪70年代中期前，米埔周围有多块淡水沼泽地，栖息着水雉（Hydrophasianus chirurgus）（Carey et al. 2001）。由于这些沼泽已消失，因此当时有人建议在《拉姆萨尔湿地管理计划》（Anon 1997）和《米埔管理计划》（世界自然基金会香港分会，2006）中增加在米埔自然保护区南端（第20~24号基围）建立若干淡水沼泽的内容（见图4-76）。

米埔的淡水沼泽是为在废弃的商业鱼塘排水，排空咸水，并用推土机或锄耕机修补堤上漏洞而建设出来的。经过雨水浇灌，池塘不久就会生出从前米埔没有的植物群落，为冬天大量植食性水鸟提供食物。这些池塘的管理目标之一是吸引两栖类动物和蜻蜓类昆虫，由于鱼类可能成为它们的天敌，所以不再往池塘里放鱼苗。从夏季池塘有大量蜻蜓类昆虫的记载看来，该管理策略是很成功的（见图4-77）。

泥滩和红树林

每年，约有5.4万只鸟来到后海湾泥滩过冬，2万~3万只候鸟在春秋迁徙时来此觅食。研究发现，泥滩上的无脊椎动物超过80种（McChesney 1997），其中约有20种是科学新发现或香港历史未曾记载过的（Lee 1993）。

滨海泥滩会因为淤泥堆积而自然升高，这时，泥滩接陆地一侧的植被（在后海湾的案例中就是红树林）就会逐渐向大海方向延伸（见图4-78）。后海湾泥滩也不例外，而且在海湾水体泥沙含量增加和海湾水流量减小的共同作用下，

Under the Mai Po Inner Deep Bay Ramsar Site Management Plan (Anon, 1997) and the Mai Po Management Plan (Young, 1999), only shrimp ponds 12-14 are to be operated as traditional gei wai ponds. The other ponds are managed for wildlife and the ecological habitats (e.g., reedbeds) that they may support.

Management of shrimp pond water levels depends on the conservation objective of that pond. For example, ponds 11, 16 and 17 are managed to provide high-tide roosting sites for migratory shorebirds during spring and autumn passage. Water levels are therefore lowered during these two seasons. During the summer when few shorebirds are present, water levels in these shrimp ponds are raised to prevent vegetation encroachment over areas of exposed mud.

Freshwater pond management

Prior to the mid-1970s, there were freshwater marshes around Mai Po that supported breeding populations of pheasant-tailed jacanas (Hydrophasianus chirurgus) (Carey et al, 2001). As these marshes have been lost, it was recommended in the Management Plan for the Ramsar Site (Anon, 1997) and the Mai Po Management Plan (WWF Hong Kong, 2006), that a series of freshwater marshes be established at the southern end of Mai Po Nature Reserve (in shrimp ponds 20-24).

The freshwater marshes at Mai Po were created by draining abandoned commercial fishponds to remove the brackish water inside and to repair any leaks in the bunds using a bulldozer and backhoe. After refilling with rain water, the ponds soon developed plant communities not previously found at Mai Po, which served as a food source for a large number of grazing and granivorous waterbirds in winter. As one of the objectives of the ponds was to attract amphibians and odonates, fish were not restocked in the ponds as they could be potential predators. This strategy appears to have been successful, with large number of odonates recorded from the ponds during the summer months.

Mudsflats & Mangroves

The Inner Deep Bay mudflat is the main feeding ground for some 54,000 wintering waterbirds and another 20,000 – 30,000 that pass through on migration in the spring and autumn. Studies have also found over 80 species of invertebrates on the mudflat (McChesney, 1997), with about 20 of those species new to science or previously unrecorded from Hong Kong (Lee, 1993).

It is a natural part of succession for coastal mudflats to silt up and increase in height. As this happens, vegetation on the landward side of the mudflats (mangroves in the case of Deep Bay) will slowly extend out towards the sea. This natural

◀ 图 4-77: 淡水生态环境的创建很成功，如今有些淡水湿地物种(如黑翅长脚鹬)已开始在米埔繁殖(提供:世界自然基金会)
Figure 4-77: Freshwater habitat creation has been very successful, with breeding populations of freshwater wetland species (such as black-winged stilt, *Himantopus himantopus*) now established at Mai Po (Photograph by Leung Wai Ki / WWF Hong Kong)

▼ 图 4-78: 由于后海湾的泥沙沉积量增加，需谨慎控制米埔的红树林生态环境，谨防红树林侵占泥滩 (摄影：易道)
Figure 4-78: Due to increased siltation in Deep Bay, mangrove habitats at Mai Po require careful control to prevent encroachment onto mudflat habitats (Photograph by EDAW)

这种沉积更加迅速突出。研究表明，自20世纪80年代中期起，泥沙沉积速率已成倍递增（北京大学，1995），现已达到约1.3~2.8cm/a（Ove Arup 2002）。这就造成红树林的幼苗逐渐向海边的泥滩上生长。为了保证有足够大的泥滩供水鸟觅食，并且使红树林不会遮挡从观鸟舍俯瞰泥滩的视线，所以必须每年清除一次这些红树林幼苗。

后海湾的泥沙沉积越来越多及随后红树林侵占泥滩，都是米埔管理中的严重问题。从长远来看，它将导致养育水鸟的大片泥滩的丧失。香港特区政府需解决这一问题，制定并实施减少后海湾泥沙沉积量且能更有效地控制红树林侵占泥滩的战略。

游客管理

为确保对米埔进行长期保护，需要通过旅游等措施提高公众对该地区重要性的认识。应该控制游人数量，将旅游活动可能引起的破坏降至最低。为协调这两个相互矛盾的目标，香港政府从1976年开始采用入区许可证制度，控制进入保护区的游人数量。保护区的入口处设有香港渔农自然护理署护理员办公室，所有游客必须出示有效的入区许可证后方能获准进入。任何人若无证进入保护区都会被处以罚款。

入区许可证都是发给要进入保护区的私人游客。世界自然基金会还组织学生团体和群众团体分别于工作日和周末参观保护区。这些游客团体主要被带到保护区东部参观，只有一些中学生团体有机会去木板浮桥和鸟舍。通常，不会将这些游客和学生团体带到南部的基围（第20~24号基围），这里是保护区受干扰较小的地区。

每年约有4万人参观米埔，其中有1/4是学生。这是政府为中小学生组织的400次特别参观活动之一。学生的参观活动由香港教育署资助，而残疾人参观则由外界慈善机构赞助（见图4-79）。

结论

米埔自然保护区是管理与维护多用途湿地的成功典范。这一个案展示出可持续发展的滨海水产养殖技术、如何结合具

siltation process has been accentuated at Deep Bay, probably due to a combination of increased sediment loads in the waters of the Bay and a reduction in water flows through the Bay. Studies have shown that the rate of sedimentation may have doubled since the mid-1980s (Peking University, 1995) and the current sedimentation rate is around 1.3 – 2.8 cm/year (Ove Arup, 2002). As a result, mangrove seedlings are colonising the mudflat, and must be removed on an annual basis. This is done to maintain an adequate area of mudflat for feeding waterbirds, and so that the mangrove trees do not obstruct the view from bird-hides looking out over the mudflats.

Increased siltation in Deep Bay and the consequent colonisation of the mudflat by mangroves is a serious management issue at Mai Po: over the long-term it could lead to the loss of large areas of mudflat feeding habitat for waterbirds. This problem needs to be addressed by the Hong Kong SAR Government, with a strategy developed and implemented to reduce sediment loads in Deep Bay, and also to manage mangrove encroachment more effectively.

Visitor management

To ensure the long-term protection of Mai Po, there is a need to promote public awareness of the site's importance through measures such as guided visits. At the same time, it is vital that disturbances caused by large numbers of visitors are minimized as far as is practicable. To balance these two conflicting goals, the Hong Kong Government introduced an entry permit system in 1976 to control access to the Reserve. There is an AFCD warden's office at the entrance to the Reserve, and all visitors must show a valid entry permit before being allowed to enter. Anyone who enters the reserve without a permit may be fined.

Entry permits are normally issued to private visitors to the reserve. The WWF also organizes visits for groups of students and public visitors to the Reserve during weekdays and weekends respectively. These visitor groups are mainly guided along the eastern, landward portion of the reserve, although certain groups of secondary school students have an opportunity of visiting the Floating Boardwalk and Hides. Generally, groups of visitors and students are not taken to the southern shrimp ponds (numbers 20-24), an area managed as a relatively undisturbed part of the reserve.

Some 40,000 people visit Mai Po annually, of which roughly a quarter are school students on one of the 400 special visits organised for primary and secondary students. The Hong Kong Education Department funds school visits while outside charities sponsor visits for the disabled (Figure 4-79).

体的保护目标维护湿地，及实现保护区
的生态价值与公共观光之间的协调。

来到米埔的许多游客，特别是来观鸟的
游客，不仅了解到保护区在生态和文化
方面的重要性，还对保护区的管理工作
本身产生了兴趣。因此，世界自然基金
会香港分会正努力增加公众对保护区管
理战略的了解，采取措施在生态环境管
理区域旁设立公告牌，说明为什么要进
行这些重要的工程以及进展情况。世界
自然基金会香港分会希望借此向游客传
达的一个主要资讯，即在保护区的管理
工作中保证"协调"的必要性。例如，
以下就是几例需要协调的关系：

1   既要保护保护区的生态意义，又要
    将公众和学生团体参观的影响减至
    最小限度。

2   基围中水鸟栖息的开放水域与塘中
    芦苇床及其他植被区之间的关系。
    在这种情况下，可能必须使用除草
    剂来控制大面积的芦苇，防止其扩
    散到开放水域。

3   长有高大树木的池塘堤岸与拥有开
    放空间的堤岸之间的关系。尽管树
    木可以为野生动物和游客遮荫，但
    如果池塘边全部生长着高大树木，
    会造成围合封闭的感觉，无法吸引
    水鸟到池塘中央的浅水中栖息。

4   红树林区与为迁徙水鸟提供觅食环
    境的泥滩之间的关系。因此，每年
    都需要将部分泥滩上的红树林幼苗
    清除掉。

随着公众对保护区工作的了解加深，他
们将更加支持米埔保护区的工作。

Summary

The Mai Po Nature Reserve is a successful example of the management and maintenance of a multipurpose wetland. As a case-study, it demonstrates sustainable coastal aquaculture techniques, and how wetland habitats can be created and maintained to achieve specific conservation goals, with the ecological value of a reserve balanced by the need for public access.

Apart from learning about the ecological and cultural importance of the Reserve, many visitors, especially birdwatchers, have taken an interest in the management of the Reserve itself. As a result, WWF Hong Kong is currently working to improve public awareness of the management strategies within the Reserve. This includes placing notices next to sites where habitat management works are taking place explaining why and how these essential works are being implemented. One of the main messages that WWF Hong Kong is trying to pass to visitors is the need to maintain a balance in the management of the Reserve. For example, there is a need to balance:

Ex.1   The ecological importance of the Reserve while minimizing disturbance from public and school group visitors;

Ex.2   The area of open water in the shrimp ponds that is used by waterbirds, with the area of reedbeds and other vegetation in the pond. In such cases, herbicides may have to be used in controlling large areas of reeds that may otherwise invade areas of open water;

Ex.3   Pond bunds supporting tall trees with bunds that have a more open landscape. While trees may provide shade for wildlife and visitors, a pond covered in its entirety by tall trees will be enclosed and less attractive for waterbirds to roost in the shallow waters in the middle of the pond; and

Ex.4   The area of mangroves and the area of open mudflats that are the primary feeding habitats for many of the migratory waterbirds that Deep Bay supports. As a result, mangrove seedlings need to be cleared from the mudflat each year.

By increasing public awareness of the Reserve's work, it is anticipated that there will be even greater support for the conservation efforts at the site.

4.3 人工处理湿地恢复与创建

4.3.1 利用湿地处理工业废水 —— 上海化学工业区自然净化系统，中国上海

背景

上海化学工业区位于上海东南的杭州湾沿岸，面积30km²，由十多处现代石油化工设施组成（见图4-80）。近期，易道应邀牵头为上海化学工业区设计了一个面积30hm²的自然污水净化系统，对已做部分处理的工业废水进行净化，使其重新为工业设施所利用，或排入杭州湾。

工业污水的独特性要求采用严密的工程方法提高水质。易道的设计方案中建议先用生物滤池去除氨氮，然后经浅水氧化塘使COD氧化，最后使用两个平行的人造湿地系统去除硝酸盐、BOD、COD及一些重金属。这个由生物滤池、浅水氧化塘和两块人造湿地组成的净化系统预计每日能够处理25000m³以上的污水。虽然该项工程的中心目标是提高水质，但设计中也引进了景观和生态概念，从而创造出具有欣赏价值的野生动植物湿地环境。因此，这一设计不仅提供了一个污水处理中心，而且还可作为当地的湿地研究中心，同时创造出一处工业区员工的休闲场所和公众的游览地。

▼ 图4-80：上海化学工业区自然净化系统位置图
Figure 4-80: Shanghai Chemical Industrial Park's Natural Treatment System Location Plan

上海
Shanghai

上海化学工业区自然净化系统
Shanghai Chemical Industrial Park's Natural Treatment System

杭州
Hangzhou

慈溪
Cixi

4.3 Water Treatment Wetland Restoration and Creation

4.3.1 Wetlands for Treatment of Industrial Wastewater: Shanghai Chemical Industrial Park's Natural Treatment System, Shanghai, China

Background

Located southeast of Shanghai on the coast of Hangzhou Bay, the 30 square kilometer Shanghai Chemical Industrial Park (SCIP) consists of more than a dozen modern petrochemical industrial facilities. EDAW recently led the design of a 30 ha Natural Wastewater Treatment System within SCIP, including a 22 ha free surface wetland which will polish partially treated industrial wastewater for reuse in the industrial facilities, and for discharge into Hangzhou Bay.

The unique characteristics of the wastewater required a strict, engineered approach to water quality improvement. EDAW's design proposed an up-front trickling filter (for ammonia removal), followed by a shallow oxidizing pond (for COD oxidation) and two parallel constructed wetland systems (for nitrate, BOD, COD, and some heavy metal removal). The treatment system has the capacity to treat up to 25,000 cubic meters per day. While improvement of water quality is the central goal, the project incorporates design from landscape and ecological teams to create an integrated aesthetic character and wildlife habitat. The design is more than a simple wastewater treatment facility, providing a research center for local academics, a recreational hub for SCIP employees, and a destination for the public.

Currently, effluent from the on-site wastewater treatment plant (WWTP) is discharged into Hangzhou Bay via an internal canal system. However, SCIP has seen the value in constructing a water treatment system that would further purify industrial wastewater, enabling its reuse within SCIP and allowing for discharge of cleaner water into Hangzhou Bay. As the design lead for the project, EDAW envisioned the use of a Natural Treatment System (NTS) for this purpose.

The SCIP project is one of the most ambitious natural treatment facilities designed for industrial wastewater in China. It will serve as a model for similar industrial operations throughout China, and as a benchmark project in the country's renewed focus on improving its environment. The project began construction in April 2006.

目前，当地污水处理厂的排放物通过内部水道系统排入杭州湾。上海化学工业区意识到建造污水净化系统的重要性：认为这样可进一步净化工业废水，使其能够在上海化学工业区重新使用，并使流入杭州湾的水更纯净。在规划上提出了"自然净化系统"的处理方案。

上海化学工业区项目是迄今为止中国为处理工业废水实施的规模最宏大的自然净化设施之一。它将为中国的类似工业项目树立典范，同时，也标志着在中国，环境改善事业再次成为关注热点。该项目2006年4月开始动工。

目的和目标

自然净化系统的主要设计目标是提高污水处理厂的排水水质，使其符合国家地面水质四级标准。然后，经净化的废水就可用于水景和河道，并且重新用于该区各项工业流程。通过水循环系统，可以进一步提高排入杭州湾的水质，减少上海化学工业区的总体耗水量从而降低处理的成本，改善上海化学工业区在环境和财务方面的运营状况。另一个重要的目标是防止净化系统产生毒性。此外，该工程还可以为野生动植物提供生境，并为休闲娱乐和科学研究提供场所。

表4-1列出设计中使用的各项水质标准。"国家地表水四类水标准"是对湿地排放物的质量要求。"污水处理厂当前净化标准"是自然净水系统进水口的最大污染物浓度（根据排放许可），"污水处理厂排放物（实际）"是自然净化系统进水口的当前水质。

Goals and Objectives

The primary design goal of the NTS is to improve water quality of the wastewater treatment plant effluent to meet Level IV of the National Standard of Surface Water Quality (NSSWQ). This polished wastewater can then be used for water features and canals, and reused in various industrial processes on site. By further improving the water quality of the effluent discharged into Hangzhou Bay, reducing SCIP's overall water consumption and thereby its water costs, the water recycling system can improve both the environmental and financial aspects of SCIP's operation. Another important requirement was to prevent toxicity in the treatment system. The project will furthermore provide a wildlife habitat and recreational and scientific research opportunities.

▼ 表4-1：符合国家地表水四类水标准的污染物和污水处理厂处理标准
国家地表水四类水标准在此作为参考值并不代表实际净水期望值
Table 4-1: Pollutants with Level IV NSSWQs and WWTP Treatment Standards

水的组成 Water Constituent	水的组成 Level IV NSSWQ (mg/L)	水的组成 WWTP Existing Treatment Standards (mg/L)	水的组成 WWTP Effluent (Measured) (mg/L)
N-NH₃	1.5	15	1.39
化学需氧量 COD	30	100	66
五日生化需氧 BOD₅	6	30	0.97
总氮 TN	1.5	无 none	31.8
总磷 TP	0.3	无 none	1.7
P-PO₄	0.3	1	1.5
溶解氧 DO	3	无 none	未知 n.a
酸碱度 pH	1.0	1.0	<1.0
高锰酸盐指数 Permanganate Index	2.0	4	<4
铜 Cu	1.5	10	<10
锌 Zn	2.0	4	<4
F⁻	0.02	0.1	<0.1
Se	1.5	10	<10
As	0.1	0.5	<0.5
Hg	0.001	0.02	<0.02
Cd	0.005	0.1	<0.1
Cr⁶⁺	0.05	0.5	<0.5
Pb	0.05	1.0	<1.0
CN	0.2	0.5	<0.5
挥发性酚 Volatile hydroxybenzene	0.01	0.4	1.77
油 Oil	0.5	无 none	0.03
阴离子表面活性剂 Anionic Surfactant	0.3	无 none	未知 n.a
硫化物 Sulfide (S)	0.5	1.0	<1.0
水中粪生大肠杆菌群 Fecal Coliform (ind./L)	20000	1000	<1000

国家地表水四类水标准不代表实际期望性能
Level IV NSSWQs do not represent actual performance expectations.

净 化 系 统 的 设 计

自然净化系统占用的土地面积共29.9hm²，呈L形分布在上海化学工业区的东北角。该区全部位于昔日养殖池塘(包括废弃的和使用中的)的围涂上。现有植被只有几种湿地植物和常见的杂乱的草本植物。养殖池塘的土壤由黏土、壤质黏土和砂土的混合物(带有缺氧症状，包括潜育化和氧化的根孔)和一些有机物组成。开工前的状况如图4-81~4-85所示。

其 他 可 选 择 的 开 发 方 案

如前所述，自然净化系统的设计主旨是提高污水处理厂排水水质，使其达到国家地表水四类水标准。为选择最合适的湿地预处理和人工湿地设计方案，规划单位仔细权衡了多种方案的利弊。

湿 地 预 处 理 方 案

氨($N-NH_4^+$)属有毒物质，尤其在pH值和温度较高的时候。它会妨碍鱼类生长，提高其死亡率，即便在$N-NH_4^+$值较低(2~4mg/L)的情况下也不例外。由于鱼是昆虫(尤其是蚊子)的主要杀手，氨含量过高会导致鱼的数量减少，进而产生虫害，这是应该避免发生的。自然净化系统进水口的氨浓度预计在15mg/L左右。因此，湿地预处理系统的主要目标就是要把这一浓度降低到2mg/L。为实现先脱氨再将水引入开放湿地，设计了几种处理方案。

方案1：曝气槽——人工湿地
用曝气槽实现硝化，然后直接排入人工湿地。

优 点

- 如果设计得当，曝气槽的工艺会很可靠；
- 可以做成漂亮的水景；
- 可以有效去除BOD。

缺 点

- 需要经过实验来得出脱氨系数和适用规模；
- 为防止盐水导致曝气器及其组成部分结垢腐蚀，需要进行专门的维护保养；
- 曝气的耗能成本高(每25000m³净化水耗电39kWh)；
- 去除了水体悬浮物，湿地就失去了微生物处理过程的动力来源——碳；
- 由于曝气槽占地面积大，水处理总量相对较低；

Table 4-1 indicates the various water quality levels used in the design. Level IV NSSWQ is the desired quality for the wetland effluent. The WWTP Existing Treatment Standard is the maximum pollutant concentration at the inflow of the NTS (according to the discharge permit), and the WWTP Effluent (Actual) is the existing water quality at the influent of the NTS.

Treatment System Design

The total land area allocated for the Natural Treatment System is 29.9 ha, constituting an L shaped lot on the northeast corner of the SCIP site. The entire site is located on reclaimed tidal lands which were occupied by aquaculture ponds (both abandoned and in-use). Existing vegetation was limited to and consisted of several wetland species as well as common grasses and weedy herbaceous plants. The soil in the aquaculture ponds consist of a mixture of clay, loamy clay, and sandy clay with indications of anoxia including gleying and oxidized root channels and some organic matter indicating wetland soils. Site conditions prior to the project are illustrated in Figures 4-81~4-85.

Alternative Development Approaches

As previously mentioned, the primary design objective of NTS is to improve water quality of the wastewater treatment plant effluent to meet Level IV of the National Standard of Surface Water Quality (NSSWQ). In order to select the most appropriate pre-wetland treatment and constructed wetland design to meet that goal, EDAW carefully weighed the benefits and drawbacks of the various alternatives.

Pre-Wetland Treatment Alternatives

Ammonia (N-NH4+) is a toxic substance, especially at higher pH and temperature values, and will inhibit growth or increase the mortality rate of fish populations, even when present at relatively low levels (2-4 mg/L). Since fish are the main source controlling insect (in particular mosquito) populations, high ammonia levels cannot be tolerated in a free surface wetland without the risk of infestation problems due to a compromised fish population. The ammonia concentration at the influent of the NTS is expected to be about 15 mg/L. Therefore the main goal of the pre-wetland treatment system is to reduce this concentration to 2 mg/L. EDAW considered several treatment options for ammonia removal prior to recommending a free surface wetland.

Alternative 1: Aerated Lagoon -> Constructed Wetland

The process is to nitrify using an aerated lagoon and discharge directly to a constructed wetland.

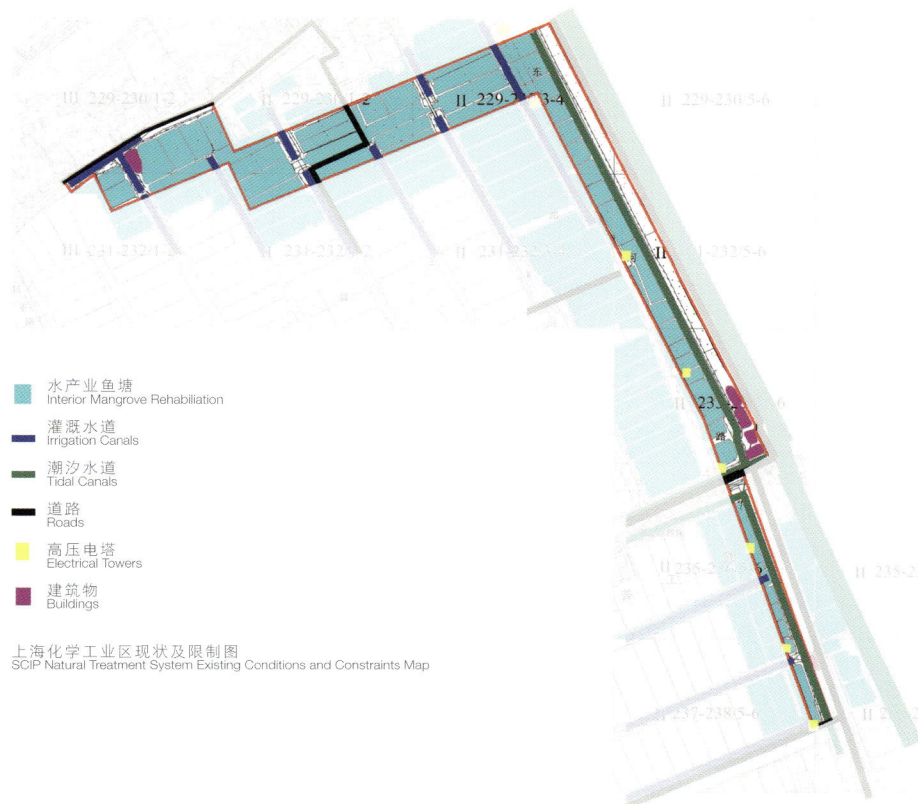

水产业鱼塘
Interior Mangrove Rehabiliation

灌溉水道
Irrigation Canals

潮汐水道
Tidal Canals

道路
Roads

高压电塔
Electrical Towers

建筑物
Buildings

上海化学工业区现状及限制图
SCIP Natural Treatment System Existing Conditions and Constraints Map

▲ 图 4-81：开工前状况（制图：易道）
Figure 4-81: Pre-project conditions site plan (Graphic: EDAW)

◀ 图 4-82：场地南端的灌溉渠（面向北）。注意巨大的送电塔（摄影：易道）
Figure 4-82 (top): The irrigation canal at the southern end of the site (looking north). Note the large electrical towers. (Photograph by EDAW)

◀ 图 4-83：场地中的一条灌溉渠（面向北朝向村庄）（摄影：易道）
Figure 4-83 (second from top): One of the irrigation channels within the site (looking north towards the village) (Photograph by EDAW)

◀ 图 4-84：养殖池塘一般以狭窄堤岸相隔（摄影：易道）
Figure 4-84 (third from top): Typical aquaculture ponds separated by a small berm (Photograph by EDAW)

◀ 图 4-85：灌溉渠边缘的典型壤质黏土。注意潜育化和氧化的根孔（摄影：易道）
Figure 4-85: Typical loamy clay soil on the edge of an irrigation ditch. Note gleying and oxidized root channels. (Photograph by EDAW)

- 可能出现虫害和难看的泡沫。

方案2：滴滤池——→人工湿地
用传统的滴滤池，或生物滤池来实现硝化，然后直接排入湿地。

优点
- 滴滤池是硝化能力已得到证实的先进技术；
- 基础设施成本不高；
- 能源需求量相对较低：每25000m³净化水耗电18kWh；
- 占地少（处理25000m³水需要0.75hm²左右），可更充分利用整个系统，处理量更大；
- 保留了水体悬浮物，为湿地提供碳源；
- 根据滴滤池表面面积使用适当的水力负荷，可避免蚊虫和臭味问题,高水力负荷可使蚊虫远离系统。

缺点
- 由于水的盐分高，可能导致滤池部件结垢并腐蚀，因而还需要这方面的维护保养；
- 滴滤池看起来不很美观；
- 滴滤池无法大量去除BOD。

方案3：改善污水处理厂的功能——→人工湿地
当前的污水处理厂可实现很高的硝化程度（约75%~80%）。但通过延长水在每个单元的平均停留时间（MCRT）并将曝气量增长5%~10%，可进一步增强处理厂的硝化能力，从而将排放物中的$N-NH_4^+$浓度控制在2~4mg/L。

优点
- 不需占用土地，因而整个场地都可用作人工湿地；
- 不用增加基础设施或增加很少就可以实现净化目标；
- 耗能低；
- 所有者和经营者都可监督管理整个水处理系统。

缺点
- 可靠性：经营污水处理厂的人不一定会把硝化作用放在第一位。经营者必须对湿地设施的排放物全权负责，才会把重点放在硝化作用上。
- 上海化学工业区达到生产能力后，产生

Benefits
- Aerated lagoon technology is reliable if appropriately designed;
- Can be made into attractive water feature;
- Effective BOD removal.

Drawbacks
- Requires some experimentation to develop removal coefficients and appropriate sizing;
- Additional maintenance requirements due to potential for saline water to cause encrustation and corrosion on aerator equipment and structural components;
- High energy costs (39 KW/25,000m3 of treated water) for aeration;
- TSS is removed and wetland is deprived of carbon source to fuel microbially facilitated removal processes;
- Total water treatment volume is lower due to large area requirement for the lagoon;
- Potential for insect infestation and unsightly foaming.

Alternative 2: Trickling Filter -> Constructed Wetland
The process is to nitrify using a conventional trickling filter, or bio-tower, and then discharge effluent directly into a wetland system.

Benefits
- Trickling filter is advanced technology with proven nitrification capabilities;
- Inexpensive infrastructure costs;
- Relatively low energy requirements – 18 KW/ 25,000 m3 of treated water;
- Small land requirements (about 0.75 has to treat 25,000 m3) allows for greater treatment volume using more of the entire system;
- TSS is not removed and will provide a carbon source for the wetland;
- Winged insects and odors can be avoided using appropriate hydraulic loading rates per surface area of trickling filter. High hydraulic loading rates flush insect larvae from the system.

Drawbacks
- Additional maintenance requirements due to potential for increased water salinity which could cause encrustation and corrosion on filter elements;
- Trickling filter may not be visually aesthetic;
- BOD is not removed in substantial amounts by trickling filter.

的排放物流速太高，可能无法实现充分硝化。

表4-2～4-4可在做出决策时应用。

基于对表4-2～4-4及易道专业团队建议的分析，方案2为实现项目目标和湿地预处理系统脱氨氮的最佳方案。因此最终方案在自然净化系统中采用了滴滤池设计。

人工湿地处理方案
对人工湿地的设计，易道考虑用两种类型：自由表面流湿地和潜流湿地。有关各自特征的描述如下：

自由表面流湿地：
- 为野生动植物提供宽阔的水面；
- 更为美观；
- 为公众提供娱乐场所；
- 具有处理大量水的能力；
- 处理BOD含量低的废水速度更快；
- 建造、运行和管理成本较低。

而潜流湿地（砂砾床）更适用于寒冷天气，但：
- 通过砂砾床的水流速度低；
- 野生动植物和美学价值低；
- 建造、运行和管理成本比自由表面流湿地高得多。

鉴于该场地条件及各项参数，易道建议将多数可用湿地面积建成自由表面流湿地，而不是潜流湿地。上海化学工业园区的水处理过程概要请参阅图4-86。

Alternative 3: Treatment Plant -> Operational Modification Constructed Wetland
The existing wastewater treatment plant achieves significant nitrification (about 75-80%). This nitrification capacity could be further improved by increasing mean cell residence time (MCRT) and increasing aeration by about 5-10% to achieve effluent with N-NH4 concentrations of 2-4 mg/L.

Benefits
- No land requirements means the entire site is available for constructed wetland;
- Treatment goals achieved with minimal or no infrastructure additions;
- Low energy costs;
- Same owner and operator would oversee the entire water treatment system.

Drawbacks
- Reliability: treatment plant operators may not place a priority on nitrification. The operator will have to be wholly responsible for the effluent of the wetland facility in order for a priority to be placed on nitrification;
- When SCIP reaches capacity, future flow rates may be too high to fully nitrify effluent.

The following matrices were developed to aid the decision making process. Each alternative was given a ranking for 1) overall water quality treatment performance (Table 4-2) and 2) operation and infrastructure requirements (Table 4-3). For water quality, alternatives were ranked based on

▼ 表4-2：水质性能表
Table 4-2: Water Quality Performance Matrix

	湿地预处理方案 Pre-wetland Treatment Alternatives		
	1	2	3
水中成份 Water Constituent	氧化槽 Oxygenated lagoon	滴滤池 Trickling Filter	污水处理厂功能改善 WWTP Op Modification
氨 Ammonia	中 Moderate (2)	高 High (3)	中 Moderate (2)
COD	差 Poor (1)	差 Poor (1)	无变化 No Change (1)
BOD	中 Moderate (2)	差 Poor (1)	无变化 No Change (1)
溶解氧 DO	中 Moderate (2)	高 High (3)	无变化 No Change (1)
平均测评 Mean Ranking	7	8	5

	处理方案 Treatment Alternatives		
	1	2	3
类别 Category	氧化槽 Oxygenated lagoon	滴滤池 Trickling Filter	污水处理厂功能改善 WWTP Op Modification
空间(公顷) Ammonia	6 (1)	0.75 (2)	无变化 No Change (3)
资本成本 Ammonia	较高 Higher (1)	中 Moderate (2)	小 Minor (3)
运行和维护成本 Ammonia	39 KW (1)	18 KW (2)	增加 5-10% increase (2)
外观效果 DO	中 Moderate (1)	低 Low (2)	改善 Enhanced (3)
湿地处理能力 Mean Ranking	25000 (1)	31000 (2)	31800 (2)
经营者可接受性 Ammonia	不适用 Not Applicable (2)	不适用 Not Applicable (3)	有问题 Questionable (1)
平均测评 Ammonia	7	13	14

▼ 表 4-4：最 后 测 评
Table 4-4: Final Ranking

	处理方案 Treatment Alternatives		
	1	2	3
	氧化槽 Oxygenated lagoon	滴滤池 Trickling Filter	改善污水处理厂的功能 WWTP Op Modification
最后测评结果(水质和运行) Total Ranking (Water Quality and Operation)	14	21	19

详细设计

易道的总体规划（见图4-87）有三大部分：滴滤池（去氨）、浅水氧化池（COD氧化）和自由表面流湿地系统(去除氮、BOD、COD和重金属元素)。

(1)滴滤池（见图4-88）

水通过四个滴滤池滤料塔进入系统。塔高4.52m，宽15m。为提高处理能力，滴滤池使用的滤料是塑胶的，而不是传统的石质滤料。水通过塔顶的分配器抽上来，再通过滤料滴下来的时候，去除氨氮。

有两套滴滤池按顺序在运行，每套由两个滤料塔组成，也是依次运行。设计没有决定兴建一个或两个更大的滴滤池滤料塔，使运行具有更大的灵活性。例如，运行的滤料塔数量可随进水口的水质不同而改变，而且运行顺序也可逆转，以防第一套滤料塔由于负荷大而堵塞。设计中还包含一个旁通管路，使污水可以根据需要直接

their ability to remove various pollutants (or supply oxygen), with a value given between 1 and 3, with 1 being the worst and 3 being the best. The same process was completed for the operations and infrastructure table. The mean ranking row at the bottom of each column shows the sum for that table. The final ranking (Table 4-4) shows the sums of each mean ranking from the water quality and operation and infrastructure tables.Based on the above matrices and recommendations of EDAW's technical staff, Alternative 2 was the preferred choice for achieving project goals and ammonia removal in the pre-wetland treatment system. The trickling filter was subsequently adopted in the NTS.

Constructed Wetland Treatment Alternatives

As for the design of the constructed wetlands, EDAW considered two types: free-surface wetlands and sub-surface wetlands. Characteristics of each are described below.

Free-surface wetlands offer:

流入COD降解池和湿地。

(2)COD降解池
污水在离开滴滤池后，即进入两个平行的COD降解池，面积各约7300m²。污水在这些平浅的砂砾池中停留3~4h，白天可以接受天然紫外线照射和与之相连的藻类提供的氧气。由于COD很难降解，易道的设计方案是先把COD放在强氧化环境中，然后再将其排入自由表面流湿地。

(3)自由表面流湿地
自由表面流湿地总面积刚过22hm²，其中湿地占18.57hm²，开放水域3.45hm²。自由表面流湿地被分成两个平行的处理系统(A和B)，每个系统由五块湿地组成。维护保养时，湿地通过流量控制阀门隔开。每块湿

- large water surface for wildlife;
- more pleasing aesthetics;
- recreational opportunities for the public;
- ability to treat large volumes of water;
- greater rates of treatment for low BOD waste; and
- lower construction, operation and management costs.

On the other hand, sub-surface wetlands (gravel beds) can be more suitable for cold climates, but have:
- low flow rate through gravel beds;
- low wildlife and aesthetic value; and
- much higher construction, operation and management costs compared to free-surface wetlands

Based on the site's conditions and parameters, EDAW proposed that the majority of the available wetland area

未经处理过的
污水
Raw Untreated
Effluent

工业排放
Industrial Discharge

物理、化学生
物处理
Physical,
Chemical and
Biological
Treatment

传统的一级和二级的
污水处理
Conventional Primary and
Secondary Wastewater
Treatment

生物处理
Biological
Treatment

滴滤池
(除氨)
Trickling Filter
(Remove Ammonia)

物理和生物
处理
Physical and
Biological
Treatment

自由表面湿地
Free Surface Wetland

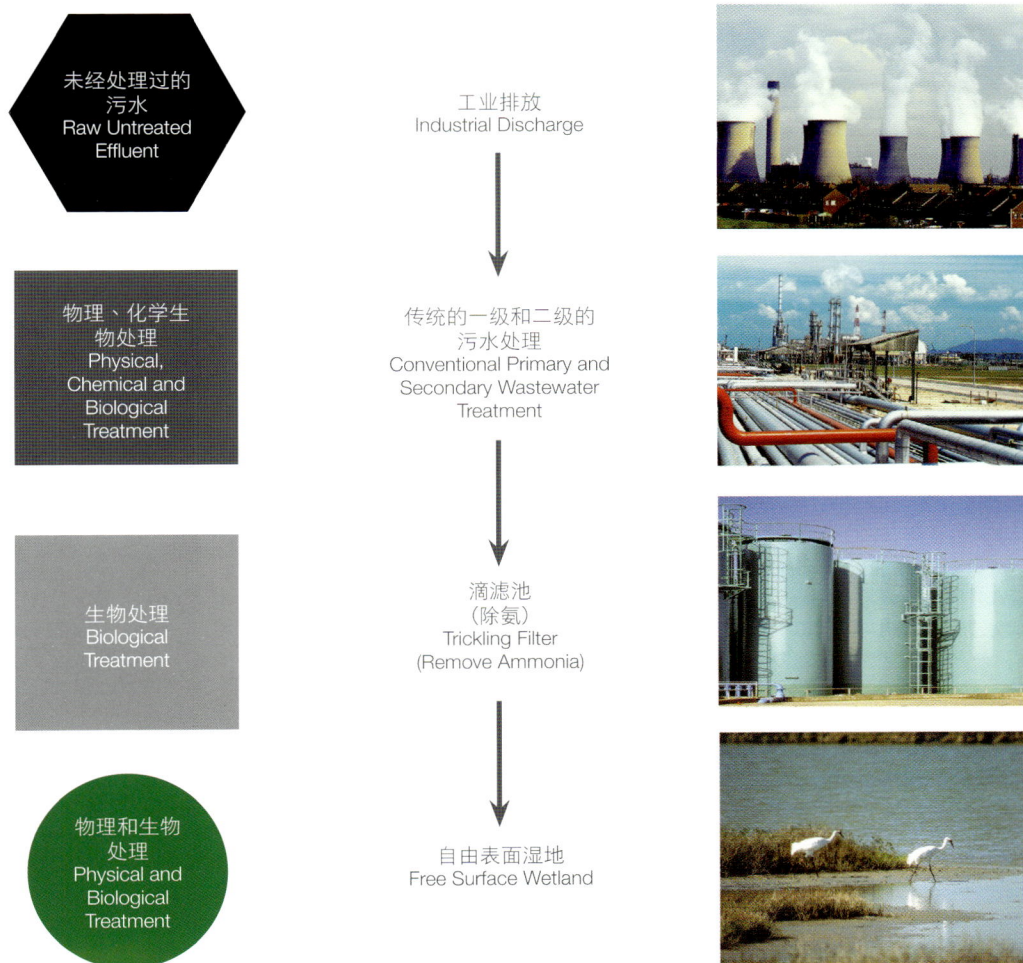

◀ 图 4-86: 上海化学工业园区的水处理过程概
要(绘制:易道，图像来源: www.photos.com)
Figure 4-86: Summary of the water treatment process
at SCIP (Graphic: EDAW. Images: www.photos.com)

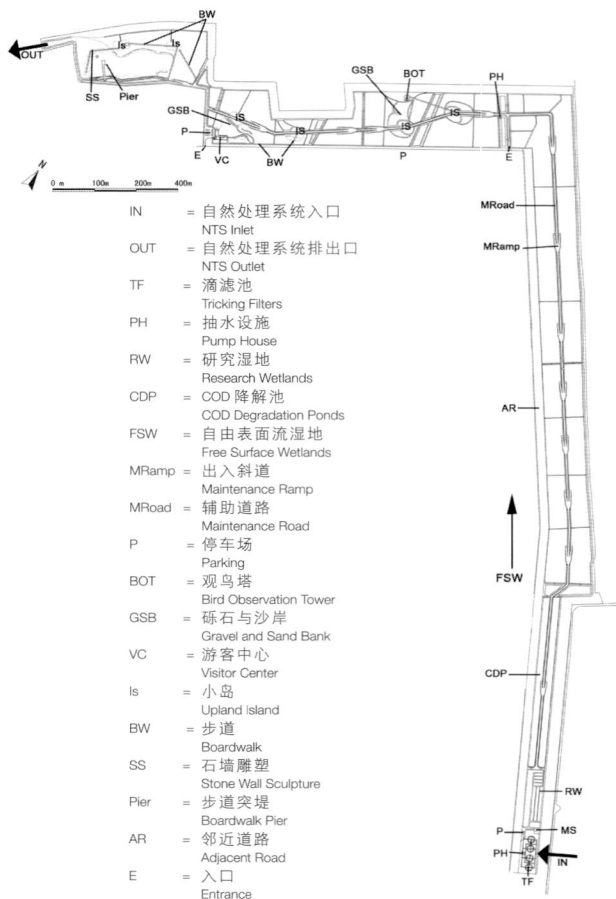

IN = 自然处理系统入口
　　　 NTS Inlet
OUT = 自然处理系统排出口
　　　 NTS Outlet
TF = 滴滤池
　　　 Tricking Filters
PH = 抽水设施
　　　 Pump House
RW = 研究湿地
　　　 Research Wetlands
CDP = COD降解池
　　　 COD Degradation Ponds
FSW = 自由表面流湿地
　　　 Free Surface Wetlands
MRamp = 出入斜道
　　　 Maintenance Ramp
MRoad = 辅助道路
　　　 Maintenance Road
P = 停车场
　　　 Parking
BOT = 观鸟塔
　　　 Bird Observation Tower
GSB = 砾石与沙岸
　　　 Gravel and Sand Bank
VC = 游客中心
　　　 Visitor Center
Is = 小岛
　　　 Upland Island
BW = 步道
　　　 Boardwalk
SS = 石墙雕塑
　　　 Stone Wall Sculpture
Pier = 步道突堤
　　　 Boardwalk Pier
AR = 邻近道路
　　　 Adjacent Road
E = 入口
　　　 Entrance

▲ 图4-87：上海化学工业区自然净化系统场地
设施特征 (Copyright EDAW)
Figure 4-87: SCIP natural treatment system site
features (Copyright EDAW)

地内的暗堤有助于保证水流在整个湿地上的流量和分配平均。这些暗堤还用来分隔不同的湿地植被，在建立湿地期间减少清除杂草的需要。共有19条堰道调节各块湿地间的流量，堰道的使用深度一般在0.5m~0.9m之间。

除了最后两个单元，整个湿地都衬砌了高密度聚乙烯衬料，以防污染地下水。

科研湿地

易道还在场地南口规划出七个单元的科研湿地，运用监测程序和适应性管理，跟踪监测具体成分(如重金属、COD和有机物)的去除率，从而为未来改善湿地性能提供有效的办法。科研湿地对于场地的长期适应性管理至关重要，可以使新方法新点子在应用于大片湿地之前在这里得到评估。同时，科研湿地也是上海工业区展示其湿地处理系统的视窗，并可与教育科研机构建立关系。

景观设计

除了这些基本处理设施，易道的设计方案还规划了许多可供野生动植物和休闲娱乐用的结构设施。面积450m²的游客中心有会议室、教育展厅、大型橱窗及小观景平台和码头。与该中心相邻的周围地区被设计成湿地植物园，上面有适于在排出的半咸水中繁衍的各种植物和其他生物。

由裸露砂石组成的小生态岛坐落在湿地植物园中一个湖泊中央，从这里可将远景尽收眼底，包括一个供水禽栖息的独立的生态环境。一条580m长的木板路从游客中心蜿蜒穿过湿地植物园。

游客从南门进园，一条2km长的公路从这里通往游客中心。在路的半中间有个岔道口，从这里有一条小路可以通往僻静的观鸟楼，俯看近处的池塘和生态岛（见图4-89）。

植物

易道在为自由表面流湿地选择植物时考虑了几个因素，分别为：耐盐碱力、对改善水质已经证实的作用以及在设定水位下的生存能力。在选择区域性及本地物种时，美学方面的考虑也占了很大成分。

基于上述标准，共为核心区域选定了四个基本湿地物种，除此以外，还为湿地植物

be composed of free-surface wetlands rather than sub-surface wetlands. A summary of the final treatment process at SCIP is shown in Figure 4-86.

Detailed Design

The main treatment components of EDAW's masterplan are:

Trickling Filter

Water enters the system through four trickling filter towers, each 4.52 meters tall by 15 meters wide. The trickling filter applies a plastic packing media rather than conventional rock packing to increase treatment capacity. The water is then pumped through distributors located at the top of each tower, and as it trickles through the media, ammonia is removed. Two sets of trickling filters are operated in series, with each set consisting of two towers, also in a series. Rather than creating one or two larger towers, the design layout was developed to allow flexibility in operation. For instance, the number of towers in operation can change depending on temporal water quality changes in the influent, and the order of operation can be reversed to prevent clogging of the first set in the series, due to higher loading rates. A bypass pipeline is also included in the design to allow wastewater to flow directly to the COD degradation pond and wetlands as needed.

COD Degradation Pond

Once it leaves these filters, water enters two parallel, COD degradation ponds, each with an area of approximately 7,300 m². Water spends roughly three to four hours in these shallow, gravel-lined ponds, where it is exposed during the day to natural UV-light and aeration from attached algae. Because much of the COD is difficult to degrade, this design would expose the COD to an intense, oxidizing environment before it enters the free surface wetland.

Free Surface Wetland

The Free Surface Wetland (FSW) has a total area of just over 22 ha, including 18.57 ha of wetlands and 3.45 ha of open water. It is divided into two parallel treatment systems (A and B), each comprised of five wetland units. Each unit can be isolated by using flow control gates to allow for maintenance. Submerged berms within each unit help maintain even water flows and distribution throughout the wetland. The berms also separate wetland vegetation types, reducing the need for weeding of unwanted growth during the establishment of the wetlands. A total of 19 weirs regulate flow between the units, which have an operating depth range between 0.5 and 0.9 meters.

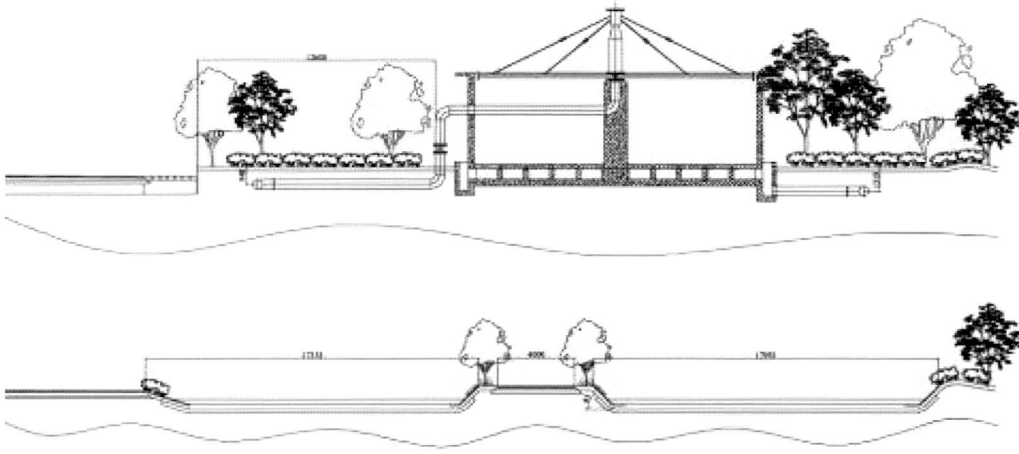

图 4-88:
Figure 4-88:

◀ 滴滤池 (Copyright EDAW)
Trickling filter (Copyright EDAW)

◀ COD降解池 (Copyright EDAW)
COD Degradation pond (Copyright EDAW)

园选择了另外25种植物。

水净化模型

易道利用专业技能,为自然净化系统设计了一个水质性能模型,结合水文、热传递和污染物去除参数,确定水处理能力和经自然净化系统处理后预期达到的水质。

这些模式共同阐释了影响水文与能量平衡的各种参数,包括降雨和水分蒸发蒸腾作用、太阳辐射、热传递、风速、反照率、湿度、环境温度及其他条件。

水质模型使易道通过在化学转化率和气候条件之间建立起关联,评估出夏季和冬季极端气候条件下的水处理能力。模型研究结果表明,在夏天最热的时候,系统处理工业废水的能力为22250m³/d,而冬天由于湿度低,微生物反应较慢,系统最大处理能力只能达到5750m³/d。

运行及监测机制

建立严格的运行及监测机制是为了确保自然净化系统的长期功用与健康。该机制要求对湿地的化学、物理和生物健康及其在处理污水方面的成功与否进行全面监测。

该机制涉及到自由表面流湿地及滴滤池的使用和保养。它旨在通过减小水、土壤和地下水中的毒性,建立起便于改革的适应性管理协议,管理或修正未来的可变条件,实现对水处理能力的监测和优化。实现这一目标靠的是用灵活的管理体制分析和改进水处理能力。

结论:工业及环境的里程碑

本设计除了系统地解决了水质改善问题外,还将净化污水的湿地变成既适合野生动植物生存又适合人们游览的一处自然美景(见图4-90)。这种结合多种设计目标的构想为上海化学工业区员工及公众提供了一个休闲娱乐中心,同时还为学术团体提供了一个湿地研究中心。

目前在国内,类似上海化学工业区自然污水净化系统少见于报导。这个项目标志着中国在创建自然污水净化系统方面一定会有更多建树。

With the exception of the last two cells, the entire wetland is lined with an impermeable High Density Polyethylene (HPDE) Liner to prevent leakage to the groundwater.

Research Wetlands

EDAW also programmed seven research wetland cells to be located near the influent pipe at the site's southern entrance. Through monitoring processes and adaptive management, these facilities would provide the means to improve future performance of the wetland by tracking removal rates of specific constituents such as heavy metals, COD, and organics. The research wetlands are essential for the long term adaptive management of the site, allowing new and innovative approaches to be assessed prior to implementation over larger site areas. They also provide a means through which SCIP can promote its wetland treatment system and form relationships with educational and research institutions.

Landscape Design

In addition to these essential treatment components, EDAW's design is programmed with a considerable number of facilities and structures with recreation and wildlife in mind. A 450 m² visitor center features a conference room, an educational display, large viewing windows and a small viewing deck and pier. The area adjacent to and surrounding the center is designed as a botanical wetland with a variety of plant species and habitats suitable for the semi-saline waters of the effluent. A small habitat island consisting of open gravel sits at the center of a lake within the botanical wetland, allowing for unimpeded views into the distance, which includes an isolated habitat for shorebirds. A 580-meter wooden plank boardwalk snakes through the botanical wetland areas from the visitor center.Visitors enter the site at its southern end, which is connected by a two-kilometer maintenance road leading to the visitor center. At the road's midway point, a turnout is provided for a trail that leads to an isolated bird observation tower overlooking an intimate pond and habitat island.

Planting

EDAW's criteria for plant species selection in the free surface wetland considered several factors, respectively: salinity-tolerance, proven performance for water quality improvement, and their ability to grow within designed water levels. Regional and local species were sought, with aesthetic considerations also playing a role in selection. Using the above criteria, four primary wetland species

were selected for core treatment areas, with an additional 25 species selected for the botanical wetlands.

Water Treatment Modeling

EDAW applied its technical expertise to engineer a water-quality performance model for the natural treatment system that incorporates hydrology, heat transfer, and pollutant removal parameters to determine treatment capacity and expected water quality of the NTS effluent. Collectively, the models account for a range of parameters affecting hydrology and energy balances including rainfall and evapotranspiration, solar radiation, heat transfer, wind speed, albedo, humidity, ambient temperature and other conditions. The modeling system allows EDAW to estimate water treatment capacity during extreme summer and winter conditions by linking chemical transformation rates to climatic conditions. The results of the model indicated that during the hottest period of the summer, the system will be able to treat 22,250 m^3/day of industrial effluent, while during the winter, due to colder temperatures and slower microbial kinetics, the system may only treat up to 5,750 m^3/day.

Operation and Monitoring Plan

A rigorous operation and monitoring program was developed to ensure the long-term function and health of the NTS. The program calls for extensive monitoring of the chemical, physical, and biological health of the wetlands and their success in treating effluent.

The plan covers the operation and maintenance of the trickling filter as well as the free surface wetland. The plan aims to optimize and monitor treatment performance, by reducing the risk of toxicity in water, soils, and groundwater, and by establishing an adaptive management protocol for modifications that could manage or offset future variable conditions. It achieves this through a flexible management framework for analyzing and improving treatment performance.

Conclusion

A Benchmark for Industry and the Environment

While the improvement of water quality is systematically addressed in this unique design, EDAW's fusion of design disciplines would allow the treatment wetlands to evolve into an attractive natural environment suitable for both wildlife habitats and visitors (Figure 4-90). The vision calls for a recreational hub for SCIP's employees and visitors, while additionally serving as a wetland research center for academic groups.

The SCIP Natural Wastewater Treatment System is a unique project for China, marking the onset of an expanding interest in creating natural water treatment systems.

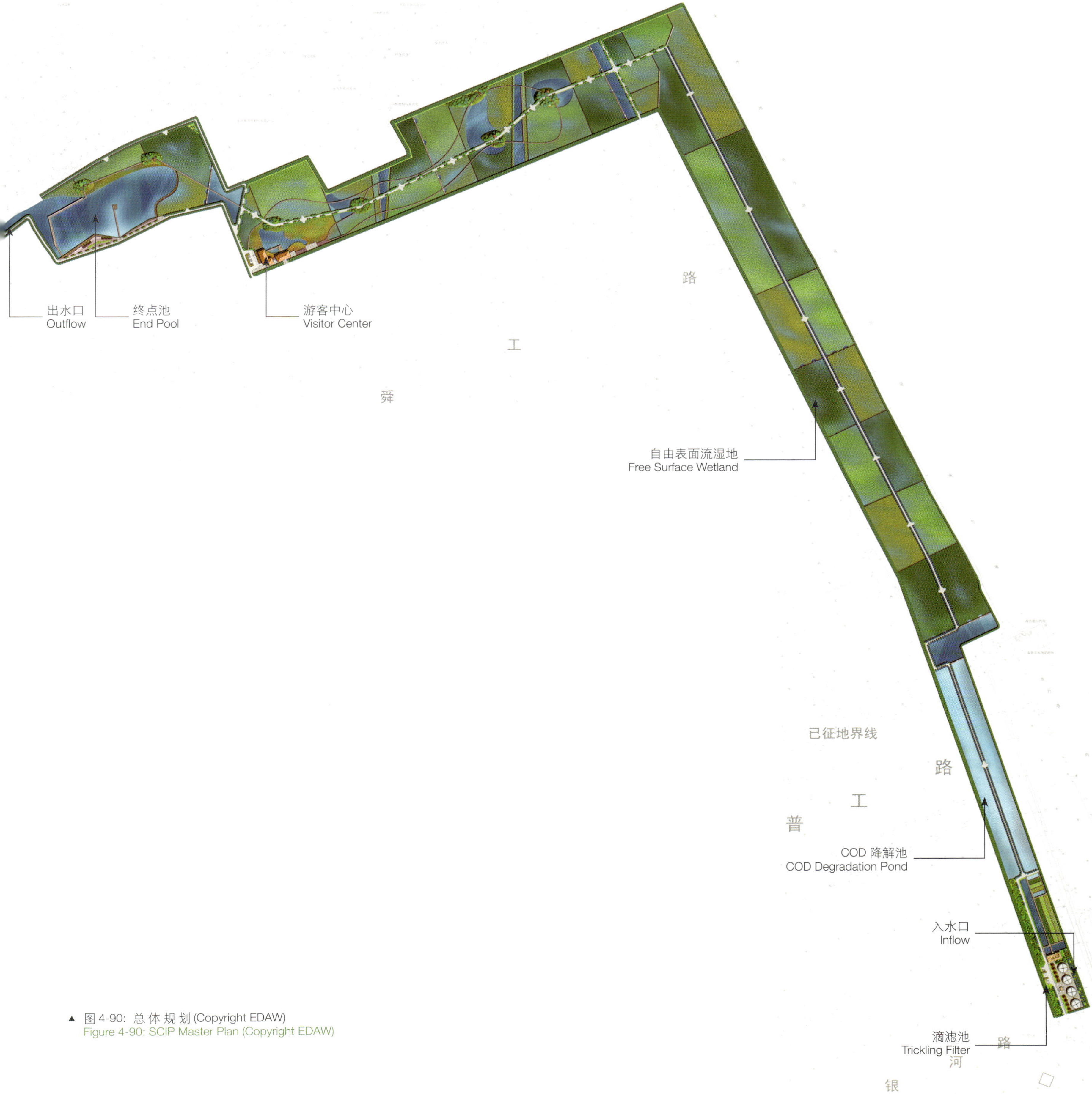

出水口
Outflow

终点池
End Pool

游客中心
Visitor Center

路

工

舜

自由表面流湿地
Free Surface Wetland

已征地界线

路

工

普

COD 降解池
COD Degradation Pond

入水口
Inflow

▲ 图 4-90: 总体规划 (Copyright EDAW)
Figure 4-90: SCIP Master Plan (Copyright EDAW)

滴滤池
Trickling Filter

路

河

银

资料提供
李荣旗
北京锦绣大地股份有限公司

Information provided by
Lee Rongqi
Beijing Glorious Land Agricultureal Co.

4.3.2 人工和自然复合湿地 —— 翠湖湿地恢复，中国北京市海淀区

背景

翠湖湿地所在地区处于近山的平原地区，西高东低，平均海拔在41~43m左右，现属于北京温榆河的上游，与上庄水库相邻，行政区划属于北京市海淀区上庄镇（见图4-91）。翠湖以南沙河为主要水源，湿地恢复的核心区约为266.67hm²，总面积约666.67hm²。

翠湖湿地（见图4-92）所处地区属暖温带半湿润季风气候，年平均气温10~12°C，1月平均气温-7~4°C，极低温度-20°C，7月平均气温25~26°C，极高温度40.8°C。无霜期190天左右，冬季冻土期100天。土壤主要为褐黄及灰色砂黏中等压缩性灰色土。翠湖湿地水资源相对丰富，虽然年平均降雨量仅约为600mm，但有南沙河作为水源补给。周边有333.33hm²水稻田，是北京市近郊最大的水稻湿地。

目的与目标

翠湖湿地以开展生态旅游、衬托城市环境作为其主要的功能定位，其外，还包括科研、科普和国际交流等活动，净化污染水体、涵养水源，保护生态与维持心理健康等功能。翠湖湿地的恢复以自然环境为基础，保持湿地生态系统的整

▶ 图4-91：翠湖湿地在北京市的位置图（绘制：易道）
Figure 4-91: Location of the Jade Lake Wetland, Beijing (Graphic: EDAW)

4.3.2 Creating a Constructed and Natural Wetland Habitat: Restoration of Jade Lake Wetland, Haidian District, Beijing, China

Background

The Jade Lake Wetland is located on a plain adjacent to mountains, descending from west to east, at an average altitude of 41 to 43 meters above sea level. It is situated on the upper reaches of the Wenyu River, lying adjacent to the Shang Zhuang reservoir, which is under the jurisdiction of Shangzhuang Town in northwest Beijing. The Nansha (South Sandy) River serves as a water source for Jade Lake. The core area for wetland restoration is approximately 267 ha out of the total lake area of 667 ha.

Jade Lake is located in a warm temperature zone within a semi-dry climate. General characteristics of the Lake and its surroundings include:

- Annual mean temperature: 10-12° C
- Average temperature in January: -7° to -4° with the lowest -20°
- A-verage temperature in July: 25° to 26° with the highest 40.8°
- Frost-free period: Around 190 days
- Frozen ground period: 100 days
- Soil type: Yellow-brown and grey, silty clay loam of medium compressibility

With the Nansha River supplying a fairly constant source of water, Jade Lake Wetland has an abundant water supply despite an annual precipitation that averages only around 60 cm. The wetland is surrounded by 333 ha of rice fields, the biggest area of cultivation of its kind near the outskirts of Beijing.

Project Purpose

The Jade Lake restoration program has been designed to meet various key goals and objectives, including:

- Serving as an eco-tourism destination and urban retreat for the public;
- Acting as a research base and center for international exchange and the promotion of related sciences;
- Serving as a treatment system to address water quality issues;
- Serving as a conservation mechanism for water resources;
- Providing a suitable environment for the establishment of self-sustaining wetland biological communities; and

体性、有机协调性，使其生态、社会以及经济效益最大化，确保公共属性，促进人与自然的协调发展。

恢复方法

翠湖湿地恢复的方法包括：采用生态工程的技术方法，重建食物链，恢复生态系统内良性的生物地球化学循环过程；采用"环境艺术"理念，使山、水、路、岛、林、草等景观表现和谐；采用生物净化、植物栽培等技术，对翠湖湿地进行综合修复与监测。

具体的修复技术包括水生植物的恢复，鸟类栖息地的建设和修复（包括建立鸟类观测系统等），食物链的构建，湿地水资源管理，人工复合湿地净化水技术。

翠湖湿地污水净化处理系统

翠湖湿地污水净化处理系统，包括地表水体自净化湿地处理系统和综合生物塘。地表水体自净化湿地处理系统是人工开挖的河道，面积约0.5hm²；综合生物塘是自然状态下形成的大规模芦苇群落，面积约5.2hm²。

设计原理主要是从处理湿地对水力学流态出发，加上导流墙和布水配水装置，同时以计流器在净化处理系统的水流路径上控制流量。由废弃鱼塘和天鹅湖进水，废弃鱼塘进水口流量为300m³/d；天鹅湖进水口流量为2000m³/d；集水池内配

– Providing a showcase for sustainable development.

The Restoration Process

The restoration of the Jade Lake Wetland employed the following processes:

– Adopting ecological engineering technologies to control nutrient cycling and establish sustainable ecological processes within the wetland habitats;
– Using the concept of environmental art to present landscape constituents such as mountain, water, road, island, woods and grass in harmony; and
– Applying technologies (such as biological water quality improvement systems and horticultural techniques) to restore and monitor the Jade Lake Wetland.

Methods employed in the Jade Lake Wetland program include the restoration of aquatic plant communities, the construction and restoration of wildlife habitats (including establishing a system of bird observation hides), the re-establishment of wetland faunal communities at different trophic levels, management of wetland water resources and utilizing constructed wetlands to improve water quality.

Wetland Treatment Systems

A key component of the wetland program is its treatment system. The system comprises a 0.5 ha water purifying channel that drains into a large (5.2 ha) treatment pool. The engineering specifications structures in the Wetland Treatment System (e.g., v-shaped weirs, levees and other structures) were based on an analysis of the hydraulic flow regime that optimizes water quality improvements (see

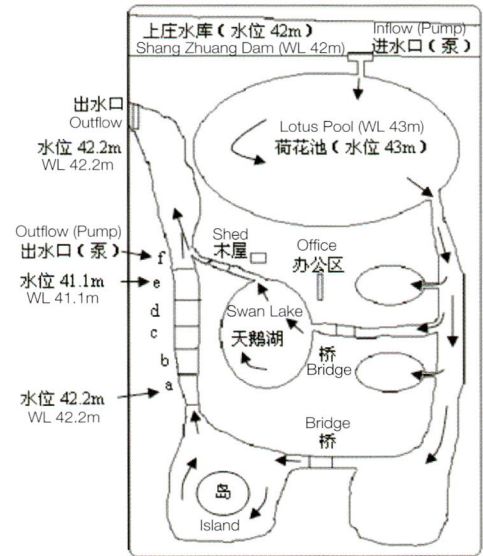

▲ 图4-92：翠湖湿地平面示意图（提供：李荣旗）
Figure 4-92: Indicative plan of the Jade Lake Wetland (Image courtesy of Li Rongqi)

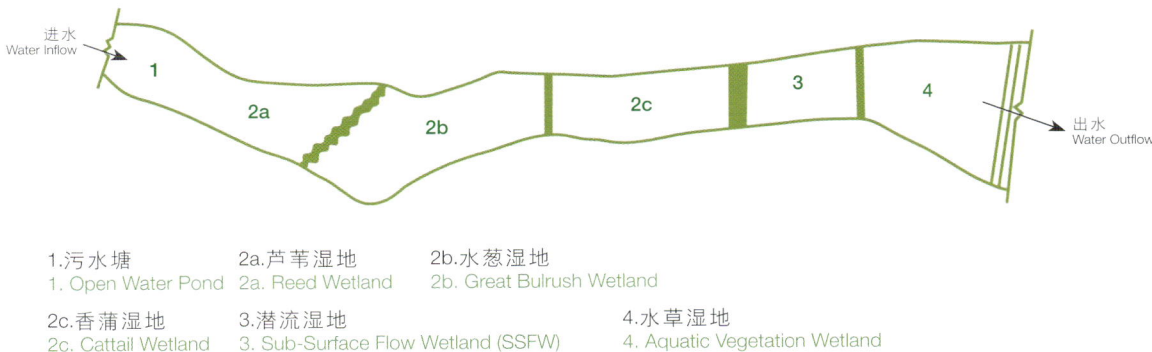

1.污水塘　　　　2a.芦苇湿地　　　　2b.水葱湿地
1. Open Water Pond　2a. Reed Wetland　2b. Great Bulrush Wetland

2c.香蒲湿地　　　　3.潜流湿地　　　　　　4.水草湿地
2c. Cattail Wetland　3. Sub-Surface Flow Wetland (SSFW)　4. Aquatic Vegetation Wetland

▲ 图4-93：地表水体自净化湿地处理系统恢复工程示意图（提供：李荣旗）
Figure 4-93: Diagram of the water treatment steps and restoration engineering at Jade Lake (Image courtesy of Li Rongqi)

▲ 图4-94: 顺水流方向(摄影: 崔丽娟)
Figure 4-94: View looking downstream (Photograph by Cui Lijuan)

▲ 图4-95: 逆水流方向(摄影: 崔丽娟)
Figure 4-95: View looking upstream (Photograph by Cui Lijuan)

置水泵2台,1台备用。地表水体自净化湿地处理系统的水流路径上均安装了3个直角三角堰流量计。(见图4-93)

地表水体自净化湿地处理系统的施工内容,主要包括河道清淤、按设计高度垫起、建布水墙(含流量测定结构和配水、排水结构)、填充湿地床基质和种植湿地植物。具体处理系统包括:

i) 污水塘前的隔断:污水塘设计水位低于废弃鱼塘内的水位,因此废弃鱼塘内的水可自流进入污水塘,并通过隔断墙内的直角三角计量堰计量进入湿地的水量。在废弃鱼塘向污水塘的进水口处建一隔断墙,并以天然石材做装饰,使其与自然环境协调,并加以计量控制流入水量,将废弃鱼塘与污水塘分开(见图4-94,4-95)。

ii) 污水塘(见图4-96):为系统的预处理单元,具有稳定水质、去除可降解颗粒物、降低污水中的有机物和氮磷物质含量的作用。以原河床体作为污水塘塘底,污水塘种植漂浮型植物凤眼莲和附着型浮叶植物睡莲、菱等。污水塘和表面流湿地及芦苇湿地间没有分隔墙,但需要建立坡度工程以利水流。

Figure 4-93). Water enters the treatment system from two sources: an abandoned fish pond (300m3/day) and a lake (2000m3/day). V-shaped weirs allow flow rates into and through the treatment system to be monitored.

Wetland Treatment System at Jade Lake
The stages of the Wetland Treatment System are described in the following sections.

i) Waterflow Design of Wetland Treatment System
The water level of the Open Water Pond was designed to be lower than that of the abandoned fish pond in order for water to flow naturally into the Wetland Treatment System. The volume flowing into the wetland can be measured by the v-shaped weir separating the fishpond and Open Water Pond. This weir was constructed with locally-sourced stone to blend in with the natural environment.

ii) Open Water Pond
The Open Water Pond is 22.5 meters long and 16.8 meters wide, and serves as a preliminary treatment unit in the Wetland Treatment System. The Open Water Pond stabilizes water quality, removes sediments from the water column, and reduces nutrient levels in water flowing into the Wetland Treatment System. The pond has been planted with water lilies (Nymphaea odorata) and water chestnuts (Trapa natans), and also supports free-floating water hyacinth (Eichhornia crassipes). After a period of retention, water flows from the pond via a gentle slope into the next stage of the Wetland Treatment System, the Surface Flow Wetland.

▼ 图4-96: 污水塘(摄影: 崔丽娟)
Figure 4-96: Open water pond (Photograph by Cui Lijuan)

▼ 图4-97: 水葱湿地(摄影: 崔丽娟)
Figure 4-97: Great bulrush wetland (Photograph by Cui Lijuan)

iii) 表面流湿地：表面流湿地建成串联的三梯级处理湿地，分别为芦苇湿地、水葱湿地和香蒲湿地。表面流湿地的水体中生长着大量具有降解有机物功能的微生物，它们以生物膜的形式附着在水生植物浸入的植株表面，与流水体充分接触中，可以高效去除有机物。芦苇等水生植物的生长过程中，也可吸收氮、磷营养物质。因此，对污水中各类污染物质都有较强的净化能力。

iv) 潜流湿地：在沿潜流湿地水流方向一侧建15cm的管道，使潜流湿地的水流可以流过，并进入水草湿地。首先在原地基上辅以砾石作为进出水滤料，然后填充介质层，最后填充厚表土作为基质层。水从湿地表面以下流过，过程中与介质层充分接触，具有良好的除磷效果。同时，芦苇的根系可以聚集硝化菌，有利于生物脱氮过程（见图4-97，4-98）。

v) 水草湿地（见图4-99）：湿地平面呈直角梯形形状，在原地面用素土垫高夯实，加上砾石滤层，再将穿孔管埋在其中，把出水导入综合生物塘，最后填充表层土壤作为植物生长基质。水草湿地是以沉水植物为特征的表面流湿地，植株浸没于水中，茎叶完全从水体中吸收营养元素，以此实现净化功能。

综合生物塘

是一个自然处理的湿地，现存的芦苇湿地，经过多年生物塘发育已经成熟，有鱼、虾等水生动物生长。在出水口处建集水池和泵水系统，以利湿地的水体循环（见图4-100）。

翠湖湿地的景观与栖息地恢复

在翠湖湿地的荷花池内种植芦苇和荷花，在岸边缓坡则采取植被自然恢复的方法，已经取得明显的景观效果。2003年9月，人工投放水生动物共5875kg，包括田螺1545kg、河蚌1650kg、河虾640kg、河蟹580kg、泥鳅600kg、花白鲢810kg、黄鳝50kg等，根据监测，湿地内的底栖动物群落已经基本恢复，能够为栖息水禽和涉禽提供充足的饵料。除了早前人工引入的鸟类外，翠湖湿地已经

iii) Surface Flow Wetland (SFW)
The Surface Flow Wetland is a three-step treatment system comprising a Reed Wetland, a Great Bul-rush (Scripus sp.) Wetland (Figure 4-97), and a Cattail (Typha sp.) Wetland. The treatment process for all three stages in the SFW is similar. The submerged stems and roots of wetland plants form an ideal substrate for mi-cro-organisms. Following a period of establishment, a bio-film supporting an abundance of microbes will form on the surfaces of the wetland plants. These microbes are capable of degrading nutrients and other pollutants in the water flowing through the SFW. Additionally, the wetland plants in the SFW are able to absorb some pollutants (such as nitrogen and phosphorus) directly from the water.

iv) Sub-surface Flow Wetland (SSFW)
The SSFW system operates under a similar principle to the SFW, with micro-organisms and vegetation acting to remove nutrients and other pollutants from water flowing through the treatment system. The structure of the SSFW, however, is very different to the SFW. The SSFW was constructed by laying a bed of gravel in the base of the wetland. A layer of planting media is placed on top of the gravel, and finally, thick topsoil is placed as the basal layer (see Figure 4-98) After a period of establishment, microbial communities will establish themselves in the gravel and planting media that will remove phosphorus from water flowing through the SSFW. Additionally, nitrifying bacteria associated with plant roots further improve water quality through biological denitrification.

v) Aquatic Vegetation Wetlands
Aquatic Vegetation Wetlands are surface flow wetlands where submerged plants absorb nutrients and other pollutants in the water directly through their stems and roots. These wetlands are trapezoidal in shape, and have a simple base of plain soil followed by a layer of gravel. A layer of topsoil is laid over this to act as a planting media for aquatic vegetation. Water drains from the Aquatic Vegetation Wetlands via a porous pipe buried in the gravel layer of the substrate.

Multipurpose Wetland
After flowing through the Wetland Treatment System, water eventually discharges into a large natural wetland habitat. This area is designed primarily as a habitat for wetland flora and fauna, but it also polishes water quality and acts as landscape feature. A catchment pool at the wetland's outlet is equipped with a pumping system to facilitate the water circulation through the wetland system.

▲ 图4-98: 潜流湿地(摄影: 崔丽娟)
Figure 4-98: (top) Subsurface flow wetland (Photograph by Cui Lijuan)

▲ 图4-99: 水草湿地（摄影: 崔丽娟）
Figure 4-99: (bottom) Aquatic vegetation wetland (Photograph by Cui Lijuan)

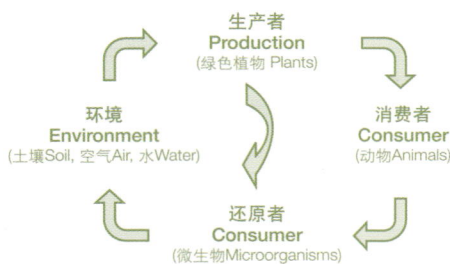

▲ 图4-100: 综合生物塘循环原理（摄影：崔丽娟）
Figure 4-100: Multipurpose wetland (Photograph by Cui Lijuan)

出现自然栖息的鸟类，尤其是包括野鸭等迁徙鸟类。

翠湖湿地的环境监测

1.翠湖湿地环境的常规监测

地面水环境质量监测系统：根据国家《地表水环境质量标准》（GB 3838－2002），在翠湖万亩（667hm²）湿地的地面水区域建立自动监站点。

地面气象观测系统：参照气象业务台站地面观测规范，翠湖湿地内建立了自动观测站点，为政府、气象部门掌握和研究湿地气候变化提供科学依据。

2.翠湖湿地自动监测与资料处理

在翠湖湿地建立水质自动监测站、空气品质监测站和图像监测站。在翠湖湿地建立监测资料处理中心，对监测的资料、图像进行编码和指标线上分析、预警系统，定时与海淀区水务局业务处理中心进行资料传输（见图4-101）。

结论

翠湖湿地水体恢复的监测与恢复效果

目前，翠湖湿地已经完成了20hm²荷花塘，20hm²的天鹅湖，33.3hm²芦苇荡，20hm²的雁鸭湖的修复。历时两年的连续监测表明，水质达到了Ⅲ类水标准，如果进一步加以改进可以达到Ⅱ类或Ⅰ类水标准（见图4-102）。

翠湖湿地生物多样性的恢复效果

经中国科学院植物研究所初步调查约有水生植物30科，100余种，具体包括莲、菱、荻、香蒲、芦苇和水葱等。

田螺、河蚌、河虾等水生动物大量增殖，小田螺、小河虾等随处可见，原有鱼类，特别是小型成鱼，如麦穗鱼、白条鱼、棒花鱼、等鱼类，由于饵料充足，繁殖速度快，鱼群随处可见。水生动物总量估计在2万千克左右。

由于翠湖湿地植被恢复良好，并有各种底栖动物和植物物种种子可为鸟类提供食物，翠湖目前已成为北京市郊的重要鸟类栖息地之一（见图4-103），逐步形成了人与自然和谐的湿地生态系统。2006年第二阶段万亩恢复工程已经进入评审阶段。

The multi-stage Wetland Treatment System constructed at Jade Lake has proved highly successful, with monitoring data from the Lake showing a steady improvement in water quality over the last two years. Water quality has now reached grade III standards, and it is predicted that with further measures being taken to improve water quality, Jade Lake is expected to meet grade II or grade I levels in the near future.

Landscape and Habitat Restoration at Jade Lake

Landscape and habitat restoration at Jade Lake has been both active and passive. Active restoration has involved the planting of reed and lotus plants in some lake areas. Sections of the lake with gently sloping banks have been restored passively by allowing a natural re-colonization by wetland vegetation. In September of 2003, various wetland fauna were introduced to the lake including aquatic snails (1.5 kg), freshwater bivalves (1.6 kg), shrimp (640 kg), crabs (580 kg), weather loaches (Misgurnus sp., 600 kg), white fish (810 kg) and finless eel (50 kg). Monitoring has shown that Jade Lake's benthic community has been essentially restored, and this newly established food source is attracting a variety of migratory water birds.

A survey carried out by the Chinese Academy of Science's Institute of Botany demonstrated that the diversity of vegetation is also increasing. Officials have recorded the presence of thirty families and over 100 species of aquatic plants in the wetland, including lotus (Nelumbo nucifera), water chestnut (Trapa natans), reeds, cattail (Typha sp.) and the great bulrush (Scripus). Aquatic invertebrates such as pond snails, freshwater mussels and shrimp populations are multiplying rapidly, with juveniles spreading throughout the wetland. Native fish are also reproducing well and schools have been observed, espe-cially of fingerling and mature fishes such as topmouth gudgeon (Pseudorasbora parva), the Chinese false gudgeon (Abbottina rivularis), and bitterling (Rhodeus sericeus). The total biomass of aquatic fauna is estimated to weigh approximately 20,000 kg.

Biological Environmental Monitoring Systems

An effective monitoring program is an essential component of any restoration project, providing valuable feedback on the relative success of restoration efforts. At Jade Lake, an extensive data collection and analysis program has been implemented to monitor changing biotic and abiotic conditions.

Regular Monitoring

1. Monitoring system for the environmental quality of surface water

Automatic water quality monitoring stations have been established over the 700 ha wetland in accordance with the Environmental Quality Standards for Surface Water (GB 3838 - 2002).

Ground Surface Meteorological Observation System

Automatic meteorological observation stations have been built within the Jade Lake Wetland. The observations provide a scientific reference of climate change for governmental and meteorological groups, as well as for researchers at the wetland.

2. Automatic Monitoring and Data Processing

At Jade Lake, automatic monitoring stations are the standard for measuring water quality, air quality and visual quality data collection. A data processing center was established to carry out encoding of monitored data and images, on-line index analysis and to act as an early-warning system (see Figure 4-101).

Conclusion

The Jade Lake Wetland restoration project has produced a robust ecosystem for benthic macroinvertebrates, wetland flora, and avifauna, making the wetland one of the most significant restoration efforts and bird habitats in the Beijing area. A well-balanced ecological system has been successfully created and continues to develop in a positive manner. Plans for the further restoration and engineering of Phase II area are currently under review.

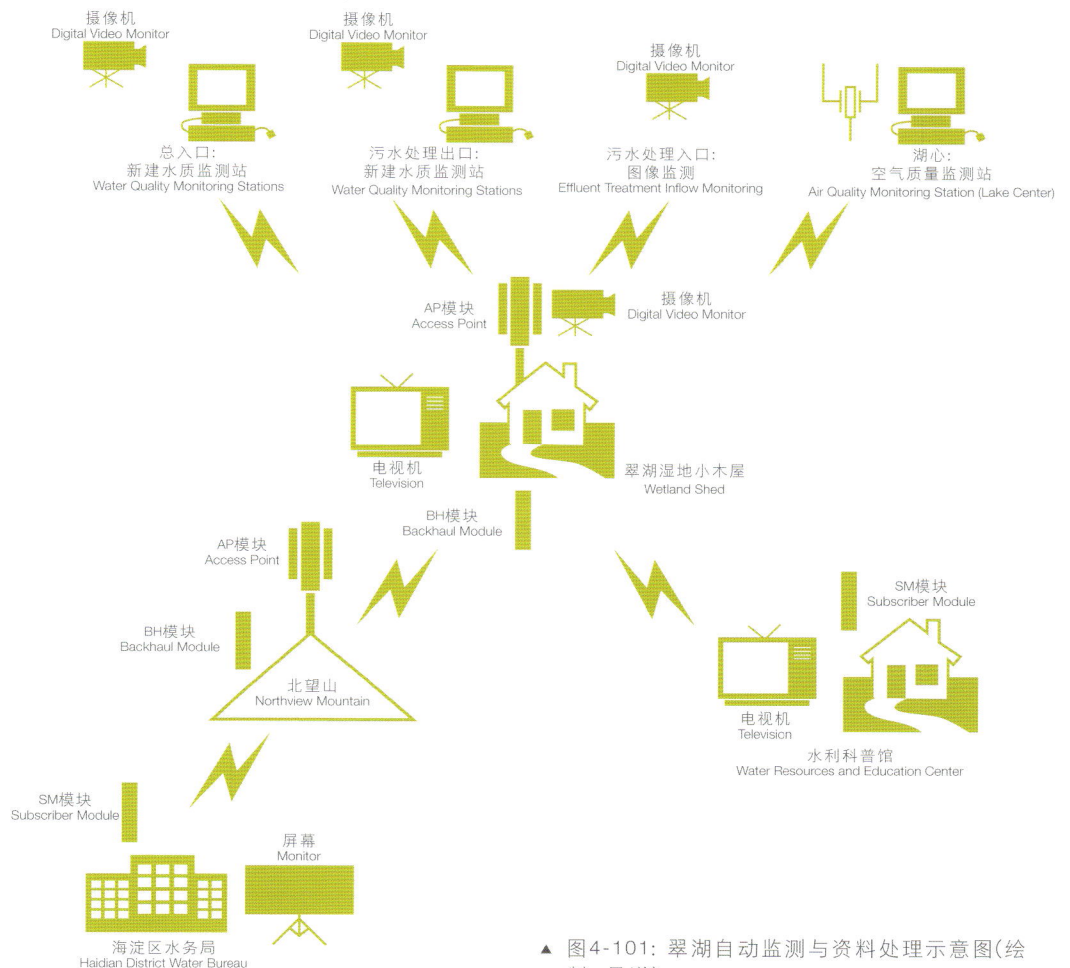

摄像机
Digital Video Monitor

摄像机
Digital Video Monitor

摄像机
Digital Video Monitor

总入口：
新建水质监测站
Water Quality Monitoring Stations

污水处理出口：
新建水质监测站
Water Quality Monitoring Stations

污水处理入口：
图像监测
Effluent Treatment Inflow Monitoring

湖心：
空气质量监测站
Air Quality Monitoring Station (Lake Center)

AP模块
Access Point

摄像机
Digital Video Monitor

电视机
Television

翠湖湿地小木屋
Wetland Shed

BH模块
Backhaul Module

AP模块
Access Point

SM模块
Subscriber Module

BH模块
Backhaul Module

北望山
Northview Mountain

电视机
Television

水利科普馆
Water Resources and Education Center

SM模块
Subscriber Module

屏幕
Monitor

海淀区水务局
Haidian District Water Bureau

▲ 图4-101：翠湖自动监测与资料处理示意图（绘制：易道）
Figure 4-101: Indicative diagram of the automatic monitoring and data processing systems of the Cui Lake Wetland project (Graphic: EDAW)

▼ 图4-102：湿地生物的恢复效果（摄影：易道）
Figure 4-102: The are many biological effects from the restoration, including an increased habitat (Photograph by EDAW)

▼ 图4-103：翠湖湿地经恢复后能为水禽提供丰富的饵料，已经成为理想的水禽栖息地（摄影：易道）
Figure 4-103: The restored wetland is ideal for waterfowl like these ducks, providing abundant foraging habitats (Photograph by EDAW)

第 五 章
Chapter 5

结 论
Conclusion

结论

今天，环境保护已成为大众舆论普遍关注的话题，政府部门、科学家、专业人士、商业团体和各种社区组织都加入到相关话题的讨论当中。过去几年以来，这些利益集团开始逐渐认同环境保护作为一门"生意"的价值，它拥有着深远的商业、文化、生态和社会价值。这些互相竞争的利益团体曾对可持续增长提出了各自不同的定义，但如今，它们开始达成共识，即可持续增长需要设立一定的标准来保护地球自然环境的健全和自然体系，而同时它又是一项可盈利的商业行为。

同样，湿地研究也经历了类似的、积极的提升。人们开始更多地关注和讨论湿地恢复；作为一门专业，湿地研究也得到了长足的发展；有越来越多的湿地项目已经落成，还有更多的正在进行之中。曾经一度被忽视和无所作为的状态都有所改观，传统的利益集团之间的争端也渐渐平息，取而代之的是各界对湿地恢复这一问题较为一致的态度和渐趋统一的解决方案。

我们已经认识到，湿地是地球上人与自然之间最重要的缓冲区之一，是地球保持自然系统平衡的一项最重要的手段，是拥有最丰富生物多样性的宝藏之一，更是减轻工业对水源和自然环境的影响的绝佳处理方案。

与此同时，湿地恢复在其起源地美国和欧洲，仍处于相当早的发展阶段，但是30年以前，《湿地公约》就极具前瞻性地认识到了湿地的重要性，并凭借先行者的优势完成了大量有益的工作：一系列基准、指导方针和法律框架纷纷出台并各司其职；工程学和相关学科中的先进技术已应用到湿地的保护和恢复中，这一切都极大地丰富了湿地研究的参考资料和相关文献。

19世纪末20世纪初，为实现其社会目标和自然保护的目标，美国积极展开了保护自然的行动，即建立美国国家公园系统。国家公园系统不仅是一个环保策略，而且向美国民众提供了可以享受自然之趣的开放空间。20世纪70年代，这种保护再次兴起，而且比以往更为积极，环境法规的里程碑——《清洁水法案》的诞生，以及随后一系列相关立法举措的实施都标志着湿地保护取得了划时代的进展。

湿地恢复的发展经历了比预期更长的历程，但毕竟迷雾渐散，阳光已现。当初利益冲突方各持己见，之后演化成商业利益与环境目标的长久对峙。现在，美国和欧洲大部分地区终于广泛达成了一致，湿地保育与恢复日趋成熟。亚洲也正沿着这个方向积极地规划可持续的未来。

许多专家早就认识到湿地的丧失对全球生态系统和生物多样性总体品质的影响，大家一致认为，应该创造出一个可持续的增长模式，在这种模式中，开发商的利益和被开发土地的环境能够被兼顾，既能够保护、加强湿地系统，更可以使其他形式的社会、文化和商贸活动能够初具规模，渐至繁荣昌盛。

人们对湿地恢复的认识以及相关法律法规为湿地恢复的发展提供了适当的机制，同时为赔偿、逆转和弥补开发所造成的影响提供了大量专业技术和法律依据。

可持续发展的中国

绿色经济的崛起并不仅适用于发达国家。由于地球的生态系统庞杂且相互联结，要维护这种复杂的依存关系需要全世界的共同努力，需要科学家、立法者、规划者、开发者和行动主义团体的共同努力。

亚洲的自然环境眼下正面临着最严重、最紧迫的威胁。几十年的繁荣增长让大自然付出了沉重的代价。中国也处于一个十字路口，现在的决策及其制定的发展道路将不仅决定自然环境的优劣，也将决定城市环境及其居民的健康。

朝前看，我们可以借鉴美国等国家湿地恢复的经验。目前，中国正积极地加入到关于自然系统价值的全球性对话之中，其中湿地恢复是一个关键部分，这让我们对前景充满信心。中国拥有令人叹为观止的迁徙或洄游物种和自然湿地，然而几百年以来对它们有意或无心的忽视，已经危及这些宝贵的遗产。

同时推动中国的经济繁荣和自然环境的健全，这是我们作为生态环境工作者孜孜以求的目标。全球的共识正在逐渐消弭分歧，同样地，《湿地恢复手册》的目的是融合有关湿地恢复的研究、工程学、科学和成功的国际案例研究。我们希望这本综合参考书能为本领域的研究和项目体系添砖加瓦，成为实用性的指南手册。

最后要强调的是，发展有合理之道，也有不当之路。合理的发展之策既要能够充分考虑到城市、人和环境三者的利益，又要兼顾可持续性和景观特征表现的多样化。真正的共识还没有完全达成，需假以时日和不懈的努力。生态问题要求原则和操作上的连贯性。在目前阶段，成功的湿地恢复项目如要考虑多方利益，各方则必须妥协，这种妥协将不但能使动植物栖息环境和人类的生存环境融洽共存，而且能推动它们之间的互利互惠，共同繁荣。

湿地恢复手册 原则·技术与案例分析

Environmentalism today is part of the general public discourse, one that includes the participation of governments, scientists, professionals, businesses and communities. Over the past several years, these interest groups are coming to view the "business" of environmentalism as a value that includes far-reaching commercial, cultural, biological and social benefits. Once defined by competing interests, a consensus is emerging today on sustainable growth, one that subscribes to a view on the necessary standards for preserving the Earth's natural integrity and natural systems, while doing so in a way that is smart and good business.

In the same light, the growing discourse and professional body of wetland restoration research, and the completed or current projects has seen a similar, positive rise in stature in every area. The time for neglect and inaction or even contention between traditionally competing interests is slowly yielding to integrated thinking and solutions.

Wetlands are recognized as one of the Earth's most critical buffers between humans and nature, as one of the Earth's critical means in helping keep nature's systems in balance, as one of the richest sources of biodiversity, and as an excellent treatment option for mitigating the effects of industries on water supplies and the natural environment.

At the same time, even within the places of its origin in the US and Europe, the practice of wetland restoration is at a relatively early stage. But with the rise of wetlands awareness more than 30 years ago, one that started and has been nurtured by the Ramsar Convention, much has been done to shape a true science and practice around the protection and restoration of wetlands: benchmarks have been established, guidelines and legal structures have been put into place, and advances in engineering and related sciences have been applied to the protection and restoration of wetlands, adding to the growing body of reference material.

At the turn of the century, the United States started its active conservation efforts with a mix of progressive social and natural goals, by establishing a National Park System as a way for its population to enjoy open spaces while embracing a conservation strategy. It was again in the 1970s that conservation efforts re-emerged in more activist forms, this time with the landmark environmental legislation of the Clean Water Act and the related legislative efforts that followed.

While it has taken longer to develop and advance, what started out as acrimony between competing interests, evolved into a lingering tension between commercial and natural aims, and is now coalescing into an emerging consensus on the business that is environmentalism in the United States and much of

Europe. Asia too is beginning to actively plan for a sustainable future along these lines.

Similarly, many experts have long recognized the impact that the loss of wetlands has had on the overall quality of the world's ecosystems and biodiversity, and in much the same way, a consensus is emerging on creating a pattern of sustainable growth that serves both the developer and the land being developed, of serving and enhancing wetland systems while allowing other social, cultural and commercial activities to take form and even flourish.

Wetland restoration awareness and regulations are providing appropriate mechanisms for development, while also providing a growing body of expertise and legislation to compensate for, reverse and offset its impacts.

A Sustainable China

The rise of green business is not limited to the developed world. Just as the Earth's remarkably complex ecosystems are inexorably linked, so now are human efforts to sustain that complexity becoming more integrated. The argument for the protection of wetlands and natural systems is today invariably a global one. Its connective tissue is the community of scientists, legislators, planners, developers and activists that share common goals.

Some of the gravest immediate threats to the natural environment today can be found in Asia. Decades of booming growth have not come without a heavy toll on our natural systems. China is truly at a crossroads, where the decisions it makes and the course it sets will determine the health of not only its natural environment, but of its urban realm and the communities that make them up.

Looking ahead, we in China can draw parallels to the wetland restoration efforts in the US. We are optimistic that China is now actively engaged in the global dialogue on the value of vital natural systems, and rehabilitating its wetlands is a critical part of that concern. While China boasts some of the world's most impressive migratory creatures and natural wetlands, centuries of either willful or unwitting neglect have compromised this inheritance.

Our aim is to act as stewards of the natural environment and to do so in a way that allows China to continue on a path that ensures both prosperity and environmental integrity. In much the same way that a global consensus is emerging, tying together once disparate, exclusive arguments, the Wetland Handbook aims to bring together the research, the engineering and science, and the successful global case studies behind wetlands restoration. We hope to offer an integrated point of

reference, a definitive guidebook that adds to the growing body of research and projects in the field.

Our message in the end is that there is a way to develop properly, a way that considers the interests of cities, people and the environment, one that embraces sustainability and the diversification of a landscape feature approach. We should not be naive and think that common ground has been reached entirely or easily, or is self-sustaining. The argument for the environment will require constancy in principle and in practice. At the same time, successful wetland restorations that consider broad interests will require compromise. But that compromise will nonetheless ensure that habitats and human developments not only coexist in harmony, but can even act at times, in ways to enhance each other.

参考文献
References

参考文献
References

第一章 Chapter 1:

1. Aber, J. D. and Jordan, W.. "Restoration ecology: An environmental middle ground". *BioScience*, 1985, 35: 399.

2. 白军红，王庆改. 中国湿地生态威胁及其对策，水土保持研究，2003，10(4):247-249.

3. Campbell, C.S. and Ogden, M.H. 1999. *Constructed Wetlands in the Sustainable Landscape*. John Wiley and Sons Inc., New York, NY.

4. 胡聃. 生态恢复设计的理论分析. 中国环境科学学会成立20周年论文选，1999:439-444.

5. Mahan, B., S. Polasky and R. Adams. 2000. "Valuing Urban Wetlands: A Property Price Approach". *Land Economics* 76: 100-113.

6. Mitsch W. J. and J. G. Gosselink. 2000. *Wetlands*, 3rd ed. Wiley, New York. 920 pp.

7. Thibodeau, FR and Ostro, BD. 1981. "An Economic Analysis of Wetland Protection". *Journal of Environmental Management*, 12: 19-30.

8. 湿地国际，湿地保护任重道远URL:http://www.wetwonder.org (16th Oct, 2002)

9. 水利部. 2004年水利统计公报URL: http://www.cws.net.cn/

10. 徐宏发，赵云龙. 崇明东滩鸟类自然保护区科学考察集. 北京：中国林业出版社，2005.

11. 水利部国际合作与科技司. 河流生态修复技术研讨会论文集. 北京：中国水利水电出版社，2005.

12. 张乔民. 我国热带生物海岸的现状及生态系统的修复与重建.海洋与湖沼，2001，32(4):454-464

13. 中国政府门户网站. "十五"期间洪涝灾害造成直接经济损失1006亿元，http://www.gov.cn/gzdt/2005-12/17/

14. 国家环境保护总局. 2005年中国环境状况公报. URL: http:// http://www.sepa.gov.cn/eic/

15. 北京市水务局. 寻梦——北护城河综合整治工程追踪，URL: http:// http://www.bjwater.gov.cn/

16. 人民日报，2005，北京：让河流自由"呼吸"（关注·城市河湖生态修复)URL: http:// www.people.com.cn/

第二章 Chapter 2:

1. Bush, Mark B. 2002. *Ecology of a Changing Planet*. New Jersey: Prentice Hall.

2. US Environmental Planning Department. 2006. "The Clean Water Act". URL: http://www.epa.gov. Accessed 17th Apr. 2006.

3. Copeland, Claudia. 1999. "Nationwide Permits for Wetlands Projects: Permit 26 and Other Issues and Controversies". URL: http://www.cnie.org

4. Daniels, Tom and Katherine. 2003. *The Environmental Planning Handbook*. Chicago: Planners Press

5. US Army Corps of Engineers. 1987. "Wetlands Delineation Manual".

6. The Ramsar Convention Secretariat, "The Ramsar Convention on Wetlands". URL: http://www.ramsar.org

7. 国家林业局. 中国湿地保护行动计划. 北京：中国林业出版社，2000.

第三章Chapter 3:

1. Mitsch W. J. and J. G. Gosselink. 2000. *Wetlands*. 3rd ed. Wiley, New York. 920 pp.

2. USEPA. 2000. "Principles for the Ecological Restoration of Aquatic Resources".

3. Lewis III, R. R. 2000. "Ecologically-based goal setting in mangrove forest and tidal marsh restoration". *Ecological Engineering* 15: 191–198.

4. 吴征镒. 中国种子植物属的分布区类型, 云南植物研究增刊IV, 1991: 1-139.

5. Simenstad C., D. Reed, and M. Ford. 2006. "When is restoration not? Incorporating landscape-scale processes to restore self-sustaining ecosystems in coastal wetland restoration." *Ecological Engineering* 26: 27–39.

6. 欧阳志云, 王如松, 赵景柱. 生态系统服务及其生态经济价值评价.应用生态学报, 1999, 10(5): 635-640.

7. 崔丽娟. 湿地价值评价研究.北京:科学出版社, 2001.

8. 陆健健. 湿地生态学, 北京：高等教育出版社, 2006.

9. 廖宝文. 深圳湾红树林恢复技术的研究.中国林业科学研究院, 2003.

10. 李玫, 廖宝文, 郑松发. 无瓣海桑海滩人工林的生态影响. 上海环境科学, 2003, 8：540－543

11. Rilov, G and Benayahu, Y. 1998. "Vertical artificial structures as an alternative habitat for coral reef fishes in disturbed environments". *Marine Environmental Research* 45: 431-451.

12. Bastian, R.K., and D.A. Hammer. 1993. "The Use of Constructed Wetlands for Wastewater Treatment and Recycling". Pages 59-68. In G.A. Moshiri (ed.), *Constructed Wetlands for Water Quality Improvement*. CRC Press, Boca Raton, FL.

13. USEPA. 1993. "Guidance Specifying Management Measures for Sources of Nonpoint Pollution in Coastal Waters". EPA-840-B-92-002, January 1993. U.S. Environmental Protection Agency, Office of Water, Washington, DC.

14. Kadlec, R.H. and R.L. Knight. 1996. *Treatment Wetlands*. Lewis Publishers, Boca Raton, FL, pp. 893.

15. Horne, A.J. 2005. "Designing Ecological Systems to Maximize Treatment, Habitat and Amenity Value". Presentation at the China Environmental Forum, 2005.

16. Knight, Robert L., Robert H. Kadlec, and Harry M. Ohlendorf. 1999. "The use of treatment wetlands for petroleum industry effluents." *Environmental Science & Technology* 33: 973-980.

17. Vrhovsek, Dani, et al. 1996. "Constructed wetland (CW) for industrial waste water treatment". *Water Research*, Vol. 30, No. 10, 2287-2292.

18. Billore, S.K., N. Singh, H.K. Ram, J.K. Sharma, Vijai P. Singh, R.M. Nelson and P. Dass. 2000. "Treatment Performance of a Molasses Based Distillery Effluent in a BioFilm-reed Bed Constructed Wetland in Central India." In 7th International Conference on Wetland Systems for Water Pollution Control, Volume III: Sections VIII-XI: Pages 1083-1600.

19. Pascoe, G.A., R.J. Blanchet, and G. Linder. 1994. "Bioavailability of metals and arsenic to small mammals at a mining waste-water contaminated wetland". *Arch. Environ. Contam. Toxicol.* 27: 44-50.

20. Kusler, J.A., and Kentula, M.E. 1990. "Executive summary". In, Kusler, J.A., and Kentula, M.E., eds. 1990. *Wetland creation and restoration-The status of the science*. Washington, D.C., Island Press, p. xvii-xxv.

21. Kentula, Mary E., Jean C. Sifneos, James W. Good, Michael Rylko, and Kathy Kunz. 1992. "Trends and patterns in Section 404 permitting requiring compensatory mitigation in Oregon and Washington, USA." *Environmental Management* 16:109-199.

22. Confer, SR and WA Niering. 1992. "Comparison of created and natural freshwater emergent wetlands in Connecticut (USA)", Connecticut College. *Wetlands Ecology and Management* 2(3): 143-156

23. Brown, M.T. 1991. "Evaluating Created Wetlands through Comparisons with Natural Wetlands". Environmental Protection Agency, Corvallis Environmental Research Lab., Corvallis, OR, Report EPA/600/3-91/058, 47 pp.

24.   Roberts, L. 1993. "Wetlands trading is a losing game, say ecologists": *Science*, v. 260, no. 5116, p. 1,890-1,892.

25.   Gearheart, Robert A. 1983. "Final Report, City of Arcata Marsh Pilot Project." City of Arcata Dept. Public Works, CA.

26.   USEPA. 1986. "Water Quality Standards for the Protection of Human Health". U.S. available on EPA website: http://www.epa.gov/waterscience/criteria/.

27.   Kelley, J.R. Jr., M. K. Laubhan, F. A. Reid, J. S. Wortham, and L. H. Fredrickson. 1993. "Options for Water-level Control in Developed Wetlands". *Waterfowl Management Handbook*. United States Department of the Interior. Noational Biological Survey. Fish and Wildlife Leaflet 13.4.8.

第四章 Chapter 4:

1.   Anon. "Development of a Comprehensive Conservation Strategy and a Management Plan in Relation to the Listing of Mai Po and Inner Deep Bay as a Wetland of International Importance under the Ramsar Convention". Aspinwall Clouston, Hong Kong (1997).

2.   N.D. Burgess, D. Ward, R. Hobbs and D. Bellamy. 1995. "Reedbeds, fens and acid bogs". In Sutherland, W.J. and D.A. Hill, 1995. *Managing Habitats for Conservation*. Cambridge University Press, Cambridge, 149-198 pp.

3.   Cary, G.J. et al. 2001. *The Avifauna of Hong Kong*. Hong Kong Bird Watching Society, Hong Kong.

4.   Hawke, C.J. and P.V. Jose. 1996. "Reedbed management for commercial and wildlife interests". RSPB, Sandy.

5.   Irving, R. and B.S. Morton. 1988. "A Geography of the Mai Po Marshes". World Wide Fund for Nature Hong Kong, Hong Kong.

6.   S.Y. Lee, "Invertebrate species new to science recorded from the Mai Po Marshes, Hong Kong". In *The Marine Biology of the South China Sea*. Vol.1. (Edited by B.S. Morton), pp. 199-210, Hong Kong University Press, Hong Kong (1993).

7.   S. McChesney, "The benthic invertebrate community of the intertidal mudflat at the Mai Po Marshes Nature Reserve, with special reference to resources for migrant shorebirds". M.Phil. Thesis, University of Hong Kong, Hong Kong (1997).

8.   Merritt, *Wetlands, Industry and Wildlife.* Wildfowl and Wetlands Trust, Slimbridge (1994).

9.   Ove Arup and Partners Hong Kong Ltd., "Agreement No. CE39/2001 Shenzhen Western Corridor Onvestigation and Planning Environmental Impact Assessment Report", Volume 1, Highway Department, Hong Kong SAR Government, Hong Kong (2002).

10.   Peking University, "Environmental Impact Assessment Study on Shenzhen River Regulation Project Final EIA Study Report", Shenzhen River Regulation Office of the Municipal Government, Shenzhen (1995).

11.   WWF Hong Kong. 1999. "Management Plan for the Mai Po Marshes Wildlife Education Centre and Nature Reserve, 2006 – 2010". WWF Hong Kong, Hong Kong.

12.   Boromthanarat, S., Chaijaroenwatana, B., Pantanahirum, W., Faiboon, A., Flos, S., and Pierre Bouret. "Ecosystem Management Pak Phanang Bay Area": Planning Study. http://www.clib.psu.ac.th/acad_41/bsom1.htm

13.   Charnsnoh, P. 1998. *Saviours of the Sea*. Trang: Yadfon Foundation.

14.   CORIN. 1991. "Coastal Management in Pak Phanang - A Historical Perspective of the Resources and Issue". Coastal Resource Institute, Prince of Songkla University. Hat Yai, Thailand.

15.   Hartwich, F., Janssen, W., and J. Tola. 2003. "Public-Private Partnerships for Agroindustrial Research: Recommendations from an Expert Consultation". ISNAR Briefing Paper No. 61.

16.   Ellison, A. M. 2000. "Mangrove restoration: Do we know enough?" *Restoration Ecology*. Vol. 8, No.3 pp.219-229.

17.   Field, C.D. 1998. "Rehabilitation of mangrove ecosystems: An overview." *Marine Pollution*, Bulletin, Vol. 37, Nos. 8±12, pp. 383-392.

18.   Lewis, R.R. 2005. "Ecological engineering for successful management and restoration of mangrove forests". *Ecological Engineering* 24 (2005) 403–418.

19.   Lewis, R.R. and M.J. Marshall. 1997. "Principles of successful restoration of shrimp aquaculture ponds back to mangrove forest". Programa/resumes de Marcuba '97, September 15/20, Palacio de Convenciones de La Habana, Cuba. Page 126.

20. Lewis, R.R. III, M.J. Phillips, B. Clough and D.J. Macintosh. 2003. "Thematic Review on Coastal Wetland Habitats and Shrimp Aquaculture". Report prepared under the World Bank, NACA, WWF and FAO Consortium Program on Shrimp Farming and the Environment. Work in Progress for Public Discussion. Published by the Consortium.

21. Aquamarkets. 2003. "Network of Aquaculture Centres in Asia-Pacific (NACA)". http://www.enacaorg/modules/wfsecion/article. php?articalid-104

22. Robertson, A.I. and M.J. Phillips. 1995. "Mangroves as filters of shrimp pond effluent: predictions and biogeochemical research needs". *Hyrdobiologia*, Vol. 295, Nos. 1-3, pp. 311-321.

23. Stevenson, N. J., R. R. Lewis, and P. R. Burbridge. 1999. "Disused shrimp ponds and mangrove rehabilitation". In: W. J. Streever (ed.). *An International Perspective on Wetland Rehabilitation*, The Netherlands: Kluwer Academic Publishers, 277-297.

24. UNEP-WCMC. 2006. "In the front line: shoreline protection and other ecosystem services from mangroves and coral reefs." UNEP-WCMC, Cambridge, UK 33pp.

25. 王国平. 保护西溪湿地，造福人民群众：关于实施西溪湿地综合保护工程的思考. 湿地公园湿地保护与可持续利用论坛交流文集，2005: 22-28.

26. 杭州市园林设计院有限公司. 杭州西溪国家湿地公园总体规划，2004.

27. 浙江省环境保护科学设计研究院，浙江大学生命科学学院. 杭州西溪湿地生态保护和修复规划，2005.

28. Peng, P.B., Zhang, J.X., Peng, B.Y., et al. 2005. "Investigation of bird resources on the West Dongting Lake Nature Reserve". *Journal of Hunan Environment Biological Polytechnic*, 11(3), pp.231-235.

附 录
Appendix

《关于特别是作为水禽栖息地的国际重要湿地公约》
（简称《湿地公约》）（翻译稿）

拉姆萨尔，1971年2月2日，经1981年12月3日的巴黎议定书修订

各缔约国，确认人与其环境相互依存；考虑到湿地的基本生态功能是作为水文状况的调节者，是某种独特植物区系和动物区系，特别是水禽赖以存活的生境；深信湿地是具有重大经济、文化、科学和娱乐价值的一种资源，一旦丧失则不可弥补；希望制止目前和今后对湿地的蚕食，乃至丧失；确认水禽在季节性迁徙时可能会超越国界，因此，应视为一种国际资源；确信具有远见的国家政策与协调一致的国际行动相结合，可以确保湿地及其动植物区系得到保护。兹议定条款如下：

第1条
1.为本公约之目的，湿地系指天然或人造、永久或暂时的静水或流水、淡水、微咸或咸水沼泽地、泥炭地或水域，包括低潮时水深不超过6m的海水区。

2.为本公约之目的，水禽是指从生态学角度依赖湿地生存的鸟类。

第2条
1.各缔约国应指定其领土内适当湿地列入《具有国际重要意义湿地名录》（下称《名录》），该《名录》由根据第8条设立的办事处保管。每块湿地的边界应在地图上精确标明和划定，可包括与湿地毗邻的河岸和海岸地区，以及位于湿地内的岛屿或低潮时水深超过6m的海洋水体，特别是具有水禽栖息地作用的岛屿或水体。

2.选择列入《名录》的湿地，应根据它们在生态学、植物学、动物学、湖沼学或水文学方面的国际重要意义来考虑。首先应列入一年四季均对水禽具有国际意义的湿地。

3.将湿地列入《名录》，并不损害其所属缔约国的专有主权。

4.每个缔约国在按照第9条规定签署本公约或交存其批准书或加入书时，应至少指定一块湿地列入《名录》。

5.任何缔约国均有权将其领土内的其他湿地增列入《名录》，扩大已列入《名录》的湿地的边界，或者出于国家利益的迫切考虑，需要取消列入《名录》湿地或缩小其边界的，应尽快将这类变动

通知负责第8条规定的常务办事处组织或政府履约单位。

6.缔约国在准备指定列入《名录》的湿地和对已列入《名录》的其领土内的湿地行使改变的权利时，应考虑履行对迁徙水禽保护、管理和合理利用的国际义务。

第3条
1.各缔约国应制订和执行规划，以促进对列入《名录》的湿地的保护，并尽可能地合理使用其领土内的湿地。

2.每个缔约国应做出安排，以便尽早获悉，由于技术发展、污染或其他人为干扰，其领土内列入《名录》的湿地生态特征已经发生变化，正在变化，或有可能发生变化。有关这类变化的情况应立即通知负责第8条规定的常务办事处的组织或政府履约机构。

第4条
1.每个缔约国应在湿地（不论是否已列入《名录》）建立自然保护区，以促进对湿地和水禽的保护，并采取充分措施予以管护。

2.当某一缔约国出于紧急的国家利益的考虑而取消列入《名录》的湿地或缩小其边界时，应尽可能弥补湿地资源的任何损失，特别应建立新的自然保护区以供水禽生存，并在同一地区或邻近地区保护原有生境的有效部分。

3.各缔约国应鼓励就湿地及其动植物区系开展研究，交换资料和出版物。

4.各缔约国应努力通过管理适当增加湿地上的水禽数目。

5.各缔约国应加强培训能胜任湿地研究、管理和看管的人员。

第5条
特别是当湿地延伸到不同缔约国领土或一条水域为数各缔约国所共有的情况下，缔约国之间应就履行《湿地公约》的义务相互协商。同时，各缔约国应努力协调和支持目前和将来关于湿地保护与动植物所制订的政策法规条例。

第6条

1.设立缔约国大会,以监督公约的履行。第8条第1段设立的公约秘书处至少每3年组织召开一次缔约国大会之例会,除非会议另有决定。在至少有1/3的缔约国提出书面要求的情况下,也可以召开特别会议。缔约国会议的每次例会均应确定举行下一次例会的时间及地点。

2.缔约国会议具有下列职权:
(a)讨论本公约的执行情况;
(b)讨论《名录》的增补和修改;
(c)审议根据第3条第2段提供的关于《名录》中所列湿地生态特性变化的资料;
(d)就保护、管理和合理使用湿地及其动植物问题,向缔约国提出一般性建议或具体建议;
(e)要求有关国际机构就涉及湿地的国际问题提出报告和提供统计资料;
(f)通过其他建议或决议,来促进本公约的执行。

3.各缔约国应保证从事湿地管理的各级负责人了解并考虑此类会议关于保护、管理和合理使用湿地及其动植物的建议。

4.缔约国会议为每次会议制定议事规则。

5.缔约国会议应制定本公约的财务条例,并定期对条例进行审议。缔约国会议应在其每次例会上以出席会议并参加表决之缔约国的2/3多数通过下一财务期的预算。

6.各缔约国应根据出席缔约国会议例会并参加表决之所有缔约国一致通过的会费额度向预算纳款。

第7条

1.参加上述会议的各缔约国代表应包括在科学、行政或其他有关方面、知识渊博、经验丰富的湿地或水禽专家。

2.出席会议的每一缔约国有一票表决权;建议、决议和决定由出席会议及参加投票的缔约国的简单多数通过,除非本公约另作其他规定。

第8条

1.国际自然及自然资源保护联盟执行本公约规定的常务办事处职责,直至全体缔约国的2/3多数指定另一个组织或政府时止。

2.常务办事处职责如下:
(a)协助召集和组织第6条规定的会议;
(b)保管《具有国际意义的湿地名录》,并接收各缔约国根据第2条第5段就列入《名录》的湿地的增补、扩大、取消或缩小所提供的资料;
(c)接收各缔约国根据第3条第2段就列入《名录》的湿地的生态特性变化所提供的资料;
(d)把对《名录》的任何修改或《名录》中所列湿地的特性变化通知所有缔约国,并为在下届会议上讨论这些事项作出安排;
(e)把会议就《名录》修改或《名录》中所列湿地的特性变化提出的建议通知有关缔约国。

第9条

1.本公约无限期开放签字。

2.联合国任何会员国、任何专门机构或国际原子能机构的会员国或国际法院规约任何当事国得依下列方式之一成为本公约缔约国:
(a)对于批准不附保留之签署;
(b)待批准之签署,继后批准;
(c)加入。

3.向联合国教育、科学及文化组织总干事(下称"保存人")交存一份批准书或加入书;批准或加入即为生效。

第10条

1.本公约在7个国家按第9条第2段方式成为本公约缔约国起4个月后生效。
2.本公约嗣后对每个缔约国应自该国对

于批准不附保留之签署或交存批准书或加入书之日起4个月后生效。

第10条 副

1.根据本条，缔约国可就《公约》修订问题召集会议，对《公约》进行修订。

2.任何缔约国均可提出修订建议。

3.所建议的任何修正案的文本及修订理由须通过根据《公约》行使常设主席团（下称"主席团"）职责的组织或政府，并由主席团随即转告所有缔约国。缔约国对文本的任何意见要在自主席团把修正案通知缔约国之日起3个月内通知主席团。主席团须于意见提交的最后一天之后立即把截至该日所收到的全部意见转告各缔约国。

4.主席团将根据1/3缔约国的书面要求召集缔约国会议，审议根据第3段提出的修正案。主席团将就会议的时间与地点同各缔约国进行协商。

5.修正案须经与会缔约国投票表决以2/3的多数通过。

6.被通过的修正案将于2/3缔约国向保管者交存接受卡之日后第4个月的第一天起对接受修正案的缔约国生效。对在2/3缔约国交存接受书之日后交存接受书的任何缔约国，修正案将于该国交存接受书之日后第4个月第一天开始生效。

第11条

1.本公约将无限期有效。

2.任何缔约国可在本公约对该国生效之日起5年后书面通知保存人退出本公约。退约应于保存人接得通知之日起4个月后生效。

第12条

1.保存人应尽快将下述事项通知所有业已签署或加入本公约的国家：

(a)《公约》之签署；

(b)本公约批准书之交存；

(c)本公约加入书之交存；

(d)本公约生效日期；

(e)退约通知。

2.本公约生效之后，保存人应根据《联合国宪章》第102条在联合国秘书处予以登记。

下列签署人经正式授权签署本公约，以昭信守。

1971年2月2日订于拉姆萨尔，原本以英文本、法文本、德文本和俄文本各式一份。所有文本具有同等效力按照签订议定书的大会最后档的要求，保存国为缔约国第二次会议提供了本公约的阿拉伯文、中文和西班牙文的正式文本。这些文本是在常务办事处的协助下与有关政府磋商后译出的。文本均交保存人，保存人则将正式副本分送所有缔约国。

与湿地公约2006-2008战略计划相关的**历届《湿地公约》缔约国大会决议和建议**

缔约国大会通过的与本《湿地公约2006-2008战略计划》的每个总目标有关的决议和建议和下。

总目标1。合理利用湿地：激励和协助所有缔约国制订、通过和使用必要而又合适的手段和措施，以确保其领土范围内所有湿地的合理利用。

关于合理利用
建议3.3:合理利用湿地
建议4.10:合理利用概念实施指导原则
决议5.6:合理利用概念实施附加指导原则
建议7.1:泥炭地合理利用和管理全球行动计划
决议8.12:加强山区湿地合理利用和保护
决议8.14:国际重要湿地和其他湿地管理规划的新指导原则
决议8.17:全球泥炭地行动指导原则
决议8.32:红树林生态系统资源合理利用
决议8.35:自然灾害尤其是干旱对湿地生态系统的影响
决议8.39:作为战略生态系统的高安第斯湿地

关于调查和评估
建议5.2:第3条（"生态特性"和"生态特性变化"）解释指导原则
建议5.3:湿地的重要特性和与湿地保护区相关的分布带必要性
决议6.1:生态特性的工作定义，描述和维持所列国际重要湿地生态特性的指导原则以及《蒙特勒记录》实施指导原则
决议7.10:湿地风险评估框架
决议7.20:湿地调查优先顺序
决议7.25:测定湿地环境品质
决议8.6:国际重要湿地调查框架
决议8.7:国际重要湿地生态特性、调查、评估和监测指南的差距和协调
决议8.8:评估和报告湿地的状态和趋势，以及《公约》第3.2条的执行
国家调查的必要性：建议1.5、建议4.6、决议5.3、决议6.12

关于政策和立法，包括影响评估和评价
建议4.4:建立湿地保护区
建议5.3:湿地的重要特性和与湿地保护相关的分布带必要性
建议6.2:环境影响评估
建议6.10:促进湿地经济评价方面的合作
决议7.6:制订和执行国家湿地政策的指导原则
决议7.7:审查促进湿地保护和合理利用的法律和制度的指导原则
决议7.16:拉姆萨尔决议和影响评估:战略

的、环境的和社会的
决议8.19:《生物多样性公约》(CBD)通过的《将与生物多样性相关的问题列入环境影响评估立法或程式并列入战略环境评估的指导原则》
国家湿地政策的必要性：建议1.5、建议3.3、建议6.9

关于湿地与可持续发展
建议6.1:保护泥炭地
建议6.7:保护和合理利用珊瑚礁和有关生态系统
建议6.8:海岸地区的战略规划
建议6.14:有毒化学品
决议6.23:拉姆萨尔和水
决议7.18:将湿地保护和合理利用列入流域管理的指导原则
决议7.21:加强潮间湿地的保护和合理利用
建议7.1:合理利用和管理泥炭地的全球行动计划
建议7.2:小岛发展中国家、岛屿湿地生态系统和《湿地公约》
决议8.1:为维持湿地生态功能分配和管理水的指导原则
决议8.2:世界大坝委员会(WCD)的报告及其与《湿地公约》的相关性
决议8.3:气候变化和湿地：影响、适应和减缓
决议8.4:海岸带综合管理(ICIM)中的湿地问题
决议8.34:农业、湿地和水资源管理
决议8.35:自然灾害尤其是干旱对湿地生态系统的影响
决议8.40:使地下水利用与湿地维持相一致的指导原则

关于湿地恢复和重建
建议4.1:湿地恢复
建议6.15:湿地的恢复
决议7.17:作为湿地保护和合理利用的国家计划内容的恢复
决议8.24:对丧失湿地生境和其他功能的弥补
决议8.16:湿地恢复的原则和指导方针

关于侵入物种
决议7.14:侵入物种和湿地
决议8.18:侵入物种和湿地

关于当地社区：本地人和文化价值
建议5.8:提高公众对湿地保护区湿地价值的认识的措施
建议5.10:1996年25周年湿地保护运动纪念
决议6.21:使当地和本地人参与国际重要湿地管理

湿地管理的指导原则
决议7.9:《公约》的延伸计划，1999-2002年
决议8.14:国际重要湿地和其他湿地管理计划的新指导原则
决议8.19:有效管理湿地时考虑湿地的文化价值的指导原则
决议8.36:作为湿地管理和合理利用之工具的《参与环境管理》(PEM)

关于私营部门参与

关于鼓励措施
决议8.15:鼓励应用合理利用原则的鼓励措施
决议8.23:作为实现湿地合理利用之工具的鼓励措施

关于交流、教育和公众意识
建议4.4:建立湿地保护区
建议4.5:教育和培训
建议5.8:提高公众对湿地保护区湿地价值的认识的措施
建议5.10:1996年25周年湿地保护运动纪念
决议6.19:教育和公众意识
决议7.9:《公约》的延伸计划，1999-2002年
决议8.31:《公约》的交流、教育和公众意识(CEPA)计划，2003-2008年

总目标2。国际重要湿地：激励和支持所有缔约国恰当地实施《国际重要湿地名录未来发展战略框架和指导原则》，包括恰当的监测和管理列出的国际重要湿地，以有助于可持续发展。
建议4.7:《湿地公约》改进应用的机制
建议4.8:国际重要湿地生态特性变化
建议5.3:湿地的重要特性和与湿地保护区相关的分布带必要性
决议5.7:国际重要湿地和其他湿地管理规划
决议5.9:应用《确定国际重要湿地的拉姆萨尔标准》
决议6.1:生态特性的工作定义，描述和维持所列国际重要湿地生态特性的指导原则以及《蒙特勒记录》实施指导原则
建议6.2:环境影响评估
建议7.11:国际重要湿地名录未来发展战略框架和指导原则
决议7.16:《湿地公约》和影响评估：战略的、环境的和社会的；以及(决议8.9)所通过的影响评估附加指南
决议7.10:湿地风险评估框架
决议7.23:涉及国际重要湿地边界定义和湿地生境弥补的问题

决议8.6:国际重要湿地调查框架
决议8.7:国际重要湿地生态特性、调查、评估和监测指南的差距和协调
决议8.8:评估和报告湿地的状态和趋势，以及《公约》第3.2条的执行
决议8.10:改进《国际重要湿地名录的战略框架和远景》的实施
决议8.11:确定和指定未被充分代表的湿地类型作为国际重要湿地的附加指南
决议8.13:加强关于国际重要湿地的资讯
决议8.14:国际重要湿地和其他湿地管理规划的新指导原则
决议8.15:促进湿地管理的《圣何塞记录》
决议8.20:解释《公约》第2.5条下"紧迫国家利益"和审议第4.2条下弥补的一般指南
决议8.21:在《拉姆萨尔资讯表》中更准地定义国际重要湿地边界
决议8.22:涉及不再符合或从来不符合指定为国际重要湿地标准的国际重要湿地的问题
决议8.33:确定、可持续管理和指定暂时工具作为国际重要湿地的附加指南
决议8.36:作为湿地管理和合理利用之工具的《参与环境管理》(PEM)
决议8.38:水禽种群估计和国际重要湿地的确定和指定
蒙特勒记录:建议4.8、决议5.4、6.1、7.12和8.8

拉姆萨尔诸询团：建议4.7、决议6.14和7.12

总目标3:国际合作：通过积极应用《湿地公约下的国际合作指导原则》，促进国际合作，尤其是为湿地保护和合理利用动员额外资金和技术援助。
决议4.4:《公约》第5条的执行
建议4.11:与国际组织的合作
建议5.4:《湿地公约》、《全球环境基金》和《生物多样性公约》之间的关系
建议5.6:非政府组织(NGOs)在《湿地公约》中的作用
决议6.9:与《生物多样性公约》的合作
决议6.10:与《全球环境基金》(GEF)及其执行机构：世界银行、联合国开发计划署(UNDP)和联合国环境规划署(UNEP)的合作
决议7.4:与其他公约的伙伴关系和合作，包括协调资讯管理基础设施
决议7.19:《湿地公约》下的国际合作指导原则
决议8.5:与其他环境公约的协同配合
决议8.30:进一步执行《公约》的地区倡议
决议8.42:大洋洲地区的小岛发展中国家
决议8.43:《湿地公约》的南美分区一级的

湿地恢复手册 原则·技术与案例分析

战略
决议 8.44:《非洲发展新伙伴关系》（NE-PAD）和《湿地公约》在非洲的实施
地中海湿地倡议：建议 5.14；建议 6.11；决议 7.22；决议 8.30
候鸟飞行路线协议：建议 3.2，决议 4.4，建议 4.12，建议 6.4；决议 8.37

总目标 4。执行能力：确保《公约》有所需的执行机制、资源和能力来实现其宗旨。
建议 5.4:《湿地公约》、《全球环境基金》和《生物多样性公约》之间的关系
建议 5.5:将湿地保护和合理利用列入多边和双边发展合作计划
建议 5.6:非政府组织(NGOs)在《湿地公约》中的作用
建议 5.7:国家委员会
决议 6.21:评估和报告湿地状态
决议 7.3:与国际组织的伙伴关系
决议 7.5:《公约》的湿地保护和合理利用小额赠款基金(SGF)及其未来运作的关键评价
决议 7.26:建立西半球湿地培训和研究的地区性拉姆萨尔中心
决议 7.28:财政和预算问题
建议 7.4:未来倡议湿地
决议 8.25:2003-2008年拉姆萨尔战略计划
决议 8.26:在 2003-2005 三年期间执行《2003-2008年战略计划》和向拉姆萨尔第 9 届缔约国大会提交国家报告
决议 8.28:科技委员会(STRP)的操作方法
决议 8.29:评价《湿地保护和合理利用拉姆萨尔小额赠款基金》(SGF)和建立拉姆萨尔捐赠基金

关于发展援助
建议 3.4:发展机构对湿地的责任
建议 3.5:《湿地公约》局对于发展机构的任务
建议 4.13:多边开发银行(MDBs)对湿地的责任
建议 5.4:《湿地公约》、全球环境基金和《生物多样化公约》之间的关系
建议 5.5:将湿地保护和合理利用列入多边和双边发展合作计划
决议 6.10:与全球环境基金(GEF)及其执行机构：世界银行、联合国开发计划署(UNDP)和联合国环境规划署(UNEP)的合作

关于培训
建议 4.5:教育和培训
建议 6.5:制订进一步的湿地管理者培训方案

关于建立中西亚湿地培训和研究的地区性拉姆萨尔中心，决议 8.41

总目标 5。成员资格：努力使所有国家都加入《公约》。
建议 1.1:扩大《公约》的成员
建议 1.2:发展加入《公约》国家
建议 3.6:发展非洲缔约国
建议 3.7:发展中美洲、加勒比和南美洲缔约国
建议 3.10:发展亚太缔约国
建议 6.18:太平洋岛屿地区湿地的保护和合理利用

湿地是功能独特的生态系统，是我国实现可持续发展进程中关系国家和区域生态安全的战略资源。保护湿地、维护湿地生态功能的正常发挥，科学管理和合理利用湿地，对于改善我国生态环境、促进经济社会可持续发展，具有重要意义。我国湿地具有面积大、类型多、生物多样性丰富等特点，为进一步加强湿地保护，由国家有关部门共同编制的《全国湿地保护工程规划》（2002—2030年)得到了国务院批准。《规划》既衔接了目前各部门已经实施的许多与湿地保护相关的规划，又在建设任务上不重叠。《规划》明确了到2030年我国湿地保护工作的指导原则、任务目标、建设布局和重点工程，对指导开展中长期湿地保护工作具有重要意义。《规划》要点如下：

一、总体目标

通过湿地及其生物多样性的保护与管理，湿地自然保护区建设、污染控制等措施，全面维护湿地生态系统的生态特性和基本功能，使我国天然湿地的下降趋势得到遏制。通过加强对水资源的合理调配和管理、对退化湿地的全面恢复和治理，使丧失的湿地面积得到较大恢复，使湿地生态系统进入一种良性状态。同时，通过湿地资源可持续利用示范以及加强湿地资源监测、宣教培训、科学研究、管理体系等方面的能力建设，全面提高我国湿地保护、管理和合理利用水准，从而使我国的湿地保护和合理利用进入良性循环，保持和最大限度地发挥湿地生态系统的各种功能和效益，实现湿地资源的可持续利用。

到2030年，使全国湿地保护区达到713个，国际重要湿地达到80个，使90%以上天然湿地得到有效保护。完成湿地恢复工程140.4万hm²，在全国范围内建成53个国家湿地保护与合理利用示范区。建立比较完善的湿地保护、管理与合理利用的法律、政策和监测科研体系。形成较为完整的湿地区保护、管理、建设体系，使我国成为湿地保护和管理的先进国家。其中从2004到2010的7年间，要划建湿地自然保护区90个，投资建设湿地保护区225个，其中重点建设国家级保护区45个，建设国际重要湿地30个，油田开发湿地保护示范区4处，富营养化湖泊生物治理3处；实施干旱区水资源调配和管理工程2项，退耕(牧)还泽(滩、草)71.5万hm²，恢复野生动物栖息地38.3万hm²；建立湿地可持续利用示范区23处，实施生态移民13769人；进行科研监测体系、宣传教育体系和保护管理体系建设。

二、建设布局和分区重点

根据全国湿地分布总的特点，考虑到不同区域明显的自然特征，尤其是与湿地形成有关的水文和地质特性、湿地功能、保护和合理利用途径的相似性、行政区域和流域的连续性及实际的可操作性，《规划》将全国湿地保护按地域划分为东北湿地区、黄河中下游湿地区、长江中下游湿地区、滨海湿地区、东南华南湿地区、云贵高原湿地区、西北干旱湿地区以及青藏高寒湿地区，共计8个湿地保护类型区域。根据因地制宜、分区施策的原则，充分考虑各区主要特点和湿地保护面临的主要问题，在总体布局的基础上，对不同的湿地区设置了不同的建设重点。

东北湿地区。位于黑龙江、吉林、辽宁及内蒙古东北部，以淡水沼泽和湖泊为主，总面积约750万hm²。三江平原、松嫩平原、辽河下游平原，大小兴安岭山地、长白山山地等是我国淡水沼泽的集中分布区。该区域湿地面临的主要问题是大规模的农业开发，使天然沼泽面积大量减少；保护能力薄弱；对湿地资源的保护和利用缺乏统一规划和协调机制。该区建设重点为，全面监测评估该天然湿地丧失和湿地生态系统功能变化情况；通过湿地保护与恢复及生态农业等方面的示范工程，建立湿地保护和合理利用示范区，提供东北地区湿地生态系统恢复和合理利用模式；加强森林沼泽、灌丛沼泽的保护；建立和完善该区域湿地保护区网路，加强国际重要湿地的保护。

黄河中下游湿地区。包括黄河中下游地区及海河流域，行政上涉及北京、天津、河北、河南、山西、陕西和山东。该区天然湿地以河流为主，伴随分布着许多沼泽、洼淀、古河道、河间带、河口三角洲等湿地。该区湿地保护的最大问题是水资源缺乏，由于上游地区的截留，河流中下游地区严重缺水，黄河中下游主河道断流严重，海河流域的很多支流已断流多年，失去了湿地的意义。该区建设重点为，加强黄河干流水资源的管理及中游地区的湿地保护，利用南水北调工程尝试性地开展湿地恢复的示范，加强该区域湿地水资源保护和合理利用，尤其是北京生态圈湿地水资源和华北平原湖区湿地的保护，缓解该地区农业及城市饮用水资源日益紧张的状况。

长江中下游湿地区。包括长江中下游地区及淮河流域，是我国淡水湖泊分布最集中和最具有代表性地区，行政上涉及湖北、湖南、江西、江苏、安徽、上海和浙江7省(市)。该区水资源丰富，农业开发历史悠久，为我国重要的粮、棉、油和水产基地，是一个巨大的自然—人工复合湿地生态系统。湿地保护面临的最大问题是围湖造田和城市化导致天然湿地面积减少，湿地功能减弱，水质污染严重，湿地生态环境退化。该区建设重点为，通过退田还林、还湖、还泽、还滩、还草及水土保持等措施，使长江中下游湖泊湿地的面积逐渐恢复，改善湿地生态环境状况；建立合理利用模式和开展污染防治及生态环境治理工程，确保这些湿地区域资源、环境和经济的可持续发展；加强保护区建设，尤其是具有国际重要意义的水禽栖息地建设，使该区域丰富的湿地生物多样性得到有效保护。

滨海湿地区。涉及我国东南滨海的11个省(区、市)，包括杭州湾以北环渤海的黄河三角洲、辽河三角洲、大沽河、莱州湾、无棣滨海、马棚口、北大港、北塘、丹东、鸭绿江口和江苏滨海的盐城、南通、连云港等湿地，杭州湾以南的钱塘江口—杭州湾、晋江口—泉州湾、珠江口河口湾和北部湾等河口与海湾湿地。该区域湿地面临的主要问题分别是油田开采、过度利用和浅海污染等，导致赤潮频发、红树林面积急剧下降、海洋生物栖息繁殖地减少、生物多样性降低。建设重点为，评估油田开采、盐田和农业开发对三角洲湿地的潜在影响和威胁，加强珍稀野生动物及其栖息地的保护，建立候鸟研究及环志基地；建立具有良性回圈和生态经济增值的湿地开发利用示范区；以生态工程为技术依托，对退化海岸湿地生态系统进行综合整治、恢复与重建；调查和评估我国的红树林资源状况，通过建立示范基地，提供不同区域红树林资源保护和合理利用模式，逐步恢复我国的红树林资源。

东南和南部湿地区。包括珠江流域绝大部分、东南及台湾诸河流域、两广诸河流域的内陆湿地。行政范围涉及福建、广东、广西、海南、台湾、香港和澳门，主要为河流、水库等类型湿地。面临的主要问题是湿地泥沙淤积、水质污染严重，生物多样性减少。该区建设重点为，加强水源地保护和流域综合治理，在河流源头区域及重要湿地区域开展植被保护和恢复措施，防止水土流失，加强湿地自然保护区建设。

云贵高原湿地区。包括云南、贵州以及川西高山区，湿地主要分布在云南、贵州、四川省的高山与高原冰(雪)蚀湖盆、高原断陷湖盆、河谷盆地及山麓缓坡等地区。面临的主要问题是

一些靠近城市的高原湖泊有机污染严重，对湿地不合理开发导致湖泊水位下降，流域缺乏综合管理，湿地生态环境退化。该区建设重点为，加强流域综合管理，保护水资源和生物多样性，进行生态恢复示范，对高原富营养化湖泊进行综合治理；通过实施宣教和培训工程，提高湿地资源及生物多样性保护公众意识，恢复和改善湿地生态环境，维护我国湿地保护方面的国际形象。

西北干旱湿地区。本区湿地可分为两个分区：一是新疆高原干旱湿地区，主要分布在天山、阿尔泰山等北疆海拔1000m以上的山间盆地和谷地及山麓平原—冲积扇缘潜水溢出地带；二是内蒙古中西部、甘肃、宁夏的干旱湿地区，主要以黄河上游河流及沿岸湿地为主。该区湿地面临的最大问题是由于干旱和上游地区的截流导致湿地大面积萎缩和干涸，原有的一些重要湿地如罗布泊、居延海等早已消失，部分地区成为"尘暴"源，荒漠干旱区的生物多样性受到严重威胁。建设重点为，加强天然湿地的保护区建设和水资源的管理与协调，采取保护和恢复措施缓解西部干旱荒漠地区由于人为和自然因素导致的湿地环境恶化、湿地面积萎缩甚至消失的趋势。通过生态措施和工程措施，遏制湿地周边区域土地沙漠化趋势，改善湿地生态环境，保证湿地的生态功能的正常发挥。

青藏高寒湿地区。分布于青海省、西藏自治区和四川省西部等，地势高亢，环境独特，高原散布着无数湖泊、沼泽，其中大部分分布在海拔3500~5500m之间。我国几条著名的江河发源于本区，长江、黄河、怒江和雅鲁藏布江等河源区都是湿地集中分布区。面临的主要问题是区域生态环境十分脆弱，草场退化、荒漠化严重，湿地面积萎缩，湿地生态环境退化，功能减退。由于该区特殊的地理位置，该区湿地保护尤其是江河源区湿地的保护涉及到长江、黄河和澜沧江中下游地区甚至全国的生态安全。该区建设重点为，加强保护区建设及植被恢复等措施，保护世界独一无二的青藏高原湿地，尤其是江河源头地区的重要湿地，发挥该地区湿地的重要储水功能，并使高原特有的珍稀野生动植物得以栖息繁衍，保护好高海拔湿地。重点在三江源头、青海湖和若尔盖沼泽地区进行湿地保护和生态示范建设。

三、重点工程
依据生态效益优先、保护与利用结合、全面规划、因地制宜等建设原则，《规划》安排

了湿地保护、湿地恢复、可持续利用示范、社区建设和能力建设等5个方面的重点建设工程。

湿地保护优先工程。一是加强自然保护区和国际重要湿地建设，建立完善的湿地自然保护区(社区)网路，包括机构、基础设施、保护管理、科研监测和宣传教育等体系及资讯交流能力、社区共管、生态旅游和资源合理利用等项建设内容。二是湿地污染控制，建立全国湿地生态环境监测和评价体系，及时监测、预测预报湿地污染和生态环境动态；开展油田开发湿地保护示范工程，加强管理与监督，控制污染物的排放量，健全环境监测网路；通过减排、收获、复壮、生物治碱等措施，开展富营养化湖泊的生物治理工程，有计划地治理已受污染的海域、湖泊和河流。

湿地恢复优先工程。一是加强水资源的调配与管理，确定全国、流域和省区水资源配置方案及水资源宏观控制指标体系和水量分配指标。在重要湿地区和重要河流流域开展水资源调配与管理工程，适当增加关键区域生态用水比例，逐步恢复原有的湿地生境。二是开展湿地恢复和综合整治工程，包括在生物多样性丰富的低产农田区实施退耕还湖(泽、滩)工程，在退化和被改造的滩涂区实施恢复与重建工程，在土地沙漠化趋势严重的湿地区实施工程退牧育林还草、封沙育林育草、休牧(轮牧)育林育草工程，已退化沼泽草地进行改良，恢复天然植被和水禽栖息地，在沿海退化红树林地区进行红树林生态恢复工程的试验示范等。

可持续利用示范优先工程。结合部门职能和行业特点，选择既有开发潜力、又有示范意义的区域和项目，多形式的开展湿地资源可持续利用示范区建设。在农区湿地主要地理区划类型为主的重要湿地，农牧渔业利用强度大，不宜建立湿地自然保护区，规划建立国家级农牧渔业综合利用管理示范区和农牧渔业可持续利用湿地管理区，主要通过加强管理，逐步引进合理利用和保护措施来逐步实现面上的恢复；在南方大江大河的三角洲地区，建立新型的人工湿地高效生态农业模式试验示范区，形成林－农－水产立体生态结构的"基塘"模式体系；在长江中游人口压力大、人地矛盾突出的地区，适当发展堤垸水产开发，根据实际情况开展湿地资源合理利用模式示范项目；在典型滨海湿地海水养殖开发区域，研究推广适用的养殖优化技

术和生态养殖技术，推进滨海湿地海水养殖产业的合理化和科学化进程。

社区建设优先工程。以保护为中心设计发展项目，通过多种形式，大力推广有利于湿地可持续利用的发展项目。因地制宜地扶持社区进行产业结构调整，鼓励开展非资源消耗性产业的发展。在一些不适合人类生活和生态脆弱的湿地区域，开展生态移民工程。

能力建设优先工程。一是国家湿地资源监测中心、湿地监测站点等湿地监测体系建设，利用"3S"技术编制全国湿地保护与合理开发利用电子地图集；二是湿地宣教培训中心、野外培训基地和人员培训等宣传教育培训体系建设；三是中国湿地研究中心、省级研究机构(省级林业科研院所)及基层研究机构和资讯化网路等科学研究、技术支撑体系建设；四是认真履行《湿地公约》等有关的国际公约，全面提高现有和新增国际重要湿地的监测、保护和管理水准，并建立全国国际重要湿地保护网路，加强国际合作与交流。

规划的实施，将极大提高国家对湿地资源的保护和管理能力，使我国天然湿地下降的趋势基本得到遏制，并充分发挥湿地调节气候、保持水土、蓄洪防旱、防风固沙和美化环境等多种功能。保护了野生动植物及其生境，使湿地野生动植物种群得到恢复和发展，为我国提供了充足的资源储备。湿地的合理利用，可以创造就业机会和发展相关产业，为社会经济提供良好的生态环境支援。工程的建设，也将提高我国履行《生物多样性公约》、《湿地公约》的能力，扩大中国湿地保护在国际上的影响。通过全面实施保护和管理工程，使我国湿地保护工作进入正规化、有序化发展的新阶段。同时又为湿地的可持续利用提供了示范，形成与当地社区协调发展、全面持久保护湿地生态系统的模式，使我国湿地资源的生态效益、经济效益和社会效益得到全面发挥，实现湿地生态系统的良性循环。

ACKNOWLEDGEMENTS

总策划 Curators

印 红 Yin Hong
国家林业局野生动植物保护司
副司长
Deputy Director, Department of Wildlife
Conservation, State Forestry Administration
of P. R. China

乔全生 Sean Chiao
易道亚洲区主席
Principal, Regional Director
EDAW Asia

主编 Editors in Chief

崔丽娟 Cui Lijuan
中国林业科学研究院林业研究所
Forestry Research Institute, Chinese
Academy of Forestry

艾思龙 Stephane Asselin
易道亚洲区生态规划总监
Principal, Regional Director of Environmental
Services
EDAW Asia

技术顾问 Editing Advisors

鲍达明 Bao Daming
国家林业局野生动植物保护司
处长
Division Chief of Department of Wildlife
Conservation, State Forestry Administration
of P. R. China

雷光春 Lei Guangchun
湿地公约秘书处亚太区域高级顾问
Senior Advisor for Asia Pacific
Ramsar Convention Secretariat

陆健健 Lu Jianjian
华东师范大学河口海岸国家重点
实验室
Key State Laboratory of Estuarine and
Coastal Research, P.R. China

盛连喜 Cheng Lianxi
东北师范大学湿地生态与植被恢复重点实
验室
Northeast Normal University, Key Laboratory of
Wetland Ecology and Vegetation
Restoration

编写人员 Authors

崔丽娟 Cui Lijuan
中国林业科学研究院林业研究所
Forestry Research Institute, Chinese Academy of
Forestry

张曼胤 Cheung Manyin
东北师范大学城市与环境科学学院
Northeast Normal University, School of Urban and
Environmental Science

何文珊 He Wenshan
易道生态规划师
Ecological Planner, EDAW Shanghai

葛礼德 David Gallacher
易道资深环境规划师
Senior Environmental Planner,
EDAW Hong Kong

熊罗恩 Rowan Roderick-Jones
易道生态工程师
Ecological Engineer, EDAW Shanghai

麦诗艳 Kimberlee Myers
易道生态规划师
Ecological Planner, EDAW Beijing

曹子俊 Hector Tso
易道公司环境规划师
Environmental Planner, EDAW Shanghai

吉井贵思 Chris Yoshii
易道亚洲区经济规划总监
Principal, Regional Director of Economics
EDAW Asia

邓伟忠 Kevin W. Teng
易道公司经济分析师
Economist, EDAW Hong Kong

柯帝文 Stephen Casale
易道亚洲区资深撰稿
Senior Writer, EDAW Asia

项 目 经 理 Project Manager

黃 巧 颖 Chloe Huang
易 道 研 究 发 展 经 理
Communication & Information Manager, EDAW
Hong Kong

美 术 编 辑 Graphic Designers

郭 嘉 华 Steve Kwok
平 面 设 计 顾 问
Graphic Consultant

郑 志 雄 Rock Cheng
亚 洲 区 平 面 设 计 经 理
Regional Graphic Manager
EDAW Asia

翻 译 校 对 Translators / Copy-editors

张 念 Jenny Zhang
EDAW Shanghai

陶 然 Ryan Tao
EDAW Shanghai

黃 锦 桂 Huang Jingui
EDAW Beijing

特 别 感 谢
Special thanks for providing projects and
support in the handbook

桂 小 杰 Gui Xiaojie
湖 南 省 林 业 厅 保 护 处
Hunan Province Forestry Administration,
Conservation Division

刘 荣 成、林 竑 斌
Liu Rongcheng, Lin Hongbin
福 建 省 惠 安 县 林 业 局
Fujian Province Forestry Administratition of Hui'an
County

李 荣 旗 Li Rongqi
北 京 锦 绣 大 地 股 份 有 限 公 司
Deputy General Manager
Beijing Glorious Land Agricultural Co.

西 溪 湿 地 公 园 管 理 办 公 室
Administration Committee of Hangzhou
Xixi National Wetland Park

世 界 自 然 基 金 会 中 国 及 香 港 分 会
WWF China/ WWF Hong Kong

Dixi Carrillo
易 道 总 部 摄 影 师
EDAW Firm Wide

易 道 萨 克 拉 门 托 办 公 室
EDAW Sacramento

copyright © URS

URS

Everglades
Florida, USA

Pak Phanang Costal
Southern Thailand

The Comprehensive Everglades Restoration Plan (CERP) involves many governmental and non-governmental organisations, and is led by the U.S. Army Corps of Engineers (USACE), Jacksonville District and the South Florida Water Management District (SFWMD).

Much of the information in this case study was sourced from the excellent CERP website maintained by USACE and SFWMD, and interested readers should visit the following website address for further details on the plan:
http://www.evergladesplan.org/index.cfm

Seth Gentzler
Project Manager
URS Corporation

Francesca Demgen
Senior Ecologist
URS Corporation

Stephane Asselin
Senior Technical Advisor
Regional Director of Environmental Service
EDAW Asia

Vatcharasinthu Choolit
Executive Director
Panya Consultants Co., Ltd.
Thailand Office

Client

Anukularmphai Apichart
President
Thailand Water Resources Association

Chinnavaso Kasemsun
Deputy Director General
National Park, Wildlife and
Plant Conservation Department and
Ministry of Natural Resources and
Environment, Thailand

Krairapanond Nawarat
Director of Strategy and Planning
Office of Natural Resources and
Policy and Planning (ONEP)
Thailand

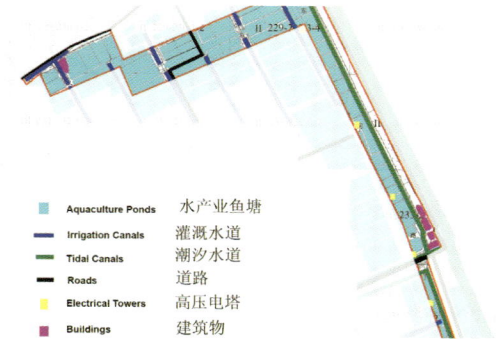

Aquaculture Ponds 水产业鱼塘
Irrigation Canals 灌溉水道
Tidal Canals 潮汐水道
Roads 道路
Electrical Towers 高压电塔
Buildings 建筑物

EDAW | AECOM

EDAW | AECOM

EDAW | AECOM

Yolo Wildlife Area
Yolo County, California, USA

Dave Feliz
Area Manager
Yolo Wildlife Area
California Department of Fish and Game

Robin Kulakow
Executive Director
Yolo Basin Foundation

Ron Unger
Project Director
EDAW Sacramento

Chris Fitzer
Project Manager
EDAW Sacramento

Curtis E. Alling, AICP
Principal in Charge
EDAW Sacramento

Client

California Department of Fish and Game

Upper Truckee River and Wetland Restoration
Lake Tahoe, California, USA

Curtis E. Alling, AICP
Principal in Charge
EDAW Sacramento

Sydney B. Coatsworth, AICP
Principal in Charge
EDAW Sacramento

Client

California Department of General Services,
Real Estate Services Division, and California
Tahoe Conservancy

Robert Sleppy
Department of General Services Real Estate Services Division

Rick Robinson
California Tahoe Conservancy

SCIP Natural Treatment System
Shanghai, China

Chi Chung Wong
Principal in Charge
EDAW Shanghai

Rowan Roderick-Jones, M.S.
Project Manger / Ecological Engineer
EDAW Shanghai

Wenshan He
Coastal Ecologist
EDAW Shanghai

Kerry McWalter, M.S.
Ecological Engineer
EDAW San Francisco

Alan Johnson
Senior Landscape Architect
EDAW Shanghai

Alexander J. Horne
Senior Wetland Specialist
Professor Emeritus of Ecological Engineering
University of California, Berkeley

Slavomir Hermanowitz
Senior Engineering Specialist
Professor of Environmental Engineering
University of California, Berkeley

Zhou Qi
Water Quality Engineer
Professor of Environmental Science and
Engineering
Tongji University, Shanghai

Client

Shanghai Chemical Industry Park
Administration Committee

图书在版编目（CIP）数据

湿地恢复手册：原则、技术与案例分析／国家林业局，易道
环境规划设计有限公司编.—北京：中国建筑工业出版社，2006
（2020.10 重印）
ISBN 978-7-112-08788-4

I.湿...　　II.①国...②易...　　III.沼泽化地－自然保
护　IV.P941．78

中国版本图书馆 CIP 数据核字(2006)第 116877 号

责任编辑：徐纺　邓卫

湿地恢复手册：原则、技术与案例分析
国家林业局，易道环境规划设计有限公司 编

中国建筑工业出版社出版、发行（北京海淀三里河路 9 号）
各地新华书店、建筑书店经销
上海恒美印务有限公司制版
临西县阅读时光印刷有限公司印刷
　　　　　*
开本：787 毫米×1092 毫米　1/8　印张：33¼　字数：879 千字
2006 年 11 月第一版　2020 年 10 月第二次印刷
印数：2201—2800 册　定价：280.00 元
ISBN 978-7-112-08788-4
　　（36234）